HIV AND AIDS: BASIC ELEMENTS AND PRIORITIES

HIV and AIDS: Basic Elements and Priorities

S. Kartikeyan
Rajiv Gandhi Medical College, Thane, India

R.N. Bharmal
Rajiv Gandhi Medical College, Thane, India

R.P. Tiwari
Nicholas Piramal India Limited, Mumbai, India

and

P.S. Bisen
Jiwaji University, Gwalior, India

 Springer

A C.I.P. catalogue record for this book is available from the Library of Congress

ISBN 978-1-4020-5788-5 (HB)
ISBN 978-1-4020-5789-2 (e-book)

Published by Springer,
P.O. Box 17, 3300 AA Dordrecht, The Netherlands.
www.springer.com

Printed on acid-free paper

Cover image courtesy of Sara Media (www.saramedia.com)

The publication of this work was supported by a grant from the Goethe-Institut

CONTENTS

PREFACE

In June 1981, the Centers for Disease Control and Prevention reported the first evidence of a new disease that would later become known as acquired immunodeficiency syndrome (AIDS). Since then, there has been an explosion of information, and volumes have been written on the clinical aspects, virology, immunology, pharmacology, therapeutics, and epidemiology of this new plague.

In the minds of the public, HIV infection is associated with sexual "minorities", "deviant" sexual behaviour, sexual promiscuity, and injecting drug users. Due to these associations, individuals who are infected with HIV face social humiliation and discrimination, which is somewhat similar to that faced by leprosy patients since biblical times. So far, HIV/AIDS is neither curable nor vaccine-preventable. Ignorance about various aspects of HIV/AIDS exists in all sections of society, irrespective of educational status. Consequently, sociology, psychology, ethics, and human rights have assumed a dominant role. HIV medicine has blossomed into a multidimensional medical specialty. Though young doctors and medical students cannot be expected to know all details, they need to be conversant with several dimensions of HIV/AIDS. As compared to their teachers, today's medical students are highly skilled in accessing electronic information. However, sources of electronic information are multiple, unregulated, and subject to perplexing malfunctions. Access to electronic sources depends on availability of electric power and functioning Internet connections. Some websites are poorly maintained.

The authors feel that a book continues to function as a dependable source of information even in the 21st century. Non-specialist doctors who are unable to undertake a detailed study due to pressures of busy medical practice may prefer a concise book with sequential arrangement of information. There are already many authoritative books on the various aspects of HIV/AIDS. Although most medical textbooks contain exhaustive details about epidemiological, diagnostic, clinical, and therapeutic aspects, the psychological, ethical, and legal aspects are either briefly touched upon or are completely ignored. Consequently, the average medical student or non-specialist doctor may remain ignorant about these dimensions. Techniques for counselling have been included because doctors have to take up the task of counselling if trained counsellors are not available in the peripheral settings.

For the above reasons, we have compiled all relevant information on HIV/AIDS in a concise form that will be a source of readily available knowledge for non-specialist doctors and medical students. All currently advocated preventive measures such as health education, condom use, safer sex practices, and treatment of sexually transmitted infections have been incorporated. Antiretroviral therapy and prospects for developing an AIDS vaccine have been focused upon. For the interested reader, a list of references is given at the end of each chapter. This book is not to be used as a substitute for a trained physician. The mention of the trade name of any product or medication does not imply its endorsement or preference to it over other products or medications made by other manufacturers. Views and opinions, if any, are those of the authors and they do not reflect the policies of the institutions to which they belong.

We thank Professor Dr. R.M. Chaturvedi, Seth GS Medical College and KEM Hospital, Mumbai and Advocate M.N. Deshmukh, Mumbai, for their valuable assistance in preparation of this book. We also thank our family members, numerous friends, and colleagues who supported our effort. We are indebted to Mr. Santosh Suryarao and Mr. Dhaval Shah of Sara Media, Thane, India for front cover image design. We wish to express our gratitude to Mr. Sachin Deshmukh, Neptune Group, Mumbai for providing us facilities to write this book.

S. Kartikeyan
R.N. Bharmal
R.P. Tiwari
P.S. Bisen

SECTION ONE

FUNDAMENTALS

CHAPTER 1

PROLOGUE

Abstract

Human immunodeficiency virus (HIV) infection ranks fourth among the world's top killers of mankind. Of the HIV-infected persons, 90 per cent belong to developing countries and over 40 per cent of those infected are women. About half of the newly infected persons belong to the age group of 15–24 years. Thus, the disease leads to loss of income to the nation, as well as to individuals and families. Sub-Saharan Africa is considered to be the *global epicentre* of the HIV epidemic, while Myanmar is the *epicentre* of the Asian epidemic. As per current projections, the number of HIV-infected persons in the Asia-Pacific region will overtake that in sub-Saharan Africa by early 21st century. International studies have shown that the indirect costs of the epidemic are 50–60 times more than the direct costs because the virus selectively affects the age groups that are involved in the national economy and socially productive activities.

Key Words

Global prevalence, HIV/AIDS statistics, Origin of HIV, Point estimate, Range estimate, Region-wise prevalence, Simian immuno-deficiency virus, Tipping point

1.1 – THE NEW "PLAGUE"

Acquired immunodeficiency syndrome (AIDS) is a calamity of the new millennium, similar to plague and smallpox that devastated mankind in the Middle Ages. Like leprosy in the bygone centuries, human immunodeficiency virus (HIV) infection is also associated with social stigma. This is a challenge that goes beyond public health, raises fundamental issues of human rights, and threatens human achievements in many areas.

HIV/AIDS is a classic example of a new and hitherto unknown disease, which has caused a worldwide epidemic. Though the epidemic is classified as "new" and does not mimic any previously known disease, studies have raised the question as to whether HIV-1 and HIV-2 are "new" agents. Africa is a known reservoir of the simian immunodeficiency virus (SIV), but out of the numerous strains, only one is closely related to HIV-1 (the strain causing the majority of AIDS cases). The known SIV strains seem to be more closely related to HIV-2, which is a common cause of AIDS in Africa (Krause, 1992).

3

Soon after the *New England Journal of Medicine* published articles on a disease with acquired cellular immune deficiency in December 1981, the scientific community soon realised that the cases reported in 1981 were not the first. In 1979, doctors in the United States had reported undiagnosed illnesses that were most likely cases of AIDS (Gottlieb, 2001), and studies showed that HIV was present in some US plasma donors as early as 1977 (Madhok *et al.*, 1985; Gottlieb, 2001). Anti-HIV antibodies were reported in a serum specimen collected in 1959 in the Belgian Congo (Nahmias *et al.*, 1986). RNA from this 1959 serum specimen was sequenced and identified as a group M strain of HIV (Zhu *et al.*, 1998). The genomic divergence of group M serotypes began sometime between 1910 and 1950 (Korber *et al.*, 2000). Comparisons of HIV-1 with SIVs isolated from chimpanzees (SIVcpz) indicate that its ancestors crossed over to humans on at least three separate occasions (Hahn *et al.*, 2000). However, it is not known when that cross-over to humans occurred (Gottlieb, 2001).

1.2 – GLOBAL SITUATION

AIDS ranks fourth among the world's top killers of mankind. Tuberculosis, respiratory tract infections, and malaria are among the top three killers. AIDS has killed more than 25 million people since it was first recognised in 1981, making it one of the most destructive epidemics in recorded history (UNAIDS/WHO, 2005). Of the HIV-infected persons, 90 per cent live in developing countries and over 40 per cent of those infected are women. Youth (15–24 years) is the fastest growing segment among the newly infected population. About half of the newly infected persons belong to the age group of 15–24 years. Each day, about 16,000 persons are newly infected, and every minute six persons below 24 years are infected. The number of *AIDS orphans* is expected to swell up to 25 million by 2010.

The number of people living with HIV worldwide as of December 2005 was 40.3 million (36.7–45.3 million). Of these, the estimated number of infected adults and children under 15 years was 38.0 million (34.5–40.6 million) and 2.3 million (2.1–2.8 million), respectively. The proportion of women infected with HIV is steadily rising. In 2005, 17.5 million (16.2–19.3 million) women were infected, one million more than in 2003. The number of people newly infected with HIV worldwide in the year 2005 was 4.9 million (4.3–6.6 million). This included 4.2 million (3.6–5.8 million) adults and 700,000 (630,000–820,000) children less than 15 years. Despite improved access to antiretroviral (ARV) treatment and care in many regions of the world, the AIDS epidemic claimed the lives of 3.1 million (2.8–3.5 million) persons in the year 2005, of whom 570,000 (510,000–670,000) were children less than 15 years. HIV/AIDS statistics pertaining to various regions of the world as of December 2005 is given below. The figures in brackets indicate *range estimates* (UNAIDS/WHO, 2005). Unlike a *point estimate*, the range estimate reflects the actual situation in the field. The upper limit of the range is usually about 20 per cent higher than the lower limit in order to take care of unaccounted number of HIV-positive persons in high-risk and other age groups.

1.3 – SUB-SAHARAN AFRICA

Sub-Saharan Africa is considered to be the *global epicentre* of the epidemic and it remains the world's most affected region. In 2005, this region had an estimated 25.8 million people living with HIV, almost one million more than in 2003. Nearly 25 per cent of AIDS-affected people in this region are in the age group of 15–49 years. This region accounts for 83 per cent of all deaths due to AIDS and 95 per cent of all AIDS orphans. It is estimated that in some countries of southern Africa, the life expectancy at birth may fall to as low as 30 years between 2005 and 2010. Two-thirds of all people living with HIV and 77 per cent of all women living with HIV live in sub-Saharan Africa. The epidemic continues to rage unabated in six southern African countries – Botswana, Lesotho, Namibia, South Africa, Swaziland, and Zimbabwe – where the prevalence of HIV infection among pregnant women is 20 per cent or higher. In Botswana and Swaziland, infection levels in pregnant women are around 30 per cent. In most of East, West, and Central Africa, HIV prevalence has remained stable for the past several years. In two East African countries (Uganda and Kenya), the prevalence of HIV infection in pregnant women is declining, probably due to behavioural changes (UNAIDS/WHO, 2005).

1.4 – THE CARIBBEAN

This region comprises seven countries – the Bahamas, Barbados, Cuba, the Dominican Republic, Haiti, Jamaica, and Trinidad and Tobago. The Caribbean is the second-most HIV-affected region in the world, second only to sub-Saharan Africa. Fifty per cent of the adults living with HIV in the region are women. An estimated 300,000 persons were living with the virus at the end of 2005. During that year, there were an estimated 30,000 (17,000–71,000) new infections and 24,000 (16,000–40,000) AIDS-related deaths. However, between 2003 and 2005, the prevalence of HIV infection in the Caribbean showed no change. Thus, this is the only region in the world where the number of people living with HIV has not increased between 2003 and 2005. The incidence of newly infected persons has decreased in urban Haiti (UNAIDS/WHO, 2005).

In the Bahamas, Haiti, and Trinidad and Tobago, more than 2 per cent of the adult population is living with HIV. AIDS is one of the leading causes of death in these countries (Noble, 2006). Between 16,000 and 24,000 persons die each year due to AIDS-related causes (UNAIDS/WHO, 2005). Heterosexual transmission, mostly associated with commercial sex, is the predominant mode of spread of HIV infection. Social precursors for the HIV epidemic include early initiation of sexual activity, predominantly young population, low rates of condom use, taboos related to sexuality and sexual activity, gender inequalities, stigmatisation, and poverty. The estimated prevalence of HIV infection in Haiti was 3.8 per cent in 2005. In the Dominican Republic, HIV prevention efforts have resulted in reduced number of sexual partners, increased use of condoms, and decreased infection rates (Noble, 2006).

In 2002, the Pan Caribbean Partnership against HIV/AIDS signed an agreement with six drug manufacturers to provide access to cheaper ARV drugs. Partly due to wide differences in drug prices, access to ARV treatment is unequal in this region. ARV therapy is made available to all those in need in Cuba, and the Bahamas and Barbados are progressing towards this objective. However at the end of 2005, just about one-third of those in need of ARV treatment were receiving it in Trinidad and Tobago. The rates were even lower in Haiti and the Dominican Republic (Noble, 2006).

1.5 – SOUTH AND SOUTH-EAST ASIA

The epicentre of the global HIV pandemic is moving from Africa to Asia. The total number of HIV-infected persons was estimated at 7.4 million (4.5–11.0 million) at the end of December 2005. Of the infected adults, 26 per cent (aged 15–49) were women. During 2005, 990,000 (480,000–2.4 million) adults and children were newly infected with the virus and an estimated 480,000 (290,000–740,000) died due to AIDS-related conditions.

Across Asia, the epidemic is fuelled by a combination of injecting drug use and commercial sex. The HIV infection prevalence rates in Asia are low in comparison with that in some continents, especially Africa. However, the populations of many Asian countries are so large that even a mere 1 per cent rise in prevalence implies addition of millions of people living with HIV. The initial driver of the HIV epidemic in most Asian countries is injecting drug use. Since large proportions of drug users are also involved in commercial sex, HIV-infected drug injectors can help build a "critical mass" of infections in sexual networks. As HIV enters commercial sex networks, its wider sexual transmission is almost certain to follow (UNAIDS/WHO, 2005).

The epidemic continues to spread in Papua New Guinea and Vietnam. Countries like Pakistan and Indonesia could be on the verge of serious epidemics (UNAIDS/WHO, 2005). Myanmar (formerly Burma) has been described as the epicentre of the Asian epidemic. As per current projections, the number of HIV-infected persons in the Asia-Pacific region will overtake that in sub-Saharan Africa by early 21st century (Chin, 1995). Many countries in the Asia-Pacific region have not initiated the strategies for prevention and control of the epidemic. Blood supply remains unsafe in many countries of the region. Cross-border trade and international migration of the workforce (particularly single migrants from Philippines, India, Sri Lanka, and Bangladesh to Middle Eastern countries) may create a favourable environment for the spread of HIV (Dwyer et al., 1997).

1.5.1 – Afghanistan

Though there are only 49 reported HIV-infected persons in the country, these statistics could be misleading because HIV screening is limited to blood donors. There are only two voluntary counselling and testing centres in the entire country – at Kabul, the capital, and Jalalabad. Afghanistan is one of the world's largest producers of opium, the raw material for heroin. Access to drugs, poverty, and lack of information increases the likelihood of widespread injecting drug use and sharing of needles. Afghanistan has the second largest number of refugees, after Palestine. As a result of two decades of armed conflict, about 3.4 million Afghans have sought refuge in other countries and an additional 200,000 persons are internally displaced. Such displaced groups have little access to HIV prevention services and are vulnerable to HIV infection due to isolation from their families and widespread poverty. Almost 70 per cent of those suffering from tuberculosis are women. Nearly half of all deaths in women of reproductive age are related to pregnancy and childbirth. Most of the population lacks access to basic health care and there is an acute shortage of trained personnel, particularly female personnel. Opportunities available for women in education, health, and employment are extremely low. Militant groups opposed to girls' education have targeted schools, particularly in rural areas (UNDP, 2006).

1.5.2 – Bangladesh

Bangladesh has a population of over 146 million and a density of 1,061 persons per square kilometre. Almost every year, the country is affected by natural calamities such as floods, tropical cyclones (about 16 in a decade), and tornadoes. The first case of HIV infection was detected in 1989. Significant under-reporting of cases is attributed to the country's limited voluntary counselling and testing (VCT) capacity, and the social stigma attached to HIV/AIDS. Only 465 cases of HIV infection were officially reported till December 2004. In 2005, the estimated number of HIV-infected persons was 11,000, with less than 500 AIDS-related deaths (UNDP, 2006). Bangladesh is among the Asian countries that still have the opportunity to prevent major epidemics. The national adult prevalence of HIV infection is well below 1 per cent. Due to early initiation of focussed HIV prevention efforts, HIV infection in female sex workers is between 0.2 and 1.5 per cent at different sentinel sites (UNAIDS/WHO, 2005).

The identified high-risk groups are commercial sex workers, men who have sex with men, migrant workers, and injecting drug users (IDUs). Sex workers in Bangladesh have the highest client turnover rate in South Asia. The estimated weekly number of clients visiting brothel-based, street-based, and hotel-based sex workers is 18, 17, and 44, respectively. Consistent condom use during commercial sex is rare. The high rates of syphilis and other sexually transmitted infections (STIs) confirm the low level of condom use (UNDP, 2006). The overall prevalence of HIV infection among IDUs in the capital city, Dhaka, was 4 per cent in 2003–2004, while it was 9 per cent in some parts of this city. Many IDUs are involved in commercial sex and among them, less than 10 per cent consistently used a condom (UNAIDS/WHO, 2005). IDUs frequently face homelessness, unemployment, and imprisonment. Illegal sale of blood by drug injectors increases the likelihood of contaminating the national blood supply.

Lack of human resources, high staff turnover, lack of training and staffing, an inefficient hierarchical programme structure, and stigma and discrimination prevalent in Bangladeshi society are among the identified obstacles that hinder the country's response to the epidemic. It has been alleged that people in positions of power often remained silent on the issue of stigma and discrimination, or aggravated the situation by misinterpreting surveillance data, or added to the perception that certain groups were responsible for spreading HIV (UNDP, 2006).

Non-governmental organisations (NGOs) involved in HIV/AIDS-related activities have formed a network and are particularly working with marginalised and hard-to-reach groups. The Shaki project of CARE works mainly with vulnerable populations. The World Bank-assisted HIV/AIDS Prevention Project (HAPP) became effective in February 2001. The new National Strategic Plan (2004–2010) aims at rapid scale-up of successful NGO programmes that focus on high-risk populations, and to strengthen government capacity in blood safety, project management, and surveillance. From 2006, students in 21,000 secondary and upper secondary schools in Bangladesh will be taught about HIV/AIDS as part of a "life skills" curriculum (UNDP, 2006).

1.5.3 – Bhutan

The reported number of HIV-infected persons in 2005 is only 76 and estimates put the number of HIV-infected persons at less than 500. Proportion of infected children is relatively high. Seven out of nine new cases among children were attributed to mother-to-child transmission (MTCT). Probable factors that could contribute to spread of HIV infection are presence of STIs, thriving commercial sex, and cross-border trade along the border with Nepal and India, population mobility, and increasing use of amphetamines. About 63 per cent of Bhutan's population is under 25 years. Lifestyle of migrant workers and truckers are conducive to commercial and casual sex. The Royal Government of Bhutan's response includes awareness campaigns, establishment of HIV sentinel surveillance system, screening of donated blood, training of all health personnel, promotion of condoms, and ARV drugs for those in need of treatment (UNDP, 2006).

1.5.4 – Cambodia

The first case of HIV infection was detected in 1991 during routine screening of donated blood. The first case of AIDS was diagnosed in late 1993. In the 1990s, the HIV epidemic focussed on the sex industry. The national adult HIV prevalence rate peaked at about 3.3 per cent in 1997–1998 and fell steeply by one-third to 1.9 per cent in 2003. This decline is attributed to possible behaviour changes and the government's policy of 100 per cent condom use in commercial sex establishments and provision of access to affordable condoms. The HIV prevalence among brothel-based and non-brothel-based sex workers decreased by half between 1999 and 2002. Condom use rates (more than 80 per cent in 2003) have been increasing steadily among commercial sex workers since the late 1990s. The potential drivers of the epidemic have changed in recent times. The main route of transmission is between spouses and from mother to child (UNAIDS/WHO, 2005; UNDP, 2006).

There are concerns regarding safety of blood supply in the country and the increasing incidence of HIV infection among pregnant women (from 0.35 per cent in 1999 to 1.48 per cent in 2002) in western Cambodia along the Thai border. This is also the only region in the country where HIV incidence among commercial sex workers has not declined. Cambodia's VCT services are largely confined to the capital, Phnom Penh. The country's weak health infrastructure is attributed to genocide, civil war, and famine in the recent past and continuing political instability and persistent poverty. Cambodia has a large commercial sex industry. Due to the country's political turmoil, there has been considerable population mobility including refugee resettlement. There is considerable rural to urban migration of young people. Most of the 140,000 workers in garment factories around Phnom Penh are migrant rural women below 30 years. Some women move between factory work and short-term sex work. Cambodia lies on drug trafficking route and illegal amphetamine laboratories have reportedly been established along the border with Thailand. Amphetamines and inhalants are commonly used drugs. At present, drug use does not seem to have a major role in facilitating risk behaviour (UNDP, 2006).

1.5.5 – Indonesia

Indonesia is an archipelago of 18,108 islands, of which about 6,000 are inhabited. It is the fourth most populous country in the world. The epidemic is unevenly distributed in this nation of 210 million people and 11 provinces (Bali, East Java, West Java, Jakarta, Papua, West Kalimantan, East Kalimantan, North Sumatra, North Sulawesi, Riau, and West Irian Jaya) out of the country's 31 provinces are severely affected. Multiple sexual partnerships are common in parts of Papua province (UNDP, 2006).

Indonesia is on the brink of a rapidly worsening HIV epidemic driven by commercial sex and injecting drug use. The epidemic is spreading to remote parts of this archipelago. During 2002–2003, HIV prevalence ranged from 66 to 93 per cent among IDUs in the capital city, Jakarta. (UNDP, 2006). Most drug injectors are young, relatively well educated and live with their families. Studies have reported that most injectors know where to get sterile needles, yet they are reluctant to carry sterile needles fearing that the police would treat this as proof that they inject drugs, which is a criminal offence. About 88 per cent of drug users use non-sterile injecting equipment. Indonesia's drug users are regularly arrested and jailed. HIV prevalence among inmates of Jakarta's Cipinang prison rose from zero per cent in 1999 to 25 per cent in early 2003 (UNAIDS/WHO, 2005; UNDP, 2006).

Condom use is infrequent or rare among sex workers working from non-brothel settings (massage parlours, clubs), as well as brothel-based sex workers, despite a decade of HIV prevention efforts. Women are reportedly averse to carrying condoms because the police who might then arrest them regard it as "proof" of commercial sex work. For controlling the growing epidemic, Indonesia may have to adapt the legal and institutional environment to facilitate effective HIV prevention strategies (UNAIDS/WHO, 2005).

1.5.6 – Iran

The first HIV case was detected in 1987. In September 2004, 7,510 HIV-infected persons were registered, of which 95.1 per cent were men. The modes of transmission are injecting drug use (57.4 per cent) and sexual activity (6.8 per cent). However, the World Health Organization (WHO) has estimated the number of HIV-infected persons to be between 22,000 and 30,000. Imprisonment appears to be the biggest risk factor for HIV infection (UNAIDS/WHO, 2005). There is a high risk of increasing HIV prevalence because the country is located on a major drug trafficking route and its north-eastern neighbour is Afghanistan – the world's biggest producer of opium. The incidence of injecting drug use is rising in Iran. The Central Asian countries to the north of Iran are also experiencing fast growing HIV epidemics. Due to the socio-cultural milieu, infection remains hidden and many infected persons to seek medical help. Though condom use has increased in recent times, the data relates only to married people (UNDP, 2006).

1.5.7 – Lao People's Democratic Republic

The overall prevalence of HIV infection in this country (formerly called Laos) is low and about two-thirds of infected persons are living in the capital city, Vientaine and Savannakhet. The prevalence of STIs is high among women who work in non-brothel settings that also provide sexual services and about 1 per cent of women in Vientaine and Savannakhet are HIV infected. Behavioural studies have reported increasing sexual activity with multiple partners among young men in Vientaine (UNAIDS/WHO, 2005).

1.5.8 – Malaysia

About 64,000 persons were reportedly living with HIV in Malaysia in late 2004, of whom, about 9,400 had developed AIDS. Most were young men aged 20–29 and three-quarters of them IDUs. Twelve per cent of cases were attributed to heterosexual transmission and less than 1 per cent to male homosexuality or bisexuality. The growing proportion of HIV infections among women (7 per cent of total cases in 2003) is due to unprotected sex with a regular partner or multiple sex partners. There are an estimated 14,000 AIDS orphans in Malaysia.

HIV prevalence of 10 per cent has been reported among sex workers in parts of Kuala Lumpur. The statistics on geographical distribution indicate the place where the infection has been diagnosed and not the place of usual residence of the infected person. Large numbers of reported cases in the states of Johor and Selangor is possibly due to high rate of detection from their relatively large prisons and drug rehabilitation centres, and the greater number of persons seeking treatment, as compared to the states in Peninsular Malaysia. The highest number of AIDS cases and AIDS-related deaths has been reported in Kuala Lumpur probably because of availability of medical treatment and support facilities in this city (UNDP, 2006).

1.5.9 – The Maldives

The nation's AIDS Control Programme was launched in 1987, before the first case of HIV infection was identified in 1991. By the end of 2003, there were 135 HIV-infected persons in the country, of which, 12 were Maldivians and the rest were foreigners. Six individuals have died of AIDS-related diseases. Numerous factors make the country susceptible to HIV epidemic. Nearly 50 per cent of the population is under 15 years and prevalence of drug use among youth is rising. The country has a highly mobile population that includes tourists, sailors, migrant workers, and students. In 1998, about 400,000 tourists visited the country, about one and half times the entire population of the Maldives. Influx of tourists and high-risk behaviour, such as unsafe sex and injecting drug use, can possibly introduce different serotypes of HIV from all over the world. However, sex tourism does not exist in the Maldives. The current high rates of divorce and remarriage may lead to infection of serial spouses, casual sex partners, and create large sexual networks capable of transmitting HIV and other STIs. The Maldivian population is dispersed over about 190 islands. In the smaller islands, persons without radio or television constitute 55 per cent and 86 per cent, respectively. These conditions hinder HIV prevention programmes, including education and condom distribution. The situation is complicated by internal displacement of population after the tsunami in 2004 (UNDP, 2006).

1.5.10 – Mongolia

The first case of HIV infection was reported in 1992. By April 2005, nine cases of HIV/AIDS had been reported. Though the numbers are small, Mongolia is vulnerable to HIV epidemic because of its young population (50 per cent below 23 years), increasing incidence of STIs and drug use, rising number of sex workers and street children, increased international and internal mobility, and low levels of condom use. Neighbouring countries (China, Kazakhstan, and Russia) are experiencing HIV epidemics, primarily driven by injecting drug use. It is feared that injecting drug use, may be followed by an HIV epidemic when the region's drug traffickers start using Mongolia's trade routes. Due to economic hardships, the commercial sex industry has proliferated in the capital, Ulaanbaatar, and in the other smaller cities. The prevalence of sexually transmitted diseases (STDs) is 30 per cent among pregnant women and up to 58 per cent among sex workers. In 2002, the incidence of syphilis and gonorrhoea per 10,000 population was 6.7 and 19.6, respectively. In the capital, donor blood from volunteers is routinely screened for blood-borne pathogens but in the other cities, screening of donor blood depends on availability of laboratory reagents and test kits (UNDP, 2006).

1.5.11 – Myanmar

Myanmar (formerly Burma) has a population of about 50 million. The country has 130 ethnic groups and is rich in natural resources. But, more than five decades of political and armed conflict combined with trade and investment restrictions imposed by the United States and the European Union in 1997 have worsened socio-economic conditions in Myanmar (UNDP, 2006).

Myanmar's first HIV-infected case was reported in 1988. The HIV epidemic is driven by injecting drug use and unprotected sex (both heterosexual and between men). Inadequate efforts to contain the spread of HIV epidemic during the 1990's led to its rapid spread initially among high-risk groups and later among the general population. Consequently, Myanmar has one of the most serious epidemics in Asia. In 2005, the estimated number of HIV-infected persons and AIDS orphans was 360,000 and 37,000, respectively with about 110,000 AIDS-related deaths. Geographically, the eastern part of the country is the worst affected. The central and delta regions have moderate rates of infection, while the western part of the country is the least affected. Epidemiologically, the country seems to be close to the "tipping point" – the point at which, the critical mass of infection becomes so great that the epidemic is self-sustaining in the general population even if high-risk behaviour is significantly reduced in the most vulnerable groups (UNDP, 2006).

Myanmar established a HIV surveillance system in 1985 and by 2000 the system was expanded to cover 27 sites nationwide. The mindset of the early years (denying severity of the epidemic) seems to be changing. HIV prevention efforts have been scaled up in recent years. In 2004, 32.6 million condoms were distributed in contrast to the earlier "rule of thumb" target of one condom per capita per year (UNDP, 2006). But, the epidemic continues to rage unabated in the low-risk populations. HIV prevalence among pregnant women estimated at 1.8 per cent in 2004. At eight sentinel sites (out of 29), HIV prevalence among pregnant women has exceeded 3 per cent. Consistently high levels of HIV prevalence (above 25 per cent since 1997) among sex workers have aggravated Myanmar's epidemic (UNAIDS/WHO, 2005).

Needle exchange programme has been established as part of harm reduction efforts. Nationally, the HIV prevalence among IDUs was 34 per cent in 2004, having decreased since 2001. But in Yangon, the nation's capital, and Mandalay, the prevalence among drug injectors in 2004 was 25 per cent and 30 per cent, respectively (UNDP, 2006). Among military recruits, the prevalence of HIV infection was 1.6 per cent in early 2004. There is a lack of national data on condom use rates during commercial sex and limited behavioural information is available. Stronger prevention efforts are needed to deal with an epidemic that already ranks as the most serious in the region (UNAIDS/WHO, 2005).

1.5.12 – Nepal

The HIV epidemic in Nepal is driven by heterosexual transmission, primarily through commercial sex and prevalence of STIs is also rising. Limited information is available about homosexual/bisexual transmission and rural-to-urban ratio of infected persons owing to inadequate surveillance data. Estimates suggest that at least 10 per cent of the 2–3 million migrant Nepalis are infected with HIV and one-third of the infections in the country are among IDUs. The prevalence consistently exceeds 5 per cent in one or more high-risk groups such as sex workers, their clients and IDUs. Every year, about 12,000 Nepali children are taken to Indian brothels and the Middle East for commercial sex work. Though HIV infection is prevalent in all regions of the country, it is concentrated in central region and the capital, Kathmandu. The national adult HIV prevalence is estimated to be 0.5 per cent in 2005 with 75,000 infected persons, 5,100 AIDS orphans, and 16,000 AIDS-related deaths. The HIV prevalence among IDUs was 38.4 per cent nationwide, while in Kathmandu it was 68 per cent. Nepal was the first developing country to establish a harm reduction programme with needle exchange for IDUs, but this programme is impeded by limited coverage. Nepal's comprehensive plan of 2002 proclaimed political commitment to halt the spread of HIV epidemic. In June 2003, satellite digital radio was established to impart information related to HIV/AIDS and gender issues to rural communities (UNDP, 2006).

Trafficking of Nepali girls and women into commercial sex work in India continues to challenge HIV control efforts. In addition, girls are also forced into traditionally institutionalised sex work practices, such as *Deuki* and *Badi*. There are between 100,000 and 200,000 internally displaced persons in Nepal due to multiple causes ranging from economic and/or political reasons to forced migration because of trafficking. About 1.5–2 million people have migrated to other countries for economic reasons and about one million Nepalis are estimated to be in India. Between 2 and 10 per cent of returning economic migrants are probably HIV-infected (UNDP, 2006).

1.5.13 – Pakistan

Though the rates reported to the country's National AIDS Control Programme (NACP) are less, the Joint United Nations programme on HIV/AIDS (UNAIDS) estimated 85,000 HIV-infected cases in 2005 with 3,000 AIDS-related deaths.

Pakistan is currently classified by WHO/UNAIDS as "low-prevalence" (adult prevalence about 0.1 per cent) but " high-risk" country for the spread of HIV infection. The HIV infection has reached epidemic proportions in Karachi (Pakistan's main trading city), Lahore, and Larkana (a small rice-growing town in Sindh province). The prevalence of STIs is high among IDUs, male sex workers, and the *hijra* (a diverse group of castrated males, transvestites, and transsexuals). The sexual network of men having sex with men (MSM) is heterogeneous, and includes the *hijra, zenanas* (transvestites who usually dress as women), and masseurs. Pakistan is a major transit and consumer country for opiates from neighbouring Afghanistan, the world's largest producer of opium. Studies have revealed that drug users were switching from smoking or "sniffing" or inhaling to injecting polydrug cocktails (UNDP, 2006).

A combination of high degree of sexual interaction between drug injectors and sex workers, their low condom use, and their limited knowledge about HIV, and a large population of displaced persons favours rapid spread of the epidemic (UNAIDS/WHO, 2005; UNDP, 2006). Of the 1.13 million refugees in the country in January 2006, those from neighbouring Afghanistan numbered nearly 1.12 million. In October 2005, a powerful earthquake in Pakistan's northern region killed more than 80,000 people. More than 3.5 million were rendered homeless. About 80 per cent of the health facilities were severely damaged or completely destroyed (UNDP, 2006).

Most infections occur between 20–44 years, with males outnumbering females by a ratio of 5:1. Sexual transmission accounted for 67.48 per cent of reported cases. It is estimated that 40 per cent of the 1.5 million annual blood donations are not screened for HIV. About four million people are employed overseas. These migrant workers are away from their homes for extended periods of time and may be exposed to risk of unprotected sex and HIV infection. Pakistan has a high rate of medical injections: about 4.5 per capita per year. Almost 94 per cent of the injections are administered with used injection equipment, and use of unsterilised needles is rampant in health facilities. An estimated 60 per cent of the country's population has access to the formal health care system, while the remaining population is dependent on traditional healers (or *hakims*) or self-described "doctors" who have little or no formal medical training (UNDP, 2006).

In 1988, the NACP was launched and HIV/AIDS was declared a notifiable disease. The Government has identified nine priority areas. However, the approach to sex workers and MSM appears to be cautious. There is no legal protection against discrimination so far and federal legislation bill for mandatory testing of blood products is awaiting Senate approval. Foreigners living in Pakistan for longer than 1 year are subject to mandatory HIV testing. Pakistan's recent Enhanced HIV/AIDS Programme aims to prevent HIV from becoming established in vulnerable populations and combating stigmatisation of those infected (UNDP, 2006).

1.5.14 – The Philippines

The Philippines is an archipelago of 7,107 islands that stretch from the south of China to the northern tip of Borneo (Indonesia). The country has more than 100 ethnic groups (UNDP, 2006). The national adult HIV prevalence has stayed low even among high-risk populations, probably due to routine screening for STIs along with other HIV prevention services for sex workers. However, condom use is rare among non-brothel-based sex workers. The rate of non-sterile needle use is reported to be 77 per cent in Cebu City and high rates have been reported in other parts of the country. This calls for efforts to close the gaps in the country's response to the epidemic (UNAIDS/WHO, 2005).

1.5.15 – Sri Lanka

The country's first case of HIV infection was reported in 1986. In 2005, the estimated number of HIV-infected persons was 5,000 with less than 500 AIDS-related deaths. At the end of 2004, the male to female ratio was 1.4:1 but the proportion of infected females is rising. Of the reported HIV infections, 86 per cent were attributed to heterosexual transmission and 11 per cent to bisexual/homosexual transmission. Since homosexual behaviour is illegal in Sri Lanka, few HIV prevention activities are targeted at this group (UNDP, 2006).

The proportion of IDUs in the country is reported to be less than 1 per cent of all drug users and so far, only one case of HIV infection has been attributed to injecting drug use. Only three cases of blood-borne infection have been reported so far. Donated blood is screened at the Central Blood Bank in the capital city, Colombo and at 56 regional blood banks. ARV therapy is provided free of charge to HIV-infected pregnant women (UNDP, 2006).

Behavioural factors that facilitate the spread of the epidemic include presence of large number of sexually active youth, low rates of condom use, internal displacement of population due to continuing ethnic strife, and increasing number of commercial sex workers (estimated at 30,000 women and girls and 15,000 boys). A significant number of sex workers operate near military camps in the strife-torn country. There are also "beach boys" and women who are involved in sex trade with tourists. Other identified vulnerable groups include women employed in free trade zones, persons seeking foreign employment, plantation workers, and the fishing community. An estimated 1.2 million Sri Lankans work in the Middle East and 79.1 per cent of unskilled migrants are women. In 2001, 48 per cent of Sri Lanka's HIV cases were among women working as housemaids abroad. High rates of unwanted pregnancies and STIs have been reported among women who comprise 80 per cent of the work force in the free trade zone at Kandy (UNDP, 2006).

1.5.16 – Thailand

Thailand has been acclaimed for its success in containing the HIV epidemic. However, coverage of HIV prevention programmes is inadequate among men who have sex with men and IDUs (UNDP, 2006). In the 1990s, a rapidly spreading HIV epidemic focussed on the sex industry. In late 2005, the estimated number of HIV-infected adults and children was 560,000 and 16,000, respectively. The prevalence of HIV infection in adults was 1.4 per cent. The number of AIDS-related deaths in 2005 was 21,000 (UNAIDS/WHO, 2005). Thailand's epidemic has metamorphosed into a more diverse epidemic in recent times. Since new HIV infections each year are occurring within marriage or steady relationships where condom use tends to be very low (UNAIDS, 2002), Thailand's much acclaimed HIV control programmes (Chapter 27) should adopt strategies to match the shifts in the epidemic (UNAIDS/WHO, 2005).

1.5.17 – Vietnam

The first case of HIV infection was reported in 1990 in Vietnam. In the 1990s, the epidemic was concentrated in a few provinces and cities. By 2004, HIV had spread to all 64 provinces and all cities. Estimates in 2005 put the number of persons living with HIV at 260,000, with 13,000 AIDS-related deaths (UNDP, 2006).

Although the overall prevalence of HIV infection is less than 0.5 per cent, the national prevalence of infection among IDUs is about 33 per cent (UNDP, 2006). In the northern coastal cities of Hai Phong and Quang Ninh, the adult prevalence was estimated at 1.1 per cent in 2003 (UNAIDS/WHO, 2005). The drug injectors are usually young (mean age: 25 years). There is a large overlap between injecting drug use and commercial sex work. Fewer than 50 per cent of drug injectors reportedly use condoms with sex workers. The average HIV prevalence among sex workers in Vietnam is about 16 per cent and the infection levels are even higher in the cities of Can Tho, Hai Phong, Ho Chi Minh City, and Hanoi. Ho Chi Minh City accounts for about one-quarter of all HIV infections in the country (UNAIDS/WHO, 2005).

Most control measures have focused on prevention and resource allocations are not adequate. Implementation and effectiveness of the current programmes are not evaluated. Due to policies to manufacture ARV drugs domestically, many persons in need of ARV treatment do not receive them. The country has only 41 laboratories in 34 provinces and cities to detect HIV infection, and laboratory skills are reportedly low. The majority of blood donations come from professional blood donors, which increases the risk of blood-borne transmission of HIV (UNDP, 2006). Studies reveal that punitive campaigns to combat "social evils" tend to drive drug injectors and sex workers beyond the scope of HIV prevention programmes and can inadvertently entrench risky behaviours (UNAIDS/WHO, 2005).

1.6 – EASTERN EUROPE AND CENTRAL ASIA

At the end of 2005, the total number of HIV-infected persons had increased by 25 per cent to 1.6 million (990,000–2.3 million) since 2003. The estimated number of AIDS-related deaths almost doubled to 62,000 (39,000–91,000) in comparison to that in 2003. During 2005, the estimated number of newly infected adults and children was 270,000 (140,000–610,000). Of the adults aged 15–49, 28 per cent were women in December 2005 in this region (UNAIDS/WHO, 2005).

Countries such as Russia and Ukraine have declining birth rates with an ageing population. The HIV epidemic will reduce the numbers of young people, which may reduce gross domestic product (GDP) and labour supply. However, the Central Asian countries have a high birth rate. An increase in AIDS-related deaths in these countries could increase the number of AIDS orphans. Throughout this region, a large number of people are IDUs. Since unauthorised possession of needles and syringes is illegal in many countries in this region, many drug users are compelled to share needles (Kirby, 2006). More than 80 per cent of HIV-infected persons in this region were less than 30 years. A large number of persons in this region are involved in commercial sex in exchange for drugs or in order to get money for drugs (UNAIDS, 2004). Many IDUs in this region are imprisoned simply because they use illegal drugs. A majority of these imprisoned drug users share needles to inject drugs in prison and there is also unprotected sex between men. This leads to spread of HIV among prisoners. In some countries of this region including Russia, substitution treatment with methadone is illegal. Thus, a combination of a thriving sex trade, high levels of injecting drug use, and unprotected sex may fuel an explosive HIV epidemic in certain parts of this region (Kirby, 2006). At the end of 2004, only a small percentage of people in high risk groups were being reached by HIV prevention programmes (UNAIDS, 2004).

1.6.1 – Eastern Europe

In the 1980s, authorities in the erstwhile communist government in Romania (Eastern Europe) thought that blood transfusions would be helpful in boosting the immune system and keeping healthy and administered blood unnecessarily to children in orphanages. Some of the transfused blood units were contaminated with HIV. Romania has the largest number of HIV-infected children in Europe and is home to the world's largest AIDS clinic for children. However, the overall prevalence of HIV infection in Romania is still low (Kirby, 2006). Only in Romania and Moldova, most people in need of ARV treatment were receiving it at the end of 2005 (WHO, 2006).

1.6.2 – Russia

Russia has the largest HIV epidemic in Europe and accounts for about two-thirds of cases in the Eastern Europe and Central Asia. At the end of 2005, there were an estimated 940,000 people living with HIV/AIDS in Russia (UNAIDS/WHO,

2006). About two million people (or 2 per cent of the population) are IDUs. An estimated 20–30 per cent of IDUs in Russia use non-sterile needles and syringes. Heterosexual transmission accounted for 20 per cent of all infections in 2003. In Russia, 9,000 HIV-infected babies were born from the start of the epidemic till the end of 2003 (UNAIDS, 2004). Only 5 per cent of the estimated 99,000 persons needing ARV treatment were receiving it at the end of 2005. This rate is to be compared with that in poor African countries – Zambia (27 per cent), Malawi (20 per cent), and Uganda (51 per cent) (WHO, 2006).

1.6.3 – Ukraine

By mid-2004, there were more than 68,000 HIV-infected persons. Tuberculosis is the leading cause of death among people living with HIV in Ukraine. Among IDUs in Ukraine, reuse of needles is widespread and condom use is low. Due to programmes to prevent MTCT of HIV infection, the percentage of HIV-infected newborns fell from 27 per cent in 2001 to 12 per cent in 2003 (UNAIDS, 2004).

1.6.4 – Central Asia

This region contains major drug trafficking routes between Russia and Europe and consequently there are many IDUs in this region. Though HIV epidemics are growing in all countries (Kazakhstan, Uzbekistan, Turkmenistan, and Kyrgyzstan) in this region, these countries are still in the early stages of the epidemic and there is still time to reverse this trend with the introduction of harm-reduction programmes for IDUs, HIV prevention programmes and improved testing and ARV treatment (Kirby, 2006).

1.6.5 – Baltic States

Though the overall numbers of HIV-infected persons are low, the epidemic is spreading in the Baltic States (Estonia, Lithuania, and Latvia). The number of HIV-infected persons more than doubled from 921 in 2000 to 2,353 in 2001. But between 2001 and 2003, Estonia and Latvia virtually halved their number of HIV cases over the 2-year period, probably due to fewer infections or changes in reporting techniques (Kirby, 2006).

1.6.6 – The Caucasus

Georgia, Armenia, and Azerbaijan have a much lower HIV prevalence rates than Russia. Some cities have very high rates of HIV infection among IDUs and an explosive epidemic is likely in these three countries. The estimated number of HIV-infected persons was 5,400 in 2005, up from 1,400 in 2003. In Baku, capital of Azerbaijan, one in four street drug users was found to be HIV-infected (UNAIDS, 2004).

1.7 – LATIN AMERICA

An estimated 1.8 million people were living with HIV/AIDS in the Latin American countries at the end of 2005. There were 200,000 new infections and 66,000 AIDS-related deaths during 2005. Of HIV-infected adults, 32 per cent are women (UNAIDS/WHO, 2005).

Though the national HIV prevalence is estimated to be at least 1 per cent in Belize, Guyana, Honduras, and Suriname, HIV infection in most Latin American countries is not generalised but concentrated in high-risk populations. Brazil, the region's most populous country, accounts for about 40 per cent of HIV-infected persons in Latin America (Noble, 2006).

In a majority of South American countries (notably Argentina, Brazil, Chile, Paraguay, Uruguay, and northern Mexico), injecting drug use and sex between men are the predominant routes for transmission of HIV. However, in countries of Central America, both heterosexual and male homosexual transmission account for most infections and injecting drug use is of minor importance (Noble, 2006).

Brazil is among the countries that produce cheaper generic ARV drugs and around 183,000 HIV-infected Brazilians were receiving treatment at the end of 2005. Argentina, Chile, Mexico, Uruguay, and Mexico are among the countries in the region with high levels of access to ARV drugs. However, in Ecuador, Paraguay, and the poorer countries of Central America, a large proportion of those in need of ARV treatment are unable to access it. Brazil has estimated that ARV treatment has contributed to 50 per cent reduction in AIDS-related deaths, while Argentina reported decreasing numbers of AIDS-related deaths between 1999 and 2004, with stabilisation for the last 2 years (Noble, 2006).

1.8 – NORTH AMERICA

The number of individuals living with HIV was estimated at 1.2 million (650,000–1.8 million) in 2005 and 25 per cent of HIV-infected adults (aged 15–49) were women. The low number of AIDS-related deaths viz. 8,000 (9,000–30,000) in 2005 is a consequence of widespread availability of ARV treatment and access to HIV care. There were between 15,000–120,000 newly infected persons in 2005.

1.8.1 – The United States

An estimated one million Americans may be infected with HIV, one-quarter of whom may be unaware of their infection (NIAID, 2005; boston.com, 2006). The epidemic is growing most rapidly among minority populations and is a leading killer of African-American males aged 25–44. AIDS affects nearly seven times more African-Americans and three times more Hispanics than whites. An increasing number of African-American women are being affected. In 2003, two-thirds of AIDS cases in the United States in both women and children were among African-Americans (NIAID, 2005). In September 2006, the Centre for Disease Control and Prevention (CDC), Atlanta, recommended that HIV screening be made routine part of medical care for all patients between the ages of 13 and 64 and to improve diagnosis of HIV infection among pregnant women. CDC recommendations strongly emphasise that HIV testing must be voluntary and undertaken only with the informed consent of the patient. Patients are to be specifically informed that they have the opportunity to decline HIV testing. CDC has recommended that written consent for HIV testing be no longer required (www.boston.com, 2006).

1.9 – EAST ASIA

In East Asia, the estimated number of HIV-infected persons was 870,000 (440,000–1.4 million) in 2005. Of HIV-infected adults aged 15–49, 18 per cent were women. In 2005, 140,000 (42,000–390,000) persons were newly infected with the virus, while 41,000 (20,000–68,000) persons died due to AIDS-related causes (UNAIDS/WHO, 2005). The situation in North Korea is not known (UNDP, 2006).

1.9.1 – China

HIV infection has been detected in all 31 provinces, autonomous regions, and municipalities, yet each area has its own distinctive epidemic pattern (UNDP, 2006). The HIV epidemic in China (particularly in the south and west), is prevalent mainly among high-risk groups such as IDUs, sex workers, former plasma donors, and their sexual partners. The majority of HIV infections have been detected in urban areas of Guangdong, Yunnan and Henan provinces, and Guangxi autonomous region. Currently, Qinghai province and Tibet autonomous region are the least affected (UNAIDS/WHO, 2005).

Among IDUs, HIV prevalence is 35–80 per cent in Xinjiang and 20 per cent in Guangdong. HIV infection levels of 10–20 per cent (rising to 60 per cent in some communities) have been found in rural areas of China, particularly in Henan, Anhui, and Shandong provinces, where rural folk sold their blood and/or plasma to supplement their meagre farm incomes (UNDP, 2006).

Studies indicate that at least half of the female drug users had also engaged in commercial sex at some stage. Most female sex workers originate from remote rural areas, are poorly educated and have little knowledge about HIV. The potential overlap between commercial sex and injecting drug use is likely to become the main driver of China's epidemic. There are also signs that the epidemic is spreading beyond the high-risk groups into the general population. HIV prevalence rate is as high as 5 per cent in some areas where HIV infection is established among drug users and sex workers. Though there is paucity of data relating to HIV transmission among men who have sex with men, some studies have reported low rates of condom use (UNAIDS/WHO, 2005). The response to the HIV/AIDS epidemic in China is described in Chapter 28.

1.9.2 – Japan

The reported annual number of HIV-infected persons peaked at 780 in 2004. Sex between men accounted for 60 per cent of new HIV infections and about one-third of the total cases in that year were among persons younger than 30 years, which seems to indicate increasing unprotected sexual intercourse among the young population (UNAIDS/WHO, 2005).

1.9.3 – Republic of Korea

The first cases of HIV infection and AIDS were reported in 1985 and 1987, respectively; 96 per cent of infections are attributed to sexual transmission, with 13 per cent occurring among women. In November 1987, the AIDS Prevention Act was enacted, to protect against discrimination and ensure confidentiality. The law requires reporting of HIV infection and stipulates mandatory HIV testing of certain subgroups in the population and donated blood. In October 1989, free-of-charge anonymous HIV testing was introduced in community health centres and quarantine stations. Mandatory HIV testing of returning sailors was banned in May 1993 (UNDP, 2006).

1.10 – OTHER REGIONS

1.10.1 – Western and Central Europe

Due to access to ARV drugs, the number of adult and child deaths due to AIDS in 2005 was less than 15,000 in 2005. However, 720,000 (570,000–890,000) persons were living with the virus in this region, and 27 per cent of adults (aged 15–49) living with HIV who were women. In 2005, 22,000 (15,000–39,000) persons were newly infected with HIV (UNAIDS/WHO, 2005).

1.10.2 – North Africa and Middle East

The Arab world has one of the lowest prevalence rates but has second fastest growing infection rate (UNDP, 2006). By the end of December 2005, 510,000 (230,000–1.4 million) persons in this region were living with HIV and 47 per cent of the adults (aged 15–49) were women. In 2005, 67,000 (35,000–200,000) of adults and children were newly infected, while 58,000 (25,000–145,000) died due to AIDS-related causes.

1.10.3 – Oceania

Oceania comprises the Pacific islands of Australasia, Polynesia, Melanesia, and Micronesia. Among adults (15–49 years) living with HIV in this region, 55 per cent were women as of December 2005. During that year, 3,600 (1,700–8,200) persons died due to AIDS-related causes and 8,200 (2,400–25,000) were newly infected with the virus. An estimated 74,000 (45,000–120,000) persons were living with HIV (UNAIDS/WHO, 2005).

Papua New Guinea: This country has the highest prevalence of HIV infection in the Pacific region and the epidemic seems largely driven by heterosexual transmission. The country has a population of 5.67 million (July 2006 estimate) and shares an island with one of Indonesia's worst affected provinces, Irian Jaya. In 2003, the estimated number of people living with HIV and AIDS-related deaths was 16,000 and 600, respectively (World Fact Book, 2003). Prevalence is over 1 per cent among pregnant women in the capital, Port Moresby, and in Goroka, and Lae. High levels of STIs could fuel an HIV epidemic in the general population (UNDP, 2006).

Sultanate of Brunei: This is an oil-rich country in Oceania with a population of 379,444 (July 2006 estimate). The adult HIV prevalence rate was less than 0.1 per cent in 2003 (World Fact Book, 2003).

Fiji: The total number of known HIV-infected cases was 171 in 2005. Though 85 per cent of transmission is through the heterosexual route, there is considerable level of homosexual and male bisexual activity in Fiji. Blood supply in Fiji is generally considered safe. The first confirmed HIV-positive person was infected through blood transfusion done abroad. Of known infections, 50 per cent have occurred in young persons aged 20–29. Risk factors that could fuel an HIV epidemic include high incidence of other STIs, teenage pregnancies, drug abuse, a mobile population, a large tourism industry, considerable levels of extramarital sexual activity particularly by men, and sexual violence (UNDP, 2006).

1.10.4 – Australia

Since the start of the HIV epidemic, until the end of 2005, 25,243 persons in Australia have been infected with HIV. There were 6,594 AIDS-related deaths and 9,759 persons were diagnosed has having AIDS. The annual incidence of HIV probably peaked around 1985. After continuously declining for 13 years, the number of new HIV infections stabilised at around 680 per year during 1998–2001. Subsequently, the number of new HIV infections rose to 826 in 2002 and 818 in 2004. The annual number of persons diagnosed with AIDS peaked in 1994 with 953 cases, declined rapidly to 206 in 1999, and increased to some extent to an estimated 239 in 2004. The decline in the number of AIDS cases since 1996 is attributed to introduction of combination ARV therapy. At the end of 2004, an estimated 14,840 persons were living with HIV in Australia (AVERT, 2006).

Among *non-indigenous* peoples, transmission occurs chiefly among MSM (63 per cent) and through heterosexual contact (20 per cent). However, heterosexual contact and male homosexual contact were reported with equal frequency among HIV-infected *indigenous* peoples. Prevalence of injecting drug use was higher among indigenous peoples (20 per cent versus 4 per cent). The proportion of HIV-infected women was also higher among indigenous peoples (33 per cent versus 10.8 per cent). However, the differences in the overall rates of HIV infection and AIDS between non-indigenous and indigenous peoples have varied little (AVERT, 2006).

The highest prevalence of HIV infection per 100,000 population until the end of 2005 was in New South Wales (209.2) followed by Victoria (107) and Capital Territory (87.6). Comparable HIV prevalence rates per 100,000 population are found in Queensland (71.7), Northern Territory (70.5), Western Australia (67.0), and South Australia (63.3). The lowest rate of 21.2 infected persons per 100,000 population was in Tasmania until the end of 2005 (AVERT, 2006).

1.11 – SITUATION IN INDIA

The presence of HIV infection was first detected in India among sex workers in Chennai in 1986. The first case of AIDS was detected in Mumbai in 1986. Since then, HIV infection has been reported from different parts of the country. Prevalence estimates are solely based on sentinel surveillance data. Since HIV/AIDS is not a notifiable disease in India, HIV testing information from the private sector is not compulsorily reported to the national information system (Fredriksson-Bass & Kanabus, 2006).

India has the second highest number of HIV-infected persons (5.21 million), after South Africa (5.5 million). But India's prevalence is still considered "low" (i.e. less than 1 per cent) because India's population is 1.1 billion, as compared to South Africa's 44 million (Stanecki, 2006; NACO, 2006). The overall estimated prevalence in previous years was 0.93 per cent (2003) and 0.92 per cent (2004), which is comparable to 0.91 per cent in 2005 (NACO, 2006). Considering India's large population, a mere 0.1 per cent rise in prevalence would increase the number of people living with the virus by over half a million (Fredriksson-Bass & Kanabus, 2006).

India has multiple HIV epidemics that are as diverse as the country's social diversity. Transmission is mainly through unprotected sex in the southern states and injecting drug use in the northeastern states. HIV prevalence of over 1 per cent has been reported among pregnant women in the industrialised western and southern states (Andhra Pradesh, Karnataka, Maharashtra, and Tamil Nadu) and in the northeastern states of Manipur, Mizoram, and Nagaland. HIV prevalence is still very low among pregnant women in the poor and densely populated northern states of Uttar Pradesh and Bihar. The small numbers of studies that have investigated the complex dimension of sexuality in India have reported HIV infection among MSM (UNAIDS/WHO, 2005). An estimated 5.21 million people were living with HIV in 2005 (NACO, 2006).

In India, *devadasis* (meaning slaves of God) are a group of women who have been historically dedicated to the service of Gods but this tradition has evolved into sanctioned prostitution. Many women from northern Karnataka's "*devadasi* belt" are sent to major cities for commercial sex work. Sex work is legal in some states of India but soliciting and brothel-keeping are penalised. Mumbai has India's largest brothel-based sex industry where sex workers are controlled by "madams", pimps, and moneylenders making HIV prevention activities more difficult (Fredriksson-Bass & Kanabus, 2006). Targeted intervention programmes in Kolkata's Sonagachi red-light area have been successful in increasing condom use among sex workers and reducing HIV prevalence from over 11 per cent in 2001 to less than 4 per cent in 2004. However, HIV prevalence among female sex workers in Mumbai has not declined below 52 per cent since 2000 (UNAIDS/WHO, 2005).

Drugs are often used in public places, though India does not have a widespread culture of professional injectors (or "street doctors") as in some other Asian countries (Fredriksson-Bass & Kanabus, 2006). In the northeastern states of Manipur, Nagaland, and Mizoram, all of which lie adjacent to the drug trafficking "Golden Triangle", HIV transmission is concentrated mainly among drug injectors and their partners, some of who are also involved in commercial sex. In Manipur, a drug injection-driven epidemic has been prevalent for at least a decade. There has been a sharp rise in HIV infections among IDUs in the southern state of Tamil Nadu (UNAIDS/WHO, 2005). There is no national policy for harm reduction but some states, such as Manipur, have adopted their own harm-reduction policies (Fredriksson-Bass & Kanabus, 2006).

India has one of the largest road networks in the world and an estimated 2–5 - million long-distance truck drivers and helpers who remain away from their families for extended periods of time and are likely to come in contact with "high-risk" sexual networks at roadside hotels (or "dhabhas"). Thus, truck drivers are crucial in spreading STIs and HIV throughout the country (Fredriksson-Bass & Kanabus, 2006).

1.12 – ECONOMIC COST OF THE EPIDEMIC

Developing countries are economically affected the most, because surveillance, care, and support programmes will consume almost half of the health budgets of these countries. The indirect costs surpass the direct costs because the epidemic selectively affects the age groups that are involved in national economy and social activities. Thus, the disease leads to loss of income to the nation, as well as to individuals and families. International studies have shown that the indirect costs are 50–60 times more than the direct costs. As compared to health-related arguments, economy-related arguments have been more persuasive in triggering government response to the HIV epidemic. More than 20 years of national development has been lost to the HIV epidemic in Africa. In 1995, it was estimated that the annual cost of the HIV epidemic to Thailand was US$ 1.6 billion and that the cost would rise in the coming years (Sahgal, 1995).

India's National Council of Applied Economic Research (NCAER) conducted a study in collaboration with UNAIDS and National AIDS Control Organization (NACO). According to the study, an increasing number of households with HIV-infected individuals would have to sell their land and borrow heavily to finance expenses on medical treatment. The NCAER report warned that the HIV/AIDS epidemic could increase health-related expenditure by both households and the State that would erode savings, crowd out investment and adversely affect economic growth. The results of the study, released in July 2006, revealed that India's economic growth could decline by 0.86 percentage points per year and per capita GDP by 0.55 percentage points. India's GDP has been growing at an estimated 8.4 per cent for the year ending March 2006 and it expects to sustain 8 per cent plus growth in the coming years and take it up to 10 per cent. The NCAER report forecast that due to the HIV/AIDS epidemic, the GDP per capita, currently at Rs. 21,000 (US$450) would fall by Rs. 7,610.61 in the next 10 years and that the growth of labour supply would slow down (Zaheer, 2006).

1.13 – CHALLENGES

Though there is ample evidence that HIV epidemic does yield to resolute and intensive interventions, the responses still do not match the extent or speed of the steadily worsening epidemic. Sustained efforts have brought down the incidence of newly infected persons in many countries. Preventive programmes initiated a few years ago are finally showing hopeful results in countries like Uganda, Kenya, and urban Haiti. Since 2003, access to ARV treatment has improved markedly. The coverage of treatment now exceeds 80 per cent in countries like Argentina, Brazil, Chile, and Cuba. However, the situation is markedly different in the poorest countries of Latin America and the Caribbean, in Eastern Europe, most of Asia, and virtually all of sub-Saharan Africa. It is optimistically estimated that one in ten Africans and one in seven Asians in need of ARV treatment were receiving it in mid-2005 (UNAIDS/WHO, 2005).

However, due to scale-up of treatment since the end of 2003, over one million HIV-infected persons in low- and middle-income countries are on ARV treatment and the estimated number of AIDS deaths averted in 2005 is between 250,000 and 350,000. The full effects of the treatment scale-up during 2005 will only be seen in 2006 and subsequent years. Though some of the treatment gaps will narrow further, this will not be at the pace required to effectively contain the epidemic (UNAIDS/WHO, 2005).

To bring the epidemic under control, the underlying factors (social and gender inequalities, social injustices, discrimination, stigma, and human rights violations) must be resolutely confronted and overcome. It is also necessary to tackle the new injustices caused by the epidemic, such as orphaning of children and depletion of human and institutional capacities. Achieving universal access to HIV prevention, treatment and care in all countries will require coordination of different approaches. Countries will need to focus on programme implementation including strengthening of human and institutional resources and initiate strategies that favour maximum possible levels of integration of services (UNAIDS/WHO, 2005).

REFERENCES

AVERT, 2006, Australia Statistics Summary. www.avert.org/ausstatsg.htm. Last updated 31 August.

Chin J., 1995, Scenarios for AIDS epidemic in Asia. Asia-Pacific Population Research Reports: 2.

Dwyer J.M., *et al.*, 1997, Challenge and response: HIV in Asia and the Pacific. In: Managing HIV (G. J. Stewart, ed.), Australasian Medical Publishing, North Sydney, Australia, pp 183–187.

Fredriksson-Bass J. and Kanabus A., 2006, HIV/AIDS in India. www.avert.org. Last updated 19 July.

Gottlieb M.S., 2001, AIDS – Past and future. N Engl J Med 344: 1788–1790.

Hahn B.H., *et al.*, 2000, AIDS as a zoonosis: scientific and public health implications. Science 287: 607–614.

Kirby M., 2006, HIV and AIDS in Eastern Europe and Central Asia. wwwavert.org. Last updated 10 July.

Korber B., *et al.*, 2000, Timing the ancestor of the HIV-1 pandemic strains. Science 288: 1789–1796.

Krause R.M., 1992, The origin of plagues – old and new. Science 257: 1073–1078.

Madhok R., *et al.*, 1985, HTLV-III antibody in sequential plasma samples from haemophiliacs 1974–1984. Lancet I: 524–525.

Nahmias A.J., *et al.*, 1986, Evidence of human infection with an HTLV-III/LAV-like virus in Central Africa, 1959. Lancet I: 1279–1280.

National AIDS Control Organization (NACO), 2006, HIV/AIDS epidemiological surveillance and estimation report for the year 2005. New Delhi: NACO, Government of India, pp 1–11.

National AIDS Control Organization (NACO). Training manual for doctors. National AIDS Control Organization, New Delhi.

National Institute of Allergy and Infectious Diseases (NIAID), 2005, HIV infection and AIDS: an overview. NIAID Fact Sheet. Bethesda: National Institutes of Health. www.niaid.nih.gov/.

Noble R., 2006, Caribbean statistics summary. www.avert.org/caribbean.htm. Last updated 31 August.

Noble R., 2006, Latin America statistics summary. www.avert.org/southamerica.htm. Last updated 31 August.

Sahgal K., 1995, Country report on HIV/AIDS in India. In: UNDP report: study on the economic implications of the HIV/AIDS epidemic in selected DMCs. United Nations Development Programme.

Stanecki K., 2006, Reuters Kaiser Daily HIV/AIDS Report. www.kaisernetwork.org. 5 July.

World Fact Book, 2003. www.sportsforum.ws/sd/factbook/geos

UNAIDS, 2002, AIDS epidemic update. Geneva: UNAIDS, p 10.

UNAIDS, 2004, Report on the global AIDS epidemic 2004. pp 39–54.

UNAIDS/WHO, 2005, Aids epidemic update. Geneva: UNAIDS/WHO, pp 1–5 (Global), 17–30 (sub-Saharan Africa), 31–44 (Asia), 45–53 (Eastern, Central Europe and Central Asia, 65–69 (Americas).

UNAIDS/WHO, 2006, 2006 Report on the Global HIV/AIDS epidemic. Geneva: UNAIDS/WHO.

United Nations Development Programme (UNDP), 2006, Asia-Pacific at a glance. www.youand aids.org

WHO, 2006, Progress on global access to anti-retroviral therapy – a report on "3 by 5" and beyond. Geneva: WHO; 28 March.

www.boston.com, 2006, CDC's revised recommendations for HIV testing of adults, adolescents, and pregnant women in health care settings. www.boston.com/yourlife/health/. 22 September.

Zaheer K., 2006, HIV may hit Indian economy. Panjim (Goa): OHeraldo, July 24.

Zhu T., *et al.*, 1998, An African HIV-1 sequence from 1959 and implications for the origin of the epidemic. Nature 391: 594–597.

CHAPTER 2

FUNDAMENTALS OF IMMUNITY

Abstract

There are three tiers of host defences against invading organisms – local defence mechanisms, acquired immunity (active and passive), and innate (natural) immunity. Systemic defence mechanisms are activated when organisms breach the local host defences. Specific immune mechanisms enable the body to recognise, destroy, and eliminate antigens that are recognised as "non-self". Thymus-derived lymphocytes (or "T-cells") and Bursa- or bone marrow-derived lymphocytes (or "B-cells") are primarily responsible for active immunity. Natural killer (NK) cells and killer T-cells destroy virus-infected cells or tumour cells by cytotoxic action. Memory cells developed during the primary immune response accelerate and enhance the secondary immune response, even if the second exposure occurs many years after the primary exposure. The *cluster of differentiation* (CD) system classifies antigens that are present on the surface of human white blood cells. Each CD is a separate molecule with a distinct molecular composition, specificity, and cellular distribution within the cells of the immune system. T4 cells (T-helper cells) "help" the action of other lymphocytes and are required for antibody production by B-cells and cytotoxic action of T8 cells (T-suppressor cells), which suppress immunoglobulin synthesis and act as a brake on immune response.

Key Words

Active immunity, Allergy, Antibody, Antibody titre, Antigen, B-lymphocytes, CD system, Cell-mediated immunity, Complement system, Helper cells, Herd immunity, Herd structure, Humoral Immunity, Hypersensitivity, Immunodeficiency, Immunoglobulin, Innate immunity, Natural immunity, Natural killer cells, Null cells, Suppressor cells, T-lymphocytes.

2.1 – INTRODUCTION

Millions of microorganisms are present in the environment. They also live on the surface of the skin and inside the body. There is a system of overlapping host defences, which provide resistance to infection by various microorganisms (Fudenberg *et al.*, 1976). An invading infectious organism stimulates multiple host defence mechanisms such as local and systemic immunity, non-specific and specific immunity, and cellular and humoral immunity. Immunity is that resistance associated with the presence of antibodies or cells that have a direct action on a pathogen or its toxin (Benenson, 1990). It includes all the physiological mechanisms that endow the host with the capacity to neutralise, eliminate, or metabolise those substances (that are recognised as "foreign" or "non-self"), with or without injury to the host's tissues. When the immune response is beneficial, it is called "immunity" and when harmful, it is termed "allergy" or "hypersensitivity".

The presence of immunity (Latin: *immunitas* = freedom from) against a pathogen confers *freedom from infection* by that pathogen (Roitt, 1997). A person is said to be immune when he or she possesses specific protective antibodies or cellular immunity as a result of previous infection or immunisation. The immune system of a person may be conditioned by previous sensitisation to respond adequately to prevent clinical or subclinical infection (Benenson, 1990). An *antigen* is a substance (usually protein, polysaccharide, or glycolipid) that is capable of eliciting a specific immune response. Antigens may be introduced in the host's body by infection or immunisation (Last, 1983). *Antibodies* are protein molecules that are produced by the host's body in response to an extraneous or "foreign" substance such as an invading microorganism. Chemicals like benzene sulphonate may induce formation of antibodies since antibody formation is specific for a particular electron cloud structure. Antibodies have the capacity to bind specifically to a foreign substance (called "antigen") that has stimulated its production. The concentration of antibodies in the serum (called "titre") can be measured in individuals (Last, 1983). The WHO (1993) recommends that the word "gamma globulin" should not be used as a synonym for "immunoglobulin". All antibodies are immunoglobulin (Ig) molecules, but not vice versa (Fudenberg *et al.*, 1976). Some Ig molecules are abnormal proteins found in some diseases like myeloma.

2.2 – HOST DEFENCE MECHANISMS

There is a stratified system of host defence against invading microorganisms – *local defence mechanisms, natural* or *innate immunity*, and *acquired immunity*. Systemic defence mechanisms are activated when the invading microorganisms breach the local defences. Specific immune mechanisms enable the body to recognise, destroy, and eliminate antigens that are recognised as "non-self".

Local Defence Mechanisms: The function of the bacterial flora is to suppress the growth of many pathogenic organisms. This suppression occurs by competition for essential nutrients or production of inhibitory substances such as colicins and acids. Other local defence mechanisms include the intact skin and mucosa; inhibition of microorganisms by the low pH generated by lactic acids, and fatty acids in sweat and sebaceous secretions; secretion of mucus, which traps microorganisms and particulate matter; ciliary movements, cough, and sneezing reflexes in the respiratory tract; washing action of tears, saliva, and urine; bactericidal components in body secretions (acid in gastric juice, spermine, and zinc in semen, lysozymes in tears, nasal secretions, and saliva); and immunoglobulin A (Roitt, 1997).

Natural or Innate Immunity: Natural (or innate) immunity is the species- or race-specific inherent resistance to a pathogen. An individual, by virtue of his or her constitutional and genetic make-up, possesses this type of immunity to infections. Prior contact with microorganisms or immunisation is not necessary for this type of immunity. For example, leprosy bacillus infects only humans and armadillo (an anteater) and not other animals. Humans are naturally immune to all plant pathogens and certain animal pathogens like canine distemper virus. Within a species, different races may show difference in susceptibility to infections. For example, some Africans are resistant to falciparum malaria due to presence of sickle cells, an abnormality of red blood cells. Non-specific responses include chemotaxis and release of mediators of inflammation.

Acquired Immunity: Acquired immunity is the resistance acquired by a host as a result of previous exposure to a pathogen, either by natural infection or by immunisation (Last, 1983). Basically, there are two types of acquired immunity – "passive" and "active". Active immunity includes "humoral" and "cellular" immunity.

2.3 – IMMUNOGLOBULINS

Ig molecules are glycoproteins containing two identical heavy polypeptide chains with molecular weight 50 kilo Daltons (kDa) and two light polypeptide chains with molecular weight 25 kDa. Each polypeptide chain contains an amino acid terminal (in variable or heterogeneous region) and a carboxy terminal (in constant region). Digestion of Ig molecule by proteolytic enzymes like papain produces two fragment antigen binding (Fab) and one fragment crystallizable (Fc) fragments. The Fc fragment lacks the ability to bind the antigen. All Ig molecules are measured in *Svedberg units* (also called "S value"). Higher the molecular weight, larger is the S value.

The normal immune system can generate an Ig for every possible antigen that the system may encounter during the life of an individual. This is done by constantly changing the amino acid sequences in the variable region of the Ig molecule. This avoids the production of millions of different Ig molecules. Ig molecules constitute 20–25 per cent of total serum proteins. Based on the structure of their *heavy* chain constant regions, Ig molecules are classified into five major classes – IgG, IgA, IgM, IgE, and IgD. Since these classes are all variants of the Ig molecule, they are termed as "isotypes". Constant regions of *light* chains also exist in isotypic forms, designated as kappa and lambda. Briefly speaking IgG protects the body fluids, IgA the body surfaces, IgM the blood stream, IgE mediates hypersensitivity, and the role of IgD is not known.

2.3.1 – Immunoglobulin G (IgG)

This class of antibodies comprises about 75 per cent of the total serum Ig. The normal serum concentration of IgG is 8–10 mg per mL. Exposure to an antigenic challenge (foreign antigen) for the first time leads to a rise in IgM titre for 2–3 days (after a latent period). Later, the IgG antibodies appear if the antigenic stimulus is strong. The required dose of antigen for eliciting IgG type of response is about 50 times more than that required for eliciting IgM response. The IgG titre reaches a peak in 7–10 days and falls gradually over weeks or months. IgG is distributed almost equally between the intra- and extravascular fluid compartments. This can easily diffuse into the extra-vascular compartment since its molecular weight is only 150 kDa (7S). It is the only immunoglobulin, which is transported across placental barrier. It is also found in breast milk. Antibodies against Gram-positive bacteria, viruses, and toxins belong to this class of Ig. IgG facilitates binding of microorganisms to phagocytic cells (Roitt, 1997).

2.3.2 – Immunoglobulin A (IgA)

IgA is the second most abundant class of Ig and comprises about 15 per cent of total serum Ig. The normal serum level is 0.6 – 4.2 mg per mL. It provides local defence at the level of mucous membranes and has a half-life of about 6–8 days. It is a 7S molecule with a molecular weight of 160 kDa. IgA is secreted by mucosa associated lymphoid tissue (MALT). It is seen selectively in saliva, tears, and seromucous secretions of respiratory, genitourinary, and gastrointestinal tracts. Antibodies against a broad range of bacteria and viruses belong to this class of Ig (Gell *et al.*, 1975). IgA does not cross placenta, but helps in protecting the newborn by its abundant presence in breast milk and colostrum. The role of IgA is to bind to the surface of bacteria or viruses and inhibit the attachment of such coated organisms to the surface of mucosal cells, thus preventing the entry of microorganisms into the tissues of the body.

2.3.3 – Immunoglobulin M (IgM)

IgM constitutes about 10 per cent of the total serum Ig, with a normal serum level of 0.5–2 mg per mL. IgM has a half-life of about 10 days. Its molecular weight is 900,000–1,000,000 Da (19S) and hence it is called the *millionaire molecule*. Because of their size, IgM antibodies are confined to the blood stream and do not cross the placenta. Hence, the presence of IgM in blood of fetus or newborn indicates intrauterine infection. It is the earliest immunoglobulin to be synthesised by the fetus, beginning at about 20 weeks of gestation. IgM forms the first line of defence against bacteraemia. The size and valency of the IgM makes it a very effective agglutinating and cytolytic agent (Roitt, 1997). On exposure to an antigenic challenge for the first time, IgM antibodies are formed after a latent period of 3–10 days. The IgM titre rises for 2–3 days and declines quickly. Since the IgM response is short-lived, its presence usually indicates acute or recent infection.

2.3.4 – Immunoglobulin E (IgE)

IgE, an 8S molecule with a molecular weight of 190 kDa, is normally present in very low concentrations in the serum. Produced mainly in the submucosa of respiratory and intestinal tracts, it has a half-life of 2–3 days. IgE has an affinity for the surface of mast cells and binds to Fc receptors on the membranes of blood basophils and tissue mast cells. Linkage of allergens with Fc receptor-bound IgE molecules results in degranulation of basophils and mast cells, leading to allergic manifestations. IgE levels are elevated in allergic conditions such as asthma, hay fever, and eczema; and when intestinal parasites are present. It does not cross the placental barrier. IgE is responsible for defence against parasitic infestations, symptoms of atopic allergy, and immediate anaphylactic reactions (Roitt, 1997).

2.3.5 – Immunoglobulin D (IgD)

IgD is mostly intravascular and has a serum concentration of 3 mg per mL. Its molecular weight is 180 kDa. It has a half-life of 2–8 days and its main role is not yet known. Since IgD is found abundantly on the surface of B-lymphocytes along with IgM, it may have a role in regulating lymphocyte activation and suppression.

2.4 – PASSIVE IMMUNITY

When antibodies produced in one host (human or animal) are administered to another host for the purpose of protection against disease, it is called "passive immunity". The body of the host does not produce its own antibodies but depends on readymade antibodies. Passive immunity may be induced *naturally* by transplacental transfer of maternal antibodies, and by transfer of maternal antibodies via breast milk and colostrum. It can also be induced *artificially* by the administration of a preparation containing protective antibodies (immunoglobulin, antitoxin, or antisera) from sensitised humans or animals (Benenson, 1990). A "biological shield" comprising maternal antibodies that are transferred across the placenta or through breast milk and colostrum protects babies born of immune mothers for duration of up to 3 months or longer. After birth, these maternal antibodies are gradually lost over a period of 6 months.

The recipient's immune system does not play any role and hence secondary response does not occur. There is no antigenic stimulus since pre-formed antibodies are transferred or administered. Its main advantage is that the protection is conferred immediately and there is no latent period. Passive immunisation is indicated for protection of a non-immune host who is faced with threat of an infection and when there is insufficient time for active immunisation to take effect. The immunity is of a short duration (few days to few months) and lasts till the passively transferred antibodies are metabolised and eliminated.

2.5 – ACTIVE IMMUNITY

Active immunity is the resistance developed in response to a stimulus by an antigen (Last, 1983). This usually lasts for months or years, and can be induced *naturally* by clinical or sub-clinical infection, or *artificially* by administering antigen (attenuated or killed) or toxin (Benenson, 1990). The duration of naturally induced active immunity varies with the type of pathogen and is usually lifelong after diseases such as measles. Thymus-derived lymphocytes (or "T-cells") and Bursa or bone marrow derived lymphocytes (or "B-cells") are primarily responsible for active immunity. T-cells promote differentiation of non-lymphoid cells, regulate production of antibodies by B-cells, and are also responsible for delayed-type hypersensitivity (DTH) and cytotoxic ("cell-killing") action. *Null cells* are non-T non-B leukocytes that do not possess the antigen-binding receptors that distinguish T- and B-cell lineages. They lack immunological memory and specificity.

Natural killer (NK) cells that belong to this group of null cells constitute 5–10 per cent of the peripheral blood lymphocytes. They are large granular lymphocytes, which bear CD16 and CD56 co-receptors. They non-specifically destroy virus-infected cells or tumour cells by cytotoxic action. Their granules contain "tumour necrosis factor-*beta*" and various "granzymes". Granzyme B functions as cytotoxin of the NK cells (Roitt, 1997).

2.6 – HUMORAL IMMUNITY

Humoral immunity is directed primarily against extracellular organisms. The humoral defences include lysozymes, complement system, acute phase proteins (C-reactive protein and mannose binding proteins), interferons (that block the replication of viruses), and antibodies. Microorganisms that penetrate the skin or mucosa are destroyed by lysozymes and by phagocytosis with intracellular digestion. The main phagocytic cells are polymorphonuclear leukocytes (neutrophils) and macrophages. In response to a stimulus by a foreign antigen, *B-cells* are activated to form plasma cells. These plasma cells produce antibodies, which act specifically against the antigens present on the surface of the microorganisms. The clonal progeny of the plasma cells produce antibody molecules with the same antigen binding specificity. Humoral immunity can be determined *in vitro* quantitatively by measuring antibody titres and qualitatively by enzyme-linked immunosorbent assay (ELISA) and fluorescent antibody absorption test.

The Complement System

The complement system is a supporting system for humoral immune response. It is a multi-enzyme cascade, which attracts the phagocytic cells to the invading microorganisms. The complement system comprises 11 proteins (10 per cent of serum globulins). Most of the proteins are heat labile. They are usually present in inactive form and need to be activated. The complement system has four subsections:

• Classical pathway, which is always activated by antigen-antibody complex.
• Properdin system (Latin: *pro* = for; *perdere* = destruction). Also called the "alternative pathway". Antibodies may activate this system.
• Amplification system for augmenting the above mentioned pathways.
• Common effector pathway.

Effects of Complement Reaction

- Amplification of humoral immune response by promoting immune adherence and cytolysis.
- Beneficial cytolysis – Holes are produced on the cell membrane. The diameter of each hole is about 100 Angstrom units (A°). 1 A° = 10^{-8} cm. This results in loss of osmotic integrity of the cell, causing cytolysis. Many Gram-negative bacteria and parasites are killed extracellularly by complement-bound eosinophils (Roitt, 1997).
- Harmful cytolysis – e.g. immunological membrane damage.
- Type II hypersensitivity reactions – e.g. acute glomerulonephritis and rheumatic fever due to streptococcal antigens, drug-induced haemolytic anaemia.
- Type III hypersensitivity reactions – e.g. Arthus reaction, erythema nodosum leprosum, and serum sickness.
- Neutralization of viruses.
- Role in blood coagulation and fibrinolysis.

2.6.1 – Primary Immune Response

When an animal or human is exposed to a foreign antigen for the first time, IgM antibodies are formed after a *latent period* (or lag phase). The latent period is of variable duration, depending on the type of antigen. For example, latent period is 2–3 weeks for diphtheria toxoid and only a few hours for pneumococcal polysaccharide vaccine. During this latent period, the antigen is recognised as "foreign". The IgM titre rises for 2–3 days, till it reaches a peak. This is followed by a steady phase when the rate of IgM synthesis equals the rate of its catabolism. Later, IgM titre declines quickly, reaching pre-exposure levels. If the antigenic stimulus is strong, IgG antibodies appear. The IgG titre reaches a peak in 7–10 days and falls gradually over weeks or months. For eliciting IgG type of response, the required dose of antigen is about 50 times more than that required for eliciting an IgM response.

During the later part of the primary immune response, antigen-containing immune complexes localise in lymphoid follicles (follicular dendritic cells) and initiate the development of germinal centres, where the memory-B cells develop. The *memory cells* (or "primed cells") educate the reticuloendothelial system and are responsible for "immunological memory". T-independent antigens cannot elicit immunological memory or a secondary response. The primary immune response is influenced by type and dose of antigen, route of administration of the antigen, presence of adjuvants, and nutritional status of the host.

2.6.2 – Secondary Immune Response

Following exposure to an antigen in a previously primed individual, B- and T-lymphocytes collaborate to initiate a *secondary response* (also called "booster" or "anamnestic" response). Due to immunological memory, there is an accelerated and enhanced response, even if the second exposure occurs many years after the primary exposure. After a brief period of production of IgM antibodies, there is a larger and more prolonged output of IgG antibodies. Protein antigens are eliminated within days or weeks. Polysaccharide antigens may persist for years. As compared to the primary immune response, the secondary response exhibits a shorter lag phase, an accelerated response due to rapid production of antibodies from already primed B-cells in the memory pool, an enhanced response due to recall of precommitted memory cells, longer duration of response, higher affinity of antibodies for antigens, and collaboration between B-cells and T-cells. A few days after the secondary response, B-memory cells migrate from mucosa-associated lymphoid tissue (MALT) to bone marrow and mature to form *plasma cells*.

2.7 – CELL-MEDIATED IMMUNITY

Cellular immunity is mediated by T-lymphocytes, which do not secrete antibodies but are responsible for recognition of "self" and "non-self". Cell-mediated immunity (CMI) is responsible for foreign body reaction, rejection of "foreign" tissue grafts (or allografts), chemically induced hypersensitivity such as contact dermatitis, immunity against some tumours, and immunity against intracellular organisms (e.g. viruses, *Mycobacteria*, *Salmonella*, *Brucella*, and fungi). CMI can be determined in vivo by skin tests such as Mantoux and Mitsuda tests, or in vitro by lymphocyte transformation test (LTT).

2.7.1 – Major Histocompatibility Complex

Each vertebrate has a major histocompatibility complex (MHC), which is responsible for rejection of transplants. Class I MHC molecules are present on virtually all the cells of the body and they signal cytotoxic T-cells. Class II MHC molecules are associated with B-cells and macrophages. They signal T-helper cells (Roitt, 1997). Macrophages, activated B-lymphocytes and dendritic cells are among the cells that have *antigen-presenting* function. They process the antigen and present it to helper T-cells and cytotoxic T-cells. The T-cell receptor can only recognise antigens being presented by a cell membrane bound protein molecule of the MHC. In a special type of immunisation, immunologically competent lymphocytes are injected. This confers *adoptive immunity*.

2.7.2 – The CD System

CD is an acronym for *cluster of differentiation*. The CD system is a system of classification of antigens that are present on the surface of human white blood cells (leukocytes). The International Leukocyte Antigen Workshops have identified 130 antigens, and have given them CD designations. Each CD is a separate molecule with a distinct molecular composition, specificity, and cellular distribution within the cells of the immune system. CD4 is a co-receptor, which is used by MHC Class II-restricted T-cells in recognising foreign antigens. In humans, the CD4 co-receptor is borne on about two-thirds of the peripheral blood T-lymphocytes, monocytes, macrophages, and most thymocytes. The normal CD4 cell count is between 950 and 1,700 cells per micro litre of blood.

The CD8 co-receptor is used by MHC Class I-restricted T-cells and is borne on most thymocytes, NK cells and about one-third of the peripheral blood T-lymphocytes. *Helper cells* (also called "T4" or "CD4 T-lymphocytes") comprise about two-thirds of the T-lymphocytes in the peripheral blood. They bear CD4 receptors on their surfaces. They are capable of recognising foreign antigens in association with MHC Class II. On activation by foreign antigens, they secrete chemicals called "cytokines". There are about 40 different types of cytokines, which have overlapping effects. Interleukin-2 (IL-2) is a cytokine, which is responsible for proliferation of T-cells. IL-4, also a cytokine, increases production of antibodies by B-cells. T4 (or CD4) cells are required for optimal production of antibodies by B-cells and for cytotoxic action of T8 cells. Thus, T4 cells "help" the action of other lymphocytes.

Suppressor cells (also called "T8", or "CD8 T-lymphocytes") comprise 20–30 per cent of the T-lymphocytes in the peripheral blood. They bear CD8 receptors on their surfaces. These cells suppress immunoglobulin synthesis and act as a brake on immune response. On activation by foreign antigens, they differentiate and proliferate with the help of IL-2, which is secreted by T4 cells. T8 cells also produce an antiviral cytokine, called CD8 cell-produced antiviral factor (CAF), which directly inhibits the replication of HIV in T4 cells, without causing their lysis.

Killer T-cells are cytotoxic cells that bear CD8 co-receptors. They can recognise foreign antigens in association with MHC Class I. These cells proliferate and differentiate in response to IL-2 secreted by T-helper cells. Killer T-cells do not require complement. Killer T-cells are responsible for antibody-dependent cell-mediated cytotoxicity (ADCC), destruction of virus-infected cells or tumour cells by forming pores on their membranes with the help of a cytolysing chemical called "perforin", and destruction of contents of target cells by releasing enzyme-filled vesicles.

All T-cells bear CD3 surface antigens and can be counted using anti-CD3 markers. Since all B-cells bear surface IgM, they can be counted using fluorescent anti-IgM markers. *Null cells* cannot be counted using either anti-CD3 or anti-IgM markers.

2.8 – MUCOSAL IMMUNE SYSTEM

Lymphoid tissue present in the gut may be *encapsulated* (e.g. liver and spleen) or *non-encapsulated* (organized or diffused lymphoid tissue).

Encapsulated Lymphoid Tissue: The inner cortical area of the *lymph node* contains B-cells, while the paracortical area contains T-cells. Lymphoid tissue in *spleen* is present in the white pulp and along the cords of Billroth in the red pulp.

Non-Encapsulated Lymphoid Tissue: Diffuse lymphoid tissue in lamina propria of gastrointestinal and respiratory tracts constitutes MALT. Organised lymphoid tissue is present in tonsils (lingual, pharyngeal, palatine), appendix, and Peyer's patches in the intestine. The Peyer's patches in the gut generate most of the mucosal activated T- and B-lymphocytes, which subsequently migrate to other mucosal sites in the gut and genital mucosa. This intermucosal movement of activated lymphocytes constitutes the "common mucosal immune system".

2.9 – VACCINATION

In vaccination, a relatively harmless antigen is used as a primary stimulus in order to impart immunological memory so that, any subsequent contact with a virulent antigen will lead to a rapid and enhanced immune response. Booster doses are given after specific time intervals to elicit a rapid and increased response on the basis of previously developed immunological memory. Humoral immunity can protect against only one type of antigen, while cellular immunity can protect against multiple antigens of the same microorganism. Hence, for generating humoral immunity against multiple serotypes of the same organism, polyvalent vaccines would be required.

2.10 – HERD IMMUNITY

It is the level of resistance of a group of individuals to an invasion and subsequent spread by an infectious agent, based on the resistance to infection, of a high proportion of individual members of the group (Last, 1983). The *Illustrated Stedman's Medical Dictionary* defines "herd immunity" as protection of a group or community, beyond that afforded by immunisation of individuals. Herd immunity is influenced by *resistance* of a large proportion of individuals in a group or community, which may be due to prevailing clinical or subclinical infection in the host species or the level of immunisation of the herd, and *herd structure*, which includes: (a) host population belonging to the herd species, (b) presence and distribution of animal hosts and insect vectors, and (c) environmental and social factors that favour or inhibit the transmission of infection from host to host. The herd structure varies constantly because of population migration, births, and deaths.

Public Health Aspects

Herd immunity may be determined by serological surveys. If the herd immunity is maintained at high levels by an immunisation programme, the number of vulnerable individuals is reduced to a small proportion of the total population (Last, 1983; Benenson, 1990). It is not necessary to achieve 100 per cent herd immunity in a population for containing an epidemic or to control a disease. However, there being no herd immunity against tetanus, 100 per cent immunisation coverage is essential for its control (Last, 1983; Benenson, 1990). In diseases such as poliomyelitis and diphtheria, high levels of herd immunity may lead to elimination of the diseases, in course of time. The eradication of smallpox was not due to high levels of herd immunity (achieved by immunisation programme) but due to elimination of the source of infection (by surveillance and containment measures).

2.11 – HYPERSENSITIVITY

Hypersensitivity leads to immuno-inflammation or immunologically mediated tissue damage. Coombs and Gell (1963) described the types of hypersensitivity reactions.

Type I (Immediate Hypersensitivity): In allergic individuals, IgE is produced in response to antigens like pollen, food, clothing, dust, and drugs. Sensitised mast cells and basophils release vasoactive amines. *Systemic anaphylaxis* occurs when a majority of the body's mast cells are sensitised. Localised immuno-inflammation can occur in the skin (urticaria), nasal mucous membrane (allergic rhinitis), or bronchial mucous membrane (extrinsic asthma).

Type II (Cytotoxicity): Types of type II hypersensitivity reactions are: (a) complement-mediated hypersensitivity, e.g. drug-induced haemolytic anaemia, immune-mediated thrombocytopenic purpura, acute glomerulonephritis, rheumatic heart disease, and Goodpasteur syndrome; (b) stimulatory hypersensitivity (or Type-V hypersensitivity) in which, the antibody binds to receptors on the cell membrane causing stimulation (*not* destruction). Hyperthyroidism due to anti-thyroid antibodies is an example; (c) ADCC, also called Type-VI hypersensitivity, which involves killer T-cell mediated cell damage. Hashimoto's disease is an example.

Type III (Immune Complex Diseases): Normally, macrophages or monocytes remove antigen–antibody complexes. In type III hypersensitivity, antigen–antibody complexes are slowly cleared from circulation or tissues. Immuno-inflammation is initiated in joints, kidney, lung, and choroid plexus. Examples of immune complex diseases are Arthus reaction, erythema nodosum leprosum, and serum sickness.

Type IV (Delayed-Type Hypersensitivity or DTH): Immuno-inflammation is initiated by T-lymphocytes that secrete lymphokines. The types of DTH include *infection-type of DTH* (caseation in tuberculosis, nerve damage in tuberculoid leprosy, and granuloma formation in chronic diseases), and *chronic dermatitis-type DTH* (contact dermatitis).

2.12 – IMMUNE DEFICIENCY

A compromised host lies open, as a form of exposed, all-purpose culture plate. He not only admits many kinds of ambient organisms with ease, but usually does so in relative silence – Paul Russel (cited by Rubin & Young, 1994).

Congenital deficiency refers to deficiency in immunity, which is present since birth and may be inherited or due to developmental defects. Defects in B-cells result in hypogammaglobulinaemias characterised by low levels of gamma globulins (antibodies) in the blood. Congenital deficiency of CMI will result in death within the first 6 months of life. However, a baby born with deficient humoral immunity may even survive up to 6 years without replacement therapy (Gell *et al.*, 1975). Individuals with T-cell defects tend to have more severe and persistent infections as compared to those with defective B-cell function. Thus, cellular immunity is more protective than humoral immunity. The complement system can also be affected by defects in function, leading to increased vulnerability to infection.

Extremes of age (infancy and old age) and pregnancy are physiological states in which the immunity is lowered. Pathological conditions causing deficiency of non-specific and specific immunity include: (a) infections such as measles, kala azar, diphtheria, and whooping cough; (b) abnormal mental states such as emotional shock and stress; (c) physical fatigue; (d) nutritional deficiencies; (e) changes in living environment; (f) malignancies; (g) metabolic disorders; and (h) corticosteroids and anti-cancer drugs (WHO, 1976). Renal disease (causing excessive protein loss) and protein-losing enteropathy can cause deficiency of Ig. Malnutrition, particularly iron deficiency, can reduce cell-mediated immune response. Recognition of opportunistic infection in an immune-deficient host requires awareness and keen observation by a clinician. A "syndrome" is a symptom complex, in which the symptoms and/or signs coexist more frequently than would be expected by chance (Last, 1983).

REFERENCES

Benenson A.S., 1990, Control of communicable diseases in man, 15th edn. Washington DC: American Public Health Association.
Fudenberg H.H., *et al.* 1976, Basic and clinical immunology. Los Altos, California: Lange Medical Publications.
Gell P.G.H., *et al.* (eds.), 1975, Clinical aspects of immunology, 3rd edn. Oxford: Blackwell Science.
Illustrat ed Stedman's Medical Dictionary, 1982, 24th edn. Baltimore: Williams & Wilkins.
Last J.M. (ed)., 1983, A dictionary of epidemiology. Oxford: Oxford University Press.
Roitt I.M., 1997, Roitt's essential immunology, 9th edn. Oxford: Blackwell Science.
Rubin R.H. and Young L.S., 1994, Clinical approach to infection in the compromised host, 3rd edn. London: Plenum Medical Book.
WHO, 1976, Public Health Papers. No. 72. Geneva: WHO.
WHO, 1993, Direction to Contributors to the Bulletin. Bull WHO 61 (1).

CHAPTER 3

HUMAN IMMUNODEFICIENCY VIRUS (HIV)

Abstract

The causative organism, HIV, is classified within Lentivirus (Latin: *lenti* = slow) sub-group of a new family of viruses, Retroviridae. During replication of these viruses, the flow of genetic information is in the opposite direction (from RNA to DNA). Hence, they are called *retroviruses*. Their unique enzyme reverse transcriptase copies the viral RNA into DNA, which is eventually inserted into the genome of the host cells. Hence, the virus persists within the host cells for years and cannot be eradicated from the host cells with any of the currently available ARV drugs. A unique two-layered envelope derived from the host cell membrane surrounds a cone-shaped protein core. HIV is a fragile virus that is easily inactivated by heat, drying, and chemical agents. There are two distinct antigenic variants of this virus: HIV-1 and HIV-2. An epidemic of HIV-2 is occurring in parallel with that of HIV-1 in India, indicating the presence of a considerable epidemic of HIV-2 outside West Africa and has implications for designing HIV vaccines for India.

Key Words

Clades, Co-receptor, Dual infection, HIV-1, HIV-2, HIV subtypes, Lentivirus, Origin of HIV, Replication, Retrovirus

3.1 – RETROVIRUSES AND HIV

Replication in most organisms involves the flow of genetic information from DNA to RNA. However, during replication of viruses belonging to Retroviridae family, the flow of genetic information is in the opposite direction (from RNA to DNA). Hence, these organisms are called retroviruses (Latin: *retro* = backwards). Retroviruses have a unique enzyme, reverse transcriptase (RNA-directed DNA polymerase), which prepares a DNA copy of the RNA genome in the host cell. This DNA copy is eventually inserted into the genome of the host cells. Hence, the virus persists within the host cells for years and cannot be eradicated from the host cells with any of the currently available ARV drugs. Based on biological properties, appearance in cell cultures and later, on sequences of nucleotides, family Retroviridae is newly classified into five groups.

Lentiviruses: Their name refers to their association with slowly progressive diseases. (Latin: *lenti* = slow). They are not associated with neoplasia. They include

39

viruses causing arthritis and encephalitis (goats), equine infectious anaemia (horses), Visna and Maedi (sheep), HIV-1 and HIV-2 infections (humans), and SIV infection in sooty mangabey monkeys.

Spumaviruses: The name spumavirus refers to foamy or vacuolated appearance of infected cells in culture. These viruses have been isolated from non-human primates, cows, and cats. Spumaviruses are not associated with human disease.

TLV-Related Oncoviruses: These are related to T-lymphocyte virus (TLV) and include bovine leukaemia virus (BLV) in cows and HTLV-1 and HTLV-2 in humans.

B- and D- Type Oncoviruses: This is a diverse group that comprises Moloney murine tumour virus (MMTV) and Rous sarcoma virus (RSV) in mice, human endogenous retrovirus, and other viruses affecting cows and non-human primates.

C-Type Oncoviruses: This includes feline leukaemia virus (FeLV) in cats, porcine endogenous retrovirus in pigs, and other viruses affecting mice and gibbon (Simmonds & Peutherer, 2006).

3.1.1 – Basic Structure of Retrovirus

All retroviruses have a diameter of about 100 nm. Those with a central condensed core and eccentric bar structure are known as type C and type D particles, respectively. All retroviruses have an outer lipoprotein envelope, which encloses a core made of other viral proteins, within which lie two molecules of viral RNA and the enzyme reverse transcriptase (RNA-dependent DNA polymerase).

3.1.2 – Human T-lymphocyte Virus (HTLV)

Adult T-cell leukaemia or lymphoma (ATL) was first recognised in Japan. The disease is a proliferative malignancy of T-cells. In 1980, the first human retrovirus was isolated from patients with ATL and was called human T-lymphocyte virus type-1 (HTLV-I). The clinical features of ATL include leukaemia, generalised lymphadenopathy, and hepatosplenomegaly with involvement of skin and bone marrow. The seroconversion stage is symptom-free. Latency of the disease lasts for many years or decades until ATL is manifested. During the latency period, high titres to *gag* proteins are detectable. T-cell proliferation is due to action of *tax* gene (Simmonds & Peutherer, 2006). Other diseases caused by HTLV-I include: (a) variant of ATL, which runs a slow course and is associated with adenopathy and splenomegaly, (b) non-Hodgkin's T-cell lymphoma, and (c) tropical spastic paraparesis, a slowly progressive myelopathy with spastic or ataxic features. HTLV-I and simian virus STLV-I are closely related. It is believed that human infection may have occurred many thousands of years ago in Africa. The presence of this virus in different parts of the world is probably due to migration of ancient peoples. The slave trade may account for foci found in the West Indies and southern United States (Simmonds & Peutherer, 2006).

HTLV-II is not linked to any particular disease, though it was first isolated from a patient with rare T-hairy cell leukaemia. Both HTLV-I and HTLV-II are cell-associated viruses and are transmitted by transfer of infected cells during sexual intercourse, breastfeeding, blood transfusion, and sharing injecting equipment. Assays are available for detecting antibodies to HTLV-I. Proviral sequences can be detected by polymerase chain reaction (PCR). PCR can also differentiate between HTLV-I and HTLV-II (Simmonds & Peutherer, 2006).

3.1.3 – Discovery of HIV

AIDS was first recognised in the United States in 1981 amongst a small cohort of young homosexuals and drug addicts in whom Kaposi's sarcoma and *Pneumocystis* pneumonia (PCP) were associated findings (Gottlieb *et al.*, 1981b; Gottlieb *et al.*, 1981c). In 1983, Luc Montagnier and his colleagues from Pasteur Institute, Paris, isolated a virus from a patient with lymphadenopathy syndrome and called it "lymphadenopathy-associated virus" (LAV) (Barre-Sinoussi *et al.*, 1983). In 1984, Robert Gallo and his colleagues from National Institutes of Health, Maryland (USA), reported that a retrovirus named "human T-lymphotrophic virus-III" (HTLV-III) caused AIDS (Gallo *et al.*, 1984; Sarangadharan *et al.*, 1984). The same virus was named "AIDS associated retro-virus" by Levy and colleagues from San Francisco, USA (Levy *et al.*, 1984). In May 1986, the International Committee on Virus Nomenclature and Taxonomy gave the HIV virus its present name. In 1986, Luc Montagnier and his associates isolated a second type of HIV (called "HIV-2") from West African patients (Clavel *et al.*, 1986).

3.2 – STRUCTURE OF HIV

HIV is an enveloped icosahedral sphere (i.e. a solid with 20 plane faces; Greek: *eikosi* = twenty; *hedra* = seat). The diameter of the virus is 80–120 nm. Two identical, non-complementary strands of RNA (the viral genome) and three enzymes (reverse transcriptase, integrase, and protease) are packaged in a cone-shaped protein core. This core is surrounded by a protein coat called "capsid". The capsid, along with the enclosed nucleic acid, is called "nucleocapsid". The capsid comprises a large number of polypeptides known as "capsomers". The functions of the capsid are to form a protective shell around the nucleic acid core, and to introduce the viral genome into host cell by adsorbing readily to host cell surfaces.

A unique two-layered envelope derived from the host cell membrane surrounds this core. This envelope is acquired by progeny viruses during release through host cell membrane by a process called "budding". The outermost envelope is made of glycoprotein subunits that are exposed as spikes (or knobs). These projecting structures are called "peplomers" (Greek: *peplos* = envelope). The surface glycoprotein 120 (gp120) is bound to the virus by a transmembrane protein: glycoprotein 41 (gp41). The gp120 and transmembrane protein (gp41) form a non-covalent complex. The gp120 binds to host cells that bear CD4 co-receptors such as lymphocytes and monocytes. A protein p18 forms the inner layer of the envelope. Various host proteins (including HLA Class I and Class II DR molecules) are incorporated in this envelope. Both gp120 and gp41 are capable of variations, which produce different strains of the virus. The number of strains increases as the HIV infection progresses.

Antibodies or cytotoxic T-lymphocytes (CTLs) recognise different regions of gp120 and gp41. Neutralising antibodies, which target gp120 and gp41, attempt to stop: (a) viral penetration of host cells, (b) binding to CD4 co-receptor, and (c) intracellular viral replication. The receptors include CD4 glycoprotein (high affinity receptor), CKR5 or CCR5 (new abbreviation: R5) chemokine receptor, and CXCR4 (new abbreviation: X4) co-receptor.

3.2.1 – Genes

Genes coding for structural proteins (the *gag, pol* and *env* genes) are common to all retroviruses, while the non-structural and regulatory genes (the *tat, rev, nef, vpr, vif*, and *vpu* genes) regulate the replication cycle of HIV by producing their characteristic proteins.

3.2.1.1 – Genes coding for structural proteins

The *gag* gene encodes for core and cell proteins and is expressed as a precursor protein (p55). This precursor protein is cleaved into three proteins (p15, p18, and p24). The major core antigen is p24, which can be detected in the serum during the early stage of infection, till the antibodies appear. The decline of anti-p24 antibodies from the circulation indicates progression of illness and is an indication for starting ARV therapy. A nucleocapsid protein (p18) codes for shell antigen (Cunningham *et al.*, 1997).

The *env* gene encodes for envelope glycoprotein gp160, which is cleaved to form gp120 (the major envelope spike antigen, which binds to CD4 co-receptor of host cell) and gp41 (the transmembrane pedicle protein, which mediates fusion of viral and host cell membranes). Antibodies to gp120 are the first to appear after HIV infection and are present in circulation till the terminal stage.

The *pol* gene is expressed as precursor protein p100, which is cleaved to form three proteins – p31, p51, and p64. This gene encodes for the following enzymes: (a) reverse transcriptase, which converts single stranded viral RNA into viral DNA duplex, (b) integrase, which integrates viral DNA duplex into host cell genome as provirus DNA, and (c) protease that cleaves core precursor polypeptide into functional core proteins.

3.2.1.2 – Non-structural and regulatory genes

While structural genes are present in all retroviruses, the non-structural and regulatory genes are specific for HIV (Cunningham *et al.*, 1997). The *tat* (transactivation) gene produces Tat protein, which activates expression of all viral genes and specifies a transactivating factor that increases synthesis of viral proteins. The *rev* (regulatory of virus) gene: the rev protein activates expression of structural and enzymatic genes. The *nef* (negative factor) gene: nef protein regulates pathogenicity and probably the latent state of HIV. The *vif* (viral infectivity factor) gene: vif protein is responsible for viral budding and infectivity of free virions. The *vpr* gene: vpr protein stimulates promoter region of HIV. The *vpu* gene exists only in HIV-1 and codes for small proteins whose function is not known. Vpu protein promotes release of budding progeny viruses from host cell. The *vpx* gene exists only in HIV-2 and performs the same functions as *vpu* gene of HIV-1. Long terminal repeat (LTR) sequences at either end give signals for enhancing and integrating to the promoter region of the virus (Simmonds & Peutherer, 2006).

3.3 – VARIANTS OF HIV

Within the host cells, viral RNA may break and recombine with RNA of other HIV subtypes, producing further inter-subtype variations. It is estimated that a change in one nucleotide base of viral RNA (called a "single point mutation") occurs about 105 times in a day in an infected individual. This mutation results in production of genetically diverse types of viruses in the same individual. The differences in the genetic sequences usually occur in the regions that code for most immunogenic viral proteins, such as gp120 and gp41. Hence, a vaccine developed from one strain of HIV may be ineffective against other strains, especially in different geographical regions. However, cross-protection has been experimentally demonstrated in monkeys vaccinated with HIV-1, HIV-2, and SIVs. Variants of HIV can be subdivided into antigenic variants, and syncytium-inducing (SI) and non-syncytium-inducing (NSI) variants.

3.3.1 – Antigenic Variants

The core and envelope antigens of HIV undergo frequent mutations. There are two distinct antigenic variants of this virus, called HIV-1 and HIV-2, both of which undergo genetic variations and differ in their envelope antigens. Their core polypeptides exhibit cross-reactivity. As the virus continues to mutate, new subtypes may be discovered in future.

HIV is genetically labile due to continuous mutations in its genome due to a high error rate of reverse transcriptase. These viruses lack mechanisms for "proof reading" and repair of errors that occur during replication. These uncorrected errors change the genetic composition during replication and the existing strain is replaced by a new antigenic variant. Thus, HIV has the ability to mutate into drug-resistant or more virulent forms. The genetic sequencing of the full genome of HIV has helped in determining the recombination between subtypes, also called "clades" (McCutchan, 2000; Piyasirisilp *et al.*, 2000; Tovanabutra *et al.*, 2001).

3.3.1.1 – HIV-1

Based on *env* or *gag* sequences, HIV-1 has three phylogenetic groups, which are further divided into subtypes (also called "clades"): (a) Group M (Major), which has 11 subtypes A to K, (b) Group O (Outlier), which has 9 subtypes and shares 55–70 per cent of genes with Group M of HIV-1, and (c) Group N (New), which is a new group (Donley *et al.*, 1993; UNAIDS, 2000). Group O is the most divergent subtype. Some ELISA kits do not detect antibodies to Group O.

HIV-1 is characterised by its genetic diversity and hyper-variability, especially in the envelope genes and to a lesser extent, in the core and regulatory genes. Variation in geographical distribution of HIV (Table 1) and routes of spread of this virus can be studied by analysing its gene sequences. Gene sequence analysis can also be used to verify links between a cohort of infected persons and a suspected source of infection (Ou *et al.*, 1992). HIV-1 subtype C and E (Asian and African strains) are more readily transmitted by the heterosexual route, while subtype B (American strain) is transmitted mainly by percutaneous, blood-borne, and homosexual routes. Globally, subtype C accounts for 47.2 per cent of all HIV infections (Osmanov *et al.*, 2002). The greatest gene sequence variation is seen in Central Africa, where HIV has been present for the longest period. Most of the subtypes of Group M are found in Africa, though subtype B is less prevalent. Rapid tests are available for diagnosing infection with multiple subtypes of HIV-1 (Phillips *et al.*, 2000). In Thailand, HIV Group M subtype A/E (E *env* gene, A *gag/pol* gene) is predominant (more than 75 per cent) and among recent seroconverters, envelope glycoprotein of HIV-1 is the genetically diverse (McCutchan *et al.*, 2000). In 2001, the first CRF01_A/B recombinant of HIV-1 was reported (Tovanabutra *et al.*, 2001). The emergence of new inter-subtype recombinant forms of HIV-1 in Central Myanmar has been reported. Subtype B predominates in Indonesia, followed by subtype E (Porter *et al.*, 1997).

Table 1. Geographic distribution of subtypes of HIV-1 (Grez *et al.*, 1993; Weniger *et al.*, 1994; Tsuchie *et al.*, 1995)

Subtype			Geographical distribution
Group M	A		Central Africa, Thailand
	B		North and South America, Western Europe, Japan, Australia, Thailand
	C		Central and Southern Africa, India, Brazil
	D		Central Africa
	E		Central Africa, Thailand, Japan, South east Asia
	F		South America, Central Africa, Europe
	G		Central Africa, Russia, Taiwan
	H		Gabon, Zaire, Central Africa
	I		Cyprus
Group O			West Africa, France

Subtypes B and C among IDUs in southern Yunnan Province of China yielded mutants CRF07_BC and CRF08_BC, which subsequently became the predominant subtypes as the HIV-1 epidemic spread to the north and east (Yu *et al.*, 2003; Su *et al.*, 2000; Beyner *et al.*, 2000). A recent outbreak of HIV-1 infection in southern China was initiated by two highly homogenous, geographically separated strains, circulating recombinant form AE, and a novel BC recombinant (Piyasirisilp *et al.*, 2000).

In India, HIV-1 subtype C predominates (91 per cent), followed by subtypes B, A, and E (Mandal *et al.*, 2000; Chakrabarti *et al.*, 2000; Gadkari *et al.*, 1998; Halani *et al.*, 2001; Sahni *et al.*, 2002; Tripathy *et al.*, 1996). Inter-subtype recombinants (A/C, B/C) have also been described in India (Lele *et al.*, 1999). HIV-2 is also circulating in India. Due to increasing international travel and population migration, increasing numbers of variants are being detected in India. Knowledge of the local subtypes would be useful in choice of appropriate vaccines for clinical trials.

3.3.1.2 – HIV-2

Though the routes of transmission and clinical manifestations are similar, there are biological differences between HIV-1 and HIV-2 (Gnann, *et al.*, 1987). This is a distinct virus but it exhibits 40–50 per cent homology with *gag* and *pol* genes of HIV-1. It has closer nucleotide similarities with some SIVs (75 per cent nucleotides are similar), as compared to HIV-1. HIV-2 has five subtypes from A to E (Donley *et al.*, 1993; UNAIDS, 2000). As compared to HIV-1, HIV-2 is less easily transmitted (i.e. less infectious), has a longer incubation period between infection and manifestation of illness, patients with HIV-2 infection have a lower viral load, and the progress of the disease is slower (Cunningham *et al.*, 1997).

Highest rates of HIV-2 infection are seen in West African countries. It is also seen in migrants or travellers from endemic areas or their sexual or injecting drug-using partners. Prevalence of HIV-2 infection in India is around 1.7–4.6 per cent (Hira *et al.*, 1996; Dattatray *et al.*, 1996). HIV-1, HIV-2 and their subtypes may be detected only if the test kit contains the prevalent subtypes in that locality. Certain subtypes are more readily transmitted by certain routes. Genetic variation in HIV-1 and HIV-2 has tremendous implications for design of HIV vaccines. A vaccine that offers protection against a serotype may not effectively protect against another serotype or inter-serotype recombinant (Excler, 2005).

3.3.1.3 – Dual Infection with HIV-1 and HIV-2

Prevalence of dual (HIV-1 and HIV-2) infection in India is about 3.3–20.1 per cent (Hira *et al.*, 1996; Dattatray *et al.*, 1996). Dual infection has also been reported among injecting drug users from Manipur in northeastern India (Singh *et al.*, 1995). Studies have revealed that epidemic of HIV-2 is occurring in parallel with that of HIV-1 in India (Rubasamen-Waigmann *et al.*, 1991; Nandi *et al.*, 1994; Singh *et al.*, 1995). These results imply the presence of a considerable epidemic of HIV-2 outside West Africa and have implications for designing HIV vaccines for India (Dore *et al.*, 1997). Dual infection is diagnosed by PCR (Cunningham *et al.*, 1997).

3.3.2 – SI and NSI Variants

Many strains of HIV have the ability to create multinucleated giant cells (called "syncytia"; singular = syncytium). These are called SI strains. The differences between SI and NSI strains are outlined in Table 2.

HIV-infected macrophages and dendritic cells can form multinucleated syncytia with uninfected T-cells, thus transmitting the virus. The gp120 on the surface of HIV binds to CD4 molecule on infected and uninfected host cells. Fusion of infected CD4 cells with CD4 protein of uninfected neighbouring cells (caused by gp120) leads to formation of multinucleated syncytia. Destruction of these multinucleated syncytia results in depletion of large numbers of uninfected CD4 cells from the circulation. This is postulated to be one of the mechanisms of destruction of CD4 cells in the lymph nodes. Other methods include killing of single CD4 cells by virus proteins and "apoptosis" or programmed cell death. Viral load should be used as a *predictor* of disease progression rather than the presence of SI or NSI variants (Cunningham *et al.*, 1997).

3.4 – TARGET CELLS AND ORGANS

HIV is a fragile virus that thrives within the cells of the immune system itself and causes a spectrum of diseases by subverting host defences. HIV can also *directly* impair function of microglial cells of the brain and epithelial cells of the gut that can result in diseases like encephalopathy and diarrhoea, respectively (Cunningham *et al.*, 1997). Three major cell types are central to the immune response – monocytes or macrophages, T-lymphocytes, and dendritic cells. After entering the blood stream, HIV infects several types of cells that bear the CD4

Table 2. Differences between SI and NSI variants (Cunningham *et al.*, 1997)

Syncytium-inducing (SI) strains	Non-syncytium-inducing (NSI) strains
• Also called *T-lymphocyte tropic* (T-tropic) strains and are rarely transmitted sexually • Infect T-lymphocytes and lymphoblastic cell lines and cannot enter macrophages because they use CXCR4 (or X4) co-receptor for entering host cells • Emerge when the CD4 cell count drops below 400 per microlitre and is associated with rapid progression to AIDS	• Also called *macrophage-tropic* (M-tropic) strains and are transmitted sexually • Can infect both macrophages and T-lymphocytes • Use CKR5 (or R5) co-receptor for entering host cells • Predominate throughout the course of the disease

co-receptor – T-helper cells, CD8 + cells, 5–10 per cent of B-lymphocytes, 10–20 per cent of monocytes, tissue macrophages, megakaryocytes, cardiac myocytes, trophoblastic cells, retinal cells, dendritic cells in peripheral blood, follicular dendritic cells of lymph nodes, glomerular cells in kidney, cells in the uterine cervix, rectal mucosal cells, astrocytes, oligodendroglia, microglia of the brain, Hofbauer cells of placenta, and Langerhans' cells of skin.

3.4.1 – Action of HIV

When the virus enters the target cell, viral RNA is transcribed by reverse transcriptase enzyme into provirus DNA. The provirus DNA is integrated into the genome of the infected host cell, causing a latent infection. The prolonged and variable incubation period of HIV infection is due to this latency. Periodically, progeny virions that are released by lysis of infected cells infect other host cells. HIV infection damages T4 cells, resulting in their depletion and reversal of T4 (helper):T8 (suppressor) ratio. HIV infection also suppresses function of infected cells without causing any structural damage, resulting in diminished cell-mediated immune response. Functions of monocytes and macrophages are also affected probably because activating factors are not secreted by T4 lymphocytes. The clinical manifestations are mainly due to immune suppression that renders the patient susceptible to life-threatening opportunistic infections and malignancies. However, dementia and other degenerative neurological lesions are probably due to direct action of HIV on the central nervous system.

In the *asymptomatic stage*, HIV resides in the following *reservoirs* – lymph nodes (where HIV adheres to follicular dendritic cells), microglial cells in the brain, and macrophages in the brain, bone marrow, or gastrointestinal tract. Infected monocytes and lymphocytes spread the infection throughout the body. HIV may enter the brain either via monocytes or through infection of endothelial cells.

3.4.2 – Cell Trophism

The gp120 spike of the viral envelope selectively binds to CD4 antigens. Conversely, antibodies to CD4 protein block the virus binding site. When HIV binds to the host cell, gp41 terminus is exposed and the host cell membrane fuses with the viral membrane. The viral core then enters the cytoplasm of the host cell. Cell fusion and virus entry requires a *co-receptor molecule* viz. CXCR4 (or X4) and CKR5 (or R5) for T-cell tropic and macrophage-tropic HIV strains, respectively.

3.4.3 – Infective Material

In an infected person, HIV may be easily isolated from peripheral blood, plasma, lymphocytes, semen, vaginal or cervical secretions, tissues, and cerebrospinal fluid (CSF). Virus isolation is less successful, if the following specimens are used – saliva, urine, breast milk, tears, and amniotic fluid. The virus is found in almost all body fluids and organs and also in infective doses in semen, vaginal/cervical secretions, and blood. Exchange of blood and body fluids from HIV infected individuals can lead to transmission of the virus to another person. Among the body fluids, the highest concentration of HIV is found in CSF. Semen contains 50 times higher concentration of the virus, as compared to cervical or vaginal secretions and blood.

3.5 – INACTIVATION

HIV in solution is inactivated by heat at 56°C within 10–20 minutes. In lyophilised protein preparations like factor VIII, it is killed at 68°C within 2 hours. Drying reduces the infectivity of the HIV. Hence, dried serum and blood are not highly infectious. Like other enveloped viruses, HIV must remain in *moist state* (or in solution) in order to be infectious. It is also susceptible to inactivation by physical and chemical agents in the moist state (Cunningham *et al.*, 1997).

Chemical disinfectants rapidly inactivate HIV in suspension but are less effective against HIV in dried body fluids (Cunningham *et al.*, 1997). HIV is inactivated by 70 per cent isopropanol (3–5 minutes), 70 per cent ethanol (3–5 minutes), 2 per cent povidone iodine (15 minutes), 4 per cent formalin (30 minutes), 2 per cent glutaraldehyde (30 minutes), household bleach (diluted) containing 1 per cent available chlorine (30 minutes), and 6 per cent hydrogen peroxide (30 minutes). For decontaminating used medical equipment, 2 per cent glutaraldehyde may be used. Lipid membrane envelope of HIV is highly susceptible to surface tension reducing action of detergents. Hence clothes and utensils may be decontaminated by washing with detergents.

3.6 – ROUTES OF TRANSMISSION

The efficacy of *sexual transmission* varies between 0.1 per cent and 1.0 per cent. The risk factors are unprotected sexual intercourse, *receptive* sex, and presence of genital lesions or other STIs. The *efficacy of transmission* is about 90–95 per cent when the following routes of transmission are involved   blood, blood products, needles, and syringes, about 50–90 per cent in case of transplantation of organs and tissues, about 3–10 per cent when the mode of transmission is by sharing needles by injecting drug users. Needle-stick injuries in health care settings have less than 0.5 per cent efficacy (NACO, Training Manual for Doctors). HIV-infected blood is a major, but easily preventable route of transmission. Since June 1989, it is mandatory to screen all blood units meant for transfusion, for HIV. Since 1 January 1990, blood collection from professional blood donors has been banned in India (NACO, Training Manual for Doctors). In spite of these precautions, the continued use of untested blood during emergency and life-threatening conditions remains a major hurdle in the prevention of parenteral transmission of HIV. The risk of transmission through sharing of infected needles and syringes is about 3–10 per cent. Sharing of infected needles and syringes is equivalent to *mini-blood transfusion*. This is a leading mode of transmission of HIV in the northeastern states of India and in many southeast Asian countries.

MTCT (or transplacental or vertical) is more likely if the women are in primary stage of infection or in terminal stage of the disease, and in those who have delivered HIV-positive babies in previous deliveries (Ramachandran, 1990). Among HIV-positive pregnant women, the risk of MTCT is 20–40 per cent. Plasma HIV load is high (more than 10,000 virus particles per mL) in primary infection and late symptomatic disease. It is low (20–40 virus particles per mL) in asymptomatic patients with normal CD4 counts. The half-life of HIV in plasma is only about 6 hours and that of newly infected blood cells, only 2 days 'mission. The risk of mortality in HIV positive babies is 50 per cent in first 2 years of life and 80 per cent in first 5 years. The physiologically lowered immunity in pregnancy, when combined with lack of spacing between pregnancies, repeated pregnancies, poor nutrition, and other infections, leads to worsened immunity and leads to rapid progression of the disease (NACO, Training Manual for Doctors). Many babies born to HIV-infected mothers show no evidence of HIV infection. Clearance of HIV infection in a perinatally infected infant has also been reported. The mechanism for clearance is not yet known (Bryson *et al.*, 1995).

3.7 – REPLICATION

Receptors are present on the surface membrane of all living cells. The receptor is compared to a lock, into which a specific key (called "ligand") will fit. HIV attaches itself to receptors on cell membrane of host cells bearing CD4 co-receptor (e.g. lymphocytes, monocytes). After infection, there is a cascade of events within the host cell. The end results are production of new viral particles, death of the host cell, and destruction of the immune system of the host (Cunningham *et al.*, 1997). Generation time (defined as "time from the release of the virus until it infects another host cell and causes release of a new generation of virions") is 2–6 days for HIV.

In an infected individual, the replication of HIV occurs rapidly and continues throughout the course of the disease, unless checked by ARV drugs. HIV-infected CD4 cells have an average life span of 2.2 days. High rate of destruction of CD4 cells leads to a decline in the CD4 cell count. The steps in HIV replication are outlined below.

Attachment to Host Cell Membrane: The surface membrane glycoprotein of HIV (gp120) locks onto the CD4 co-receptor. After binding, gp120 interacts with a second co-receptor – either CXCR4 (X4) or CKR5 (R5) – embedded in the host cell membrane, exposing the fusion peptide of gp41.

Entry of Viral RNA into Host Cell: Tight attachment to the host cell receptors enables fusion of viral membrane with host cell membrane. This fusion is mediated by transmembrane glycoprotein gp41. Following fusion, the contents of the virus are emptied into the cytoplasm of the host cell.

Reverse Transcription of Viral RNA into Complementary DNA: Reverse transcriptase reads the sequence of viral RNA that has entered the host cell and transcribes the sequence into a complementary DNA sequence. Thus, reverse transcriptase converts single stranded viral RNA into viral DNA duplex. In case there are errors in reading the viral RNA sequence, the viral progeny may have molecular differences in their surface membrane and enzymes, which may lead to production of drug-resistant and immunological escape mutants in each cycle of viral replication.

Integration of Viral DNA Duplex into Host Cell DNA: The viral RNA duplex is integrated into the DNA of the host cell. This process is facilitated by the viral enzyme integrase. The integrated DNA is called a "provirus".

Transcription of Viral DNA to Viral RNA: If the infected host cells are not activated, the viral DNA remains dormant. If activated, the viral DNA is transcripted, resulting in the production of multiple copies of viral RNA.

Cleavage of Core Precursor Polypeptides: The viral enzyme protease cleaves the long polypeptide chains of viral core precursor into its individual components. Cleavage makes the viral enzymes functional. The genes in the RNA of HIV produce viral envelope, core, and enzymes.

Assembly and Budding: The viral RNA, core protein, envelope, and enzymes are assembled to form new budding viruses. These budding viruses are released from the surface of the host cell. During the release, the viruses carry with them a piece of the host cell membrane containing viral surface proteins, which will bind to receptors on other host cells (Cunningham *et al.*, 1997).

3.8 – ORIGIN OF HIV

HIV/AIDS is a classic example of a new and hitherto unknown disease, which has caused a worldwide epidemic. Although the AIDS epidemic is "new" and it does not mimic any previously known disease, studies have raised the question as to whether HIV-1 and HIV-2 are "new" agents. Wild African monkeys remain asymptomatic despite chronic infection with "SIVs" which are similar to HIV-1 and HIV-2. However, transmission of SIVs to captive macaques leads to AIDS-like diseases, suggesting that cross-species transmission may change the virulence of these viruses. It is postulated that HIV-1 and HIV-2 crossed the species barrier from monkey viruses, in Africa, many centuries ago (Krause, 1992).

Africa is a known reservoir of SIV, but out of the numerous strains, only one is closely related to HIV-1 (the strain causing the majority of AIDS cases). The known strains of SIV seem to be more closely related to HIV-2, which is a common cause of AIDS in Africa (Krause, 1992). Computer programmes have been used to compare the genetic sequences of viruses. It is likely that the AIDS viruses have been in existence for at least a century, (or perhaps longer in Central Africa), causing sporadic human infection. Mathematical models indicate that there may have been a passage of 100 years between the emergence of AIDS at its source and the outbreak of the epidemic in the 1980s. It is also probable that the epidemic gained momentum as a result of long distance travel, changing cultural patterns, urbanisation, and human migration (Krause, 1992).

Similar historical circumstances have been noted for the spread of bubonic plague and tuberculosis. Old social habits allowed people to coexist with the plague bacillus in the Burma-Unnan region long before the Mongol invasion. The lifestyle of the inhabitants did not eliminate the plague bacillus, but curtailed its spread. However, social disruption following the Mongol invasion favoured the spread of the bacillus. The socio-economic changes triggered by the industrial revolution promoted the outbreak of tuberculosis. The social forces unleashed by the changing patterns of sexual behaviour, the ongoing epidemic of STDs, and the use of injectable drugs could have promoted the spread of AIDS (Krause, 1992).

On December 10, 1981, the *New England Journal of Medicine* published three consecutive articles on a disease with acquired cellular immune deficiency by Gottlieb *et al.* (1981a), Siegal *et al.* (1981), and Masur *et al.* (1981), Scientists soon realised the cases reported in 1981 were not the first. In 1979, doctors in the East and West coasts of the United States had reported undiagnosed illness that were most likely cases of AIDS (Gottlieb, 2001). Studies of European recipients of Factor VIII imported from the United States showed that the virus was present in some US plasma donors as early as 1977 (Madhok *et al.*, 1985; Gottlieb, 2001). Anti-HIV antibodies were reported in a serum specimen obtained in 1959 at Leopoldville (now Kinshasa) in the Belgian Congo (Nahmias *et al.*, 1986). RNA from this 1959 serum specimen was sequenced and identified as a group M strain of HIV (Zhu *et al.*, 1998). Bette Korber from Los Alamos National Laboratory in New Mexico, USA, traced the disease to a single viral ancestor that could have emerged sometime between 1910 and 1950 due to genomic divergence of group M serotypes (Korber *et al.*, 2000). Comparisons of HIV-1 with SIVcpz indicate that its ancestors crossed over to humans on at least three separate occasions (Hahn *et al.*, 2000). However, it is not known when that cross-over to humans occurred (Gottlieb, 2001).

3.9 – IMMUNITY TO HIV?

During the early years of the HIV epidemic, researchers observed that some HIV-infected individuals took a long time to progress to AIDS, while some individuals did not get infected at all, in spite of repeated exposures to HIV. A mutation in a gene that produces CKR5 (or R5) co-receptor is common among people of Western European descent. The frequency of heterozygotes (those possessing only *one* mutant gene, inherited from one parent) is about 20 per cent, while that for homozygotes (those possessing *one pair* of the mutant gene, inherited from both parents) is about 1 per cent.

Homozygotes do not have CKR5 co-receptors on their cells and are virtually immune or highly resistant to HIV infection in spite of multiple exposures to the virus. Heterozygotes of similar ancestry have increased immunity, which confers limited protection against infection and slows the progression to AIDS. (Liu *et al.*, 1996). This mutant gene is seen mainly in North Europeans and to a lesser extent, in South Europeans. It is rare in other ethnic groups. In 1998, it was estimated that this gene mutation occurred about 700 years ago. Initially, it was thought that plague (a disease that had wiped out 40 per cent of the European population in the 14th century) might have caused this mutation. But, recently smallpox has been implicated for the following reasons:

• During the Middle Ages, smallpox was constantly present while plague came and went in waves. Constant presence of a disease in a population can cause gene mutation.

• Plague affected all ages, while smallpox affected mainly children, who were not immune.

A "protective" gene is more likely to survive through generations if the survivors of an epidemic have a long reproductive span. Hence, it is more likely that the mutation would have occurred in children who survived smallpox. The debate on the origins of the observed immunity to HIV is likely to continue. Studies have been conducted in persons exposed to HIV perinatally (Rowland-Jones *et al.*, 1993), sexually (Rowland-Jones *et al.*, 1995), or occupationally (Clerici *et al.*, 1994), but have not been infected. According to Rowland-Jones *et al.* (1993, 1995), exposed infants and adults, who remain uninfected; have high levels of HIV-specific CTLs.

REFERENCES

Barre-Sinoussi F., Chermann J.C., Rey F., *et al.*, 1983, Isolation of lymphocytotrophic virus from a patient at risk for acquired immuno deficiency syndrome (AIDS). Science 220: 868–871.

Beyner C., Razak M.H., Lisam K., *et al.*, 2000, Overland heroin trafficking routes and HIV-1 spread in south and south-east Asia. AIDS 14: 75–83.

Bryson Y.J., Peng S., Wei L.S., *et al.*, 1995, Clearance of HIV infection in a perinatally infected infant. N Engl J Med 332: 833–838.

Chakrabarti S., Panda S., Chatterjee A., *et al.*, 2000, HIV-1 subtypes in injecting drug users & their non-injecting wives in Manipur, India. Indian J Med Res 111: 189–194.

Clavel F., Guelard F., Brun-Vezinet F., *et al.*, 1986, Isolation of new human retrovirus from West African patients with AIDS in West Africa. Science 233: 343–346.

Clerici M., Levin J.M., Kessler H.A., *et al.*, 1994, HIV-specific T-helper activity in sero-negative health care workers exposed to contaminated blood. JAMA 271: 42–46.

Cunningham A.L., Dwyer D.E., Mills J., and Montagnier L., 1997, Structure and function of HIV. In: Managing HIV (G.J. Stewart, ed.). North Sydney: Australasia Medical Publishing.

Dattatray S., Maniar J.K., Kurimura T., and Gurmally E., 1996, Natural history of dual infection with HIV-1 and HIV-2. XI International Conference on AIDS, Vancouver, Canada, July 7–12. Abstract No. MOB-1209.

Donley C., Heisenring W., Sandberg S., *et al.*, 1993, Comparison of transmission rates of HIV-1 and HIV-2 in a cohort of prostitutes in Senegal. Bull Math Bio 55: 731–741.

Dore G.J., Kaldor J.M., Ungchusak K., Mertens T.E., 1997, Epidemiology of HIV and AIDS in the Asia-Pacific region. In: Managing HIV (G.J. Stewart, ed.). North Sydney: Australasian Medical Publishing, pp 188–192.

Excler J.L., 2005, AIDS vaccine development: perspectives, challenges, and hopes. Indian J Med Res 121(4): 568–581.

Gadkari D.A., Moore D., Sheppard H.W., *et al.*, 1998, Transmission of genetically diverse strains of HIV-1 in Pune, India. Indian J Med Res 107: 1–9.

Gallo R.C., Salahuddin S.Z., Popovic M., *et al.*, 1984, Frequent detection and isolation of cytopathic retrovirus (HTLV-III) from patients with AIDS and at risk for AIDS. Science 224: 497–500.

Gnann J.W., McCormick J.B., Mitchell S., *et al.*, 1987, Synthetic peptide immunoassay distinguishes HIV Type 1 and HIV Type 2 infections. Science 237: 1346–1349.

Gottlieb M.S., 2001, AIDS – past and future. N Engl J Med 344: 1788–1790.

Gottlieb M.S., *et al.*, 1981a, *Pneumocystis carinii* pneumonia and mucosal candidiasis in previously healthy homosexual men: evidence of a new acquired cellular immunodeficiency. N Engl J Med 305: 1425–1431.

Gottlieb M.S., Schanker H.M., Fan P.T., *et al.*, 1981b, *Pneumocystis* pneumonia – Los Angeles. Mobr Mort Wkly Rep 30: 250–252.

Gottlieb M.S., Scroff R., Schanker H.M., *et al.*, 1981c, *Pneumocystis carinii* pneumonia and mucosal candidiasis in previously healthy homosexual men: evidence of a new acquired cellular immunodeficiency. N Engl J Med 305: 1425–1431.

Grez M., Dietrich U., Maniar J.K., *et al.*, 1993, High prevalence of HIV-1 and HIV-2 mixed infections in India. IX International Conference on AIDS, Berlin, June 6–11. Abstract No. PO-A11-0177.

Hahn B.H., *et al.*, 2000, AIDS as a zoonosis: scientific and public health implications. Science 287: 607–614.

Halani N., Wang B., Ge Y.C., *et al.*, 2001, Changing epidemiology of HIV type-1 infection in India: evidence of subtype B introduction in Bombay from a common source. AIDS Res Hum Retroviruses 17: 637–642.

Hira S., Oberoi C., Gharpure H., and Dupont H., 1996, Clinical profile of persons with single and dual HIV-1/2 infections in Bombay. XI International Conference on AIDS, Vancouver, Canada, 7–12 July. Abstract No. MOB-1210.

Holmes E.C., *et al.*, 1993, Molecular investigation of HIV infection of an HIV-infected surgeon. J Infect Dis 167: 1411–1414.

Korber B., *et al.*, 2000, Timing the ancestor of the HIV-1 pandemic strains. Science 288: 1789–1796.

Krause R.M., 1992, The origin of plagues: old and new. Science 257: 1073–1078.

Lele K.S., Bollinger R.C., Paranjape R.S., *et al.*, 1999, Full length human immunodeficiency virus type-1 genomes from subtype coinfected converters in India, with evidence of intersubtype recombination. J Virol 73: 152–160.

Levy J.A., Hoffman A.D., Kramer S.M., *et al.*, 1984, Isolation of lymphocytopathic retroviruses from San Francisco patients with AIDS. Science 225: 840–842.

Liu R., Paxton W.A., Choe S., *et al.*, 1996, Homozygous defect in HIV-1 co-receptor accounts for resistance in some multiple-exposed individuals to HIV-1 infection. Cell 86: 367–377.

Madhok R., *et al.*, 1985, HTLV-III antibody in sequential plasma samples from haemophiliacs 1974–1984. Lancet I: 524–525.

Mandal D., Jana S., Panda S., *et al.*, 2000, Distribution of HIV-1 subtypes in female sex workers of Calcutta, India. Indian J Med Res 112: 165–172.

Masur H., *et al.*, 1981, An outbreak of community-acquired Pneumocystis carinii pneumonia: initial manifestation of cellular immune dysfunction. N Engl J Med 305: 1431–1438.

McCutchan F.E., 2000, Understanding the genetic diversity of HIV-1.AIDS 14 (Suppl 3): 531–534.

McCutchan F.E., Viputtigul K., De Souza M.S., *et al.*, 2000, Diversity of envelope glycoprotein from human immunodeficiency virus type-1 of recent seroconverters in Thailand. AIDS Res Hum Retroviruses 16: 801–805.

Nahmias A.J., *et al.*, 1986, Evidence of human infection with an HTLV-III/LAV-like virus in Central Africa, 1959. Lancet I: 1279–1280.

Nandi J., Kamat H., Bhavalkar V., and Banerjee K., 1994, Detection of human immunodeficiency virus antibody among homosexual men in Bombay. Sexual Trans Dis 21: 235–236.

National AIDS Control Organisation (NACO). Training manual for doctors. New Delhi: Government of India.

Osmanov S., Pattou C., Walker N., *et al.*, 2002, WHO-UNAIDS network for HIV isolation and characterization. Estimated global distribution and regional spread of HIV-1 genetic sub-types in the year 2000. J Aids 29: 184–190.

Ou C-Y., Ciesielski C.A., Myers G., *et al.*, 1992, Molecular epidemiology of HIV transmission in a dental practice. Science 256: 1165–1171.

Phillips S., Granada T.C., Pau C-P., *et al.*, 2000, Diagnosis of human immunodeficiency virus type 1 infection with different sub-types using rapid tests. Clin Diag Lab Immunol 7: 698–699.

Piyasirisilp S., McCutchan F.E., Carr J.K., *et al.*, 2000, A recent outbreak of human immunodeficiency virus type-1 infection in southern China was initiated by two highly homogenous, geographically separated strains, circulating recombinant form AE and a novel BC recombinant. J Virol 74: 11286–11295.

Porter K.R., Mascola J.R., Hupudio H., *et al.*, 1997, Genetic, antigenic, and serologic characterization of human immunodeficiency virus type-1 from Indonesia. J AIDS 14: 1–6.

Ramachandran P., 1990, HIV infection in women. ICMR Bulletin 20 (11&12): 111–119.

Rowland-Jones S.L., Nixon D.F., Ariyoshi K., *et al.*, 1993, HIV-specific cytotoxic T-cell activity in an HIV-exposed but uninfected infant. Lancet 341: 860–861.

Rowland-Jones S.L., Sutton J., Ariyoshi K., *et al.*, 1995, HIV-specific cytotoxic T-cell activity in HIV-exposed but uninfected Gambian women. Nature Med 1: 59–64.

Rubasamen-Waigmann H., Briesen H.V., Maniar J.K., *et al.*, 1991, Spread of HIV-2 in India (letter). Lancet 337: 550–551.

Sahni A.K., Prasad V.V., and Seth P., 2002, Genomic diversity of human immunodeficiency virus type-1 in India. Int J STD AIDS 13: 115–118.

Sarangadharan M.G., Popovic M., Bruch L., *et al.*, 1984, Retroviruses (HTLV-III) in the serum of patients with AIDS. Science 224: 506–508.

Siegal F.P., *et al.*, 1981, Severe acquired immunodeficiency in male homosexuals manifested by chronic perianal ulcerative herpes simplex lesions. N Engl J Med 305: 1439–1444.

Simmonds P. and Peutherer J.F., 2006, Retroviruses. In: Medical microbiology (D. Greenwood, R.C.B. Slack, and J.F. Peutherer, eds.). 16th edn. New Delhi: Elsevier, pp 527–538.

Singh N.B., Panda S., Naik T.N., *et al.*, 1995, HIV-2 strikes injecting drug users (IDUs) in India. Jour Infect 31: 49–50.

Su L., Graf M., Zhang Y., *et al.*, 2000, Characterization of a virtually full-length human immunodeficiency virus type-1 genome of a prevalent intersubtype (C/B') recombinant strain in China. J Virol 74: 11367–11376.

Tovanabutra S., Polonis V., De Souza M., *et al.*, 2001, First CRF01_A/B recombinant of HIV-1 is found in Thailand. AIDS 15: 1063–1065.

Tripathy S., Renjifo B., Wang W.K., *et al.*, 1996, Envelope glycoprotein 120 sequences of primary HIV type-1 isolates from Pune and New Delhi, India. AIDS Res Hum Retroviruses 12: 1199–1202.

Tsuchie H., Saraswathy T.S., Sinniah M., *et al.*, 1995, HIV-1 variants in South and South-east Asia. Jour STD AIDS 117–120.

UNAIDS, 2000, AIDS epidemic update. Geneva: UNAIDS. www.unaids.org

Weniger B.G., Takebe Y., Ou C-Y., and Yamazaki S., 1994, Molecular epidemiology of HIV in Asia. AIDS 8 (Suppl 2): S13-S28.

Yu X.F., Wang X., Mao P., *et al.*, 2003, Characterization of HIV type-1 heterosexual transmission in Yunnan, China. AIDS Res Hum Retroviruses 14: 1051–1055.

Zhu T., *et al.*, 1998, An African HIV-1 sequence from 1959 and implications for the origin of the epidemic. Nature 391: 594–597.

CHAPTER 4

SOCIAL AND ENVIRONMENTAL FACTORS

Abstract

Multiple factors act as social precursors for spread of HIV infection. Lack of family life education and ignorance about sexual matters even among educated individuals are also contributory factors. Wars and civil disturbances affect the physical and social security of people, resulting in increased incidence of rape and forced sexual activities. There is a shortage of essential commodities, condoms, and HIV kits. Community leaders may refuse to accept the presence of HIV-infected persons in their communities since the disease is associated with "deviant" sexual behaviour and injecting drug use. Self-appointed "guardians of public morals" may oppose programmes that promote sex education and safer sex. Some religious groups can be mobilised to provide care and support for HIV-infected persons.

Key Words

Alcoholism, Armed conflicts, Behavioural factors, Biological factors, Drug use, Family life education, High-risk groups, Jakarta Declaration, Marginalised groups, Men having sex with men, Migration, Myths, Natural calamities, Peer pressure, Religious groups, Social precursors, Socio-political environment, Women having sex with women, Women's status.

4.1 – SOCIAL FACTORS

The *social precursors* responsible for rapid spread of HIV in India include: (a) social taboo regarding open discussion of sexual matters and learning about sex and sexuality, (b) family pressure to give birth to a male child (heir) and implicit threat to a marriage when a woman is unable to conceive, (c) social acceptability and high prevalence of domestic violence against women, (d) double standards of morality for men and women, and (e) low social status of women (Solomon & Ganesh, 2002). In many cultures, discussion on sexual lifestyle of the client, an integral component of HIV counselling is considered taboo. Persons with high-risk behaviour are regarded as "deviant" (Solomon & Ganesh, 2002).

4.1.1 – Myths

In some parts of the world, it is believed that having sex with a virgin will cure HIV infection. Consequently, many young girls have been infected with HIV. A study in Texas, USA, found that about 30 per cent of persons of Latin American or African descent believed that HIV was a government conspiracy to kill minorities (Fact Sheet 158, 2006).

4.1.2 – Lack of Family Life Education

Young adolescents need accurate information about sexuality so that they can make informed decisions. Programmes addressing that need have been called "family life education", "sexuality education", "family life skills", "reproductive health education", or "responsible parenthood education" in some countries (WHO, 1985). Some programmes provide only biological information, while others include issues such as gender sensitivity and self-esteem.

Family life education for adolescents can result in behaviour changes such as delay in first intercourse or increase in protected sexual intercourse. Almost all of the 250 programmes evaluated and reviewed by the US National Campaign to prevent Teen Pregnancy did not lead to an increase in frequency of sexual activity (Kirby, 2001). HIV prevention programmes were more likely to show a decrease in number of sexual partners and an increase in use of condoms, while sex education programmes had more impact on use of contraceptives by sexually active youth. A review of 47 programmes from developed and developing countries found that sex education programmes had greater impact on behaviour if the courses were imparted before youth became sexually active than after (Grunseit et al., 1997). This finding underscores the importance of starting sexuality education at an early age. A broad-based curriculum for family life education should include structure and function of reproductive system, changes during adolescence, sex and sexuality, factors causing marital harmony and disharmony, sexual health problems, and STIs.

The lack of family life education and ignorance about sexual matters, even among educated individuals, is another factor that contributes to the spread of HIV infection. Family life education helps in initiating intervention measures.

4.1.3 – Vested Interests

In ancient India, sexually explicit sculptures adorned temples. Rituals covered marriage, sexual intercourse (nuptial nights), pregnancy, and childbirth. But, in present-day Indian society, talking about sex is taboo (Solomon & Ganesh, 2002). Programmes to promote correct scientific information regarding sex, sexuality, and safer sexual behaviour may be opposed by self-appointed "guardians of public morals" and other vested interests having illusory fears about increase in promiscuity (Vas & de Souza, 1991). So far, only half-hearted attempts have been made in India to introduce education on sexual and reproductive issues in schools (Solomon & Ganesh, 2002).

4.1.4 – High-risk Groups

These groups include commercial sex workers, promiscuous individuals, persons with STIs, recipients of multiple blood transfusions, single migrant males, individuals in certain occupations (transport workers, sailors), asylum inmates, refugees, prisoners, and marginalised groups in society (NACO, Training Manual for Doctors; UNAIDS, 2000). *Marginalised groups* in society include illegal immigrants, refugees, MSM, commercial sex workers, the *hijra* (a diverse group of castrated males or eunuchs, transvestites, and transsexuals in India and Pakistan), IDUs, and criminals. Their behaviour may be perceived to be socially "deviant" by the rest of the society. They may be unable, or reluctant to utilise the available health services since their presence in the country may be illegal, or because their behaviour is against prevalent social norms, or against the law of the land (NACO, Training Manual for Doctors).

4.1.5 – Womens' Status

The success of interventions depends on educational status of women and level of women's empowerment (i.e. role of women in decision-making at family and community levels, and extent of their economic independence). In societies where women have a low status, women with ulcerative STIs would be either reluctant to seek treatment, or unable to seek treatment because their health-seeking behaviour depends on the cooperation of their menfolk. Due to extreme family pressure to bear children, women often choose to conceive in spite of likelihood of HIV infection rather than being childless and HIV-seronegative (Solomon & Ganesh, 2002). Women are often blamed for infecting their husbands or for not controlling their partners' urges to have sex with other women (Fredriksson-Bass & Kanabus, 2006). In male-dominated societies, there is a thin line between sexual violence and sexual subjugation (De Bruyn, 1992) and women cannot question their husbands about their extramarital encounters, negotiate condom use, or refuse to have sexual intercourse. Female sex workers cannot negotiate safer sexual practices, such as condom use with their customers due to multiple problems such as peer pressure, poor education, and poverty.

4.2 – BIOLOGICAL FACTORS

- AGE – Individuals aged between 20 and 49 are the most affected by HIV infection. Men over the age of 50, who tend to be promiscuous, may also be vulnerable.
- GENDER – Due to anatomical differences in genitalia, many women with other STIs usually do not experience painful or visible clinical symptoms and therefore, may not be aware of the infection.
- HIGHER RISK IN "RECEPTIVE" PARTNER – Larger surface area (vaginal mucosa in heterosexual and anal mucosa in homosexual intercourse) of the "receptive" partner is exposed to HIV. The concentration of HIV in semen is high. The risk is higher during menstruation.
- ADOLESCENT FEMALES – The secretion of vaginal mucus in adolescent girls is less as compared to that in adult women. Consequently, the natural barrier to infection is reduced significantly.
- WOMEN WITH SEXUALLY TRANSMITTED INFECTIONS (STIs) – Ulcerative STIs give rise to few symptoms. Hence, cases among women are not detected.

- IMMUNITY: In individuals with a normal immune system, the CD4 cells (T-helper cells) outnumber the CD8 cells (T-suppressor cells). The normal CD4:CD8 ratio in adults varies from 1.2 to 3.5. However, in HIV positive persons, this ratio is reversed since HIV destroys CD4 cells (NACO, Training Manual for Doctors).

4.3 – BEHAVIOURAL FACTORS

4.3.1 – Risky Behaviour

Use of drugs or alcohol by the male partner may result in faulty use of condoms (due to impaired judgement) and failure to use condoms. Single migrant males and travellers are vulnerable to *high-risk* behaviour. Married males who are temporarily away from their wives are also likely to have multiple sexual partners. More educated individuals have better access to information and are more likely to make well-informed decisions regarding their lifestyle and health. In addition, educated people usually have better jobs and better access to money and other resources, which can help to support healthier lives. In general, as the level of education increases, the prevalence of some types of risky behaviour decreases, while that of other kinds of risky behaviour may increase. Better-educated girls tend to have sexual intercourse later, while the tendency is the reverse in case of boys (UNAIDS, 2000). Poverty and the transmission of STIs or HIV go hand-in-hand. Poverty forces women to opt for commercial sex, despite the knowledge of the risk of acquiring STIs or HIV.

The high-risk behaviour that causes transmission of HIV takes place in private. MSM and bisexual men account for majority cases of HIV/AIDS in North America, Europe, and Australia. However, in many countries of Asia and Africa, the epidemic is driven by heterosexual transmission.

4.3.2 – Personal Hygiene

Sharing of towels may result in non-sexual contact transmission of STIs, such as gonorrhoea. While washing after defecation, some individuals tend to wash from anal region towards the genitalia. This causes transfer of bacterial flora such as *Escherichia coli* and *Streptococcus faecalis* from the lower gut to the external genitalia, where they act as pathogens. Due to anatomical differences, women are more vulnerable to this type of infection. Breach in continuity of mucosa following infection facilitates entry of organisms causing STI, including HIV.

4.3.3 – Migration and Displacement of Population

Displacement of population may be due to economic and/or political reasons, armed conflict, or trafficking in women and girls (UNDP, 2006). The rate of migration of population, along with that of urbanisation, is also related to the socio-economic conditions of the population. There is a large migration of males in the reproductive age groups, from some states of the country to other states, in search of employment. Migration and population mobility *per se*, are not risk factors for HIV infection. The risky bahaviours and situations encountered during mobility or migration increase vulnerability and risk of HIV infection (UNAIDS, 2001). There were more than 19.9 million refugees and internally displaced persons worldwide as on 1 January 2005, of which 4.86 million and 6.9 million were in Africa and Asia, respectively (Ramachandran & Gardner, 2005). Afghanistan has the second largest number of refugees in the world, after Palestine. About 3.4 million Afghans have sought refuge in other countries and an additional 200,000 persons are internally displaced. Displaced people have little access to HIV prevention services and are vulnerable to HIV infection due to isolation from their families and widespread poverty (UNDP, 2006).

4.3.4 – Traditionally Institutionalised Sex Work

In India, *devadasis* (meaning slaves of God) are a group of women who have been historically dedicated to the service of gods but in recent times, this has evolved into sanctioned prostitution. Many women from northern Karnataka's *"devadasi* belt" are sent to major cities for commercial sex work (Fredriksson-Bass & Kanabus, 2006). In Nepal, girls are forced into traditionally institutionalised sex work practices such as *Deuki* and *Badi* (UNDP, 2006).

4.3.5 – Sexual Exposure

Higher difference in the ages of the sexual partners increases the risk of HIV transmission. In some parts of sub-Saharan Africa, young women from poor families enter into sexual relationships (and sometimes cohabit) with ageing wealthy men, for varying periods, in exchange for non-monetary benefits like expensive clothes, jewellery, and better lifestyle. Such men are called "sugar daddies". Anal sex is more risky due to the absence of natural lubricant, lack of elasticity, and comparatively thinner mucosa. This makes peno-anal sex act more traumatic. Thus, the risk of acquiring infection is higher in peno-anal sex than in peno-vaginal sex. Similarly, the risk of transmission in peno-vaginal sex is more than that in peno-oral sex (NACO, Training Manual for Doctors). The risk of transmission is the lowest in any type of sexual activity where quality lubricated condoms are used correctly. Thus, using condoms in any type of sexual act (that involves exchange of body fluids) can prevent HIV transmission. Unprotected sex during menstruation, with an HIV infected male increases the risk of acquiring infection by the female.

4.3.6 – Other Sexually Transmitted Infections

The risk of transmission of HIV is enhanced by the presence of other STIs. Both *ulcerative* and *non-ulcerative* STIs in any of the sexual partners are known to enhance risk of transmission by 10 and 5 times, respectively (Wasserheit,

1992; Laga *et al.*, 1994). HIV has been isolated from genital secretions, tissue, and mononuclear cells in patients with STIs. The lesions in the *genital ulcer syndrome* (chancroid, herpes genitalis, syphilis, and granuloma venereum) disrupt the integrity of the mucosa and provide a raw area through which the entry of HIV is facilitated. HIV has been isolated from genital ulcers (Kreiss *et al.*, 1989). Similarly, in *genital discharge syndrome* (gonorrhoea), there is a higher concentration of HIV in the discharge.

4.3.7 – Women having Sex with Women

Surveys of behavioural risk factors have been conducted on women having sex with women (WSW) but these surveys differ in definition of WSW, location for recruitment, and intake criteria. Therefore, the findings of these surveys cannot be generalised to all WSW (CDC, 2006). As of March 2005, there were no confirmed reports of female-to-female transmission of HIV in the United States. It is possible that sexual behaviour was not specifically asked for or revealed, or some women may have declined to reveal information on having sex with other women. Some women may have other behavioural risk factors such as injecting drug use and unprotected vaginal intercourse with MSM or men who inject drugs. Health care providers need to remember that sexual identity does not necessarily predict behaviour and that some women who identify themselves as WSW or "lesbian" may be at risk for HIV infection through unprotected sex with men (CDC, 2006).

Vaginal secretions and menstrual blood are potentially infectious and mucous membrane exposure to these secretions has the potential to cause HIV infection. The potential for transmission is greater during early and terminal stages of HIV infection when the viral load in the blood is high. WSW need to know their own and their partner's HIV sero-status so that uninfected women can change their behaviours and reduce their risk of becoming infected. While for women who are already infected, this knowledge would help them seek early treatment and avoid infecting others. Condoms should be used correctly and consistently during every sexual contact with men or when using sex toys. Sex toys should not be shared. Till date, no barrier method for use during oral sex has been found effective. However, natural rubber latex sheets, dental dams, condoms that have been spread open, or plastic wraps may offer some protection from contact with body fluids during oral sex and thus may reduce the likelihood of HIV transmission (CDC, 2006). Use of alcohol or drugs before or during sexual activity increases the risk of ignoring safer sex guidelines (Fact Sheet 151, 2006).

4.4 – SOCIO-POLITICAL FACTORS

4.4.1 – Political Situation

Community leaders may refuse to accept the presence of HIV infection in their communities in which they live, since the disease is associated with "deviant" sexual behaviour and injecting drug use. Many fears and misconceptions associated with HIV epidemic are not supported by scientific facts and have tended to confuse relevant issues. In some countries, the urgency to contain the HIV epidemic has led to rather hasty political decisions such as compulsory HIV testing of foreign students and immigrants, with scant attention given to ethical considerations. It is unethical for lawmakers to consider abrogation of fundamental rights of HIV-infected individuals on the basis of concern for common good, especially when specific measures for containing other communicable diseases, such as vaccination, isolation, and quarantine are not applicable to control of HIV infection (Vas & de Souza, 1991).

4.4.2 – Urbanisation

Increasing urbanisation leads to disparity in availability of job opportunities and standards of living, thus causing selective migration of rural males to cities. Factors that cause high-risk behaviour among single migrant males include alcoholism and drug addictions, lack of access to counselling and contraceptive services, lack of awareness, lack of traditional constraints in urban areas, long working hours and lack of entertainment, peer pressure, poverty, and relative isolation from their families or single living that provides anonymity. Subsequently, the infected males spread HIV infection among women in rural areas (UNAIDS, 2001).

4.5 – ARMED CONFLICTS AND NATURAL CALAMITIES

There were more than 19.9 million refugees and internally displaced persons worldwide as on 1 January 2005, of which 4.86 million and 6.9 million were in Africa and Asia, respectively. As of mid-2005, about 45 countries (28 in Africa and 12 in Asia) faced crises related to armed conflicts or natural disasters (Ramachandran & Gardner, 2005). Most of the crisis situations have occurred in developing countries, where health care systems can be quickly overwhelmed by the added burden. Consequently, prevention and treatment of STIs including HIV, family planning, and reproductive health services are neglected (Palmer, 1998; UNHCR, 1999).

Irrespective of their status, people who have been displaced by armed conflicts or natural disasters have similar needs – food, shelter, security, and basic health care (Ramachandran & Gardner, 2005). People living in refugee camps are usually better off, as compared to displaced persons who are dispersed within local communities (Creel, 2003; McGinn, 2000). Relief agencies usually concentrate on providing these basic needs and also emphasise on preventing outbreaks of epidemics of infectious diseases in refugee camps. Where refugees are dispersed, their status and needs are unknown and it is difficult for relief agencies to meet their emergency needs (Creel, 2003; McGinn, 2000). The responsibility for protection of people displaced within their own country is *not* defined by any international treaty, as is the case for international refugees (Deng, 2000; UNFPA, 2002).

4.5.1 – Sexual and Gender-based Violence

Gender-based violence refers to acts of violence committed against females because they are female and against males because they are male (Vann, 2002). This includes a wide range of violent acts – sex trafficking, forced prostitution, sexual exploitation, sexual harassment, and harmful traditional practices, such as forced marriages in many parts of Asia and Africa, and female genital mutilation in parts of Africa. Most frequently the victims of sexual and gender-based violence are women and girls, though men and boys may also be victims (WHO, 2000). Sexual violence may occur during escape from place of conflict, in refugee camps, and during repatriation (UNHCR, 1999). The causative factors include breakdown of family and social norms, ethnic tensions, loss of security, overcrowding, predominantly male camp leadership who do not see preventing gender-based violence as a high priority, and psychological trauma (UNHCR, 2003; Ward & Brewer, 2004; WCRWC, 2003).

4.5.2 – Factors Favouring Exposure to Unsafe Sex

- Poverty, powerlessness, food insecurity, and displacement (Spiegel, 2004)
- Disruption of supplies of condoms and family planning services (McGinn *et al.*, 2004)
- Weakening or total collapse of social support networks (Creel, 2003; Krause *et al.*, 2000)
- Increased incidence of rape and other forms of sexual and gender-based violence (Ramachandran & Gardner, 2005)
- Risk of exposure to forced sex in exchange for food, shelter, and protection especially for adolescent girls (UNHCR, 1999; UNFPA, 2001)
- Even in situations where free condoms are available, unsafe sexual behaviour has been found to increase among adolescents (UNHCR, 2004)

While rape and other forms of sexual and gender-based violence take place in all societies at all times, their incidence escalates to a great extent during armed conflicts and natural disasters. The use of rape as a weapon of war has been documented in Algeria, Bangladesh, Bosnia and Herzegovina, Indonesia, Liberia, Rwanda, and Uganda (Ramachandran & Gardner, 2005). The UNHCR has documented some instances where peacemakers and aid workers have been the perpetrators, exchanging food for sex by threatening to withhold food rations (UNHCR, 1995).

In Liberia, the prevalence of HIV infection was estimated at about 8 per cent before the civil war. During the civil war, women and girls were abducted to act as sex slaves for soldiers and there was widespread sexual violence. After the war, screenings for STIs showed that 93 per cent of male combatants and 83 per cent of female combatants had at least one STI. Based on projected estimates from these high rates of STIs, it was estimated that the prevalence of HIV was much higher than before the war (Ramachandran & Gardner, 2005).

4.6 – RELIGION AND HIV/AIDS

In many cultures, religions provide important ethical guidelines for living and for coping with life's events. Religions are thriving in the midst of modernisation because they also provide an anchor in a time of rapid social changes (Tan, 2000). According to religious conservatives, the only way to prevent HIV/AIDS is to return to the demands of religion and faith (Mas'udi, 2000). Some religions have always exerted social control, especially in the area of sexuality. The HIV/AIDS epidemic has posed new challenges to many religions and intensified their indecisive attitudes to sexuality. Ambivalent attitudes continue to exist, often creating hurdles for HIV prevention and care programmes. Religious stigma affects HIV-infected persons and their families, who may be left to fend for themselves. On the other hand, many religions especially in ancient times have respected or even celebrated sexuality, either for reproduction or for eroticism (Tan, 2000). The sculptures in Khajuraho in northern India confirm this fact.

Religious prejudices, in combination with misconceptions about HIV/AIDS may be fatal for HIV-infected persons. The HIV epidemic has also caused ethical dilemmas. Though many orthodox believers contend that condom promotion programmes may encourage sexual transgressions, progressive religious thinkers view the programme as a method of saving lives. Some progressive religious thinkers have also questioned the religious norms that bolster gender inequalities, thus contributing to women's vulnerability to HIV (Tan, 2000).

Some religious groups object to HIV prevention programmes that distribute condoms without educating people on the importance of monogamous sexual relationships. HIV prevention programmes need to deal with people's beliefs and attitudes. For example, if women consider the risk of HIV infection as a part of *karma* (deeds in previous births), then HIV education programmes may not be very effective (Tan, 2000).

4.6.1 – Tapping Religious Groups

Since many religious groups have extraordinary resources and run their own educational institutions, these groups can be tapped for HIV prevention programmes. Though some religious groups may be reluctant to promote condoms, they may be mobilised to provide care and support (Tan, 2000).

In some Asian countries, Christian missionaries provide care for terminally ill AIDS patients (Tan, 2000). In Thailand, Buddhist monks conduct home visits to talk to HIV-infected persons. Besides social, spiritual, and emotional support, monks also provide HIV-infected individuals their basic needs such as food, clothing, and soap (Bayoneta-Leis, 2000). A Buddhist monk's initiative has helped in establishing "Friends for Life", a hospice for people living with HIV/AIDS on the outskirts of Chiang Mai, Thailand (Manning, 1995). The Buddhist AIDS Project (www.buddhistaidsproject.org) is a non-profit project of the Buddhist Peace Fellowship. The project provides information on HIV/AIDS and alternative health care and is also involved in community service projects in Thailand and Cambodia. The First HIV/AIDS Association of South East Asian Nations (ASEAN) Regional Workshop of Islamic religious leaders held in November–December 1998 prepared the "Jakarta Declaration", which explored the rationale for involvement of Muslims in the regional response to HIV/AIDS (Mas'udi, 2000).

REFERENCES

Bayoneta-Leis N.D., 2000, Buddhist monks: responding to HIV/AIDS. AIDS Action. Asia–Pacific Edition. 47: 4–5.

Centers for Disease Control and Prevention (CDC), 2006, Fact sheet: HIV/AIDS among women who have sex with women, pp 1–2. www.cdc.gov

Creel L., 2003, Meeting the reproductive health needs of displaced people (policy brief). Washington DC: Population Reference Bureau, p. 4. www.dec.org/pdf

De Bruyn M., 1992, Women and AIDS in developing countries. Soc Sci Med 34: 249–262.

Deng F.M., 2000, Introductory note to guiding principles on internal displacement. United Nations Office for Coordination of Humanitarian Affairs (OCHA). www.reliefweb.int/ocha

Fredriksson-Bass J. and Kanabus A., 2006, HIV/AIDS in India. www.avert.org. Last updated 19 July.

Grunseit A., et al., 1997, Sexuality education and young people's sexual behaviour: a review of studies. J Adolesc Res 12(4): 421–453.

Kirby D., 2001, Emerging answers: Research findings on programs to reduce teen pregnancy. Washington DC: National Campaign to Prevent Teen Pregnancy.

Krause S.K., Jones R.K., and Purdin S.J., 2000, Programmatic responses to refugees' reproductive health needs. Int Fam Plan Perspect 26(4): 181–187.

Kreiss J.K., Coombs R., Plummer F.A., et al., 1989, Isolation of HIV from genital ulcers in Nairobi prostitutes. J Infect Dis 160: 380–384.

Laga M., Diallo M.O., Buve A., 1994, Inter-relationship of sexually transmitted diseases and HIV – Where are we now? AIDS Suppl 1: S119-S124.

Manning R., 1995, Friends for life. AIDS Action. Asia-Pacific Edition 25: 11.

Mas'udi M.F., 2000, HIV/AIDS: Between two paradigms. AIDS Action. Asia-Pacific Edition 47: 6.

McGinn T., 2000, Reproductive health of war-affected populations: What do we know? Int Fam Plan perspect 26(4): 174–180.

McGinn T., Casey S., Purdin S., and Marsh M., 2004, Reproductive health for conflict-affected people: policies, research and programmes. Humanitarian Practice Network. www.rhrc.org/pdf/

NACO. Training manual for doctors. New Delhi: Government of India.

New Mexico AIDS Education and Training Center, 2006, Fact Sheet 151. Safer sex guidelines. University of New Mexico Health Sciences Center. www.aidsinfonet.org. Revised 18 July.

New Mexico AIDS Education and Training Center, 2006, Fact Sheet 158. AIDS myths and misunderstandings. University of New Mexico Health Sciences Center. www.aidsinfonet.org. Revised 18 April.

Palmer C., 1998, Reproductive health for displaced populations. London: Overseas Development Institute (ODI). www.odihpn.org

Ramachandran D. and Gardner R., 2005, How providers can meet reproductive health needs in crisis situations. Population Reports Series J. No.53. Baltimore: Johns Hopkins Bloomberg School of Public Health, pp 3–19.

Solomon S., and Ganesh A.K., 2002, HIV in India. Topics in HIV medicine. International AIDS Society – USA. 10(3): 19–24.

Spiegel P., 2004, UNHCR, HIV/AIDS, and refugees: lessons learned. Forced Migration Review 19: 21–23.

Tan M.L., 2000, Religion and HIV/AIDS. AIDS Action. Asia-Pacific Edition 47: 1–3.

UNAIDS, 2000, Report on the global HIV/AIDS epidemic, pp 37–51.

UNAIDS, 2001, Population mobility and AIDS. Technical update. Geneva: UNAIDS, p. 5.

UNFPA, 2001, Reproductive health for communities in crisis: UNFPA emergency response. New York: UNFPA, pp 38. www.unfpa.org/

UNFPA, 2002, Impact of conflict on women and girls: A UNFPA strategy for gender mainstreaming in areas of conflict and reconstruction. New York: UNFPA, pp 140. www.unfpa.org/

UNHCR, 1995, Sexual violence against refugees: guidelines for prevention and response. Geneva: UNHCR, pp 102. www.reliefweb.int/

UNHCR, 1999, Reproductive health in refugee situations. An inter-agency field manual. New York: UNHCR. www.unfpa.org/

UNHCR, 2003, Sexual and gender-based violence against refugees, returnees and internally displaced persons: Guidelines for prevention and response. Geneva: UNHCR, pp 158. www.rhrc.org/resources/gbv/gl_sgbv03.html

UNHCR, 2004, Reproductive health services for refugees and internally displaced persons: Report of an inter-agency global evaluation. Geneva: UNHCR, pp 261. www.rhrc.org/resources/lawg/

United Nations Development Programme (UNDP), 2006, Asia-Pacific at a glance. www.youandaids.org

Vann B., 2002, Gender-based violence: Emerging issues in programs serving displaced populations. Arlington, VA: Reproductive Health for Refugees Consortium, pp 144.

Vas C.J. and de Souza E.J., 1991, Ethical concerns in AIDS. Mumbai: FIAMC Biomedical Ethics Centre.

Ward J. and Brewer J., 2004, Gender-based violence in conflict-afflicted settings: Overview of a multi-country research project. Forced Migration Review, 19 January 2004, pp 26–28.

Wasserheit J., 1992, Epidemiological synergy: Inter-relationship between HIV and other STDs. Sex Trans Dis 19: 61–77.

WHO, 1985, Reproductive Health and the Law. WHO Chronicle.

WHO, 2000, Reproductive health during conflict and displacement: a guide for programme managers. Geneva: WHO, pp 175.

Women's Commission for Refugee Women and Children (WCRWC), 2003, Sexual violence in refugee crises: A synopsis of the UNHCR guidelines for prevention and response. New York: WCRWC. www.women'scommission.org/projects/P&P/guidelines/

CHAPTER 5

IMMUNOPATHOLOGY

Abstract

During the phase of acute infection, most patients manifest clinical symptoms of viral infection (called "seroconversion illness"). Many HIV-infected individuals remain asymptomatic in spite of continuing damage to their immune system. When the CD4 count falls markedly, AIDS-defining opportunistic infections usually appear. Patients with advanced HIV infection may develop neurological disorders and malignancies. Autoimmune antibodies may be produced against platelets, lymphocytes, neutrophils, and myelin. Antibody-mediated drug allergies are more frequent. Autoimmune disorders may occur in the early stages since the immune system is relatively effective. The origin of many autoimmune disorders is different, as compared to that of similar conditions in HIV-negative individuals.

Key Words

Apoptosis, Autoimmune disorders, Candidiasis, CD4 lymphocytes, CD8 lymphocytes, Herpes zoster, Hypersensitivity reactions, Kaposi's sarcoma, Malignancies, Molluscum contagiosum, Non-Hodgkin's lymphoma, Oncogenic viruses, Opportunistic infections, *Pneumocystis* pneumonia, Primary infection, Seroconversion illness

5.1 – PRIMARY INFECTION

The gp 120 on the outermost envelope of HIV binds to cells bearing CD4 co-receptor (T-helper cells, monocytes, and macrophages). Soon after infection, HIV reaches the regional lymph nodes and stimulates both cellular and humoral immune responses. More and more cells bearing CD4 co-receptor get infected due to the flow of lymphocytes to the lymph nodes. Within a few days, there is leukopenia (decrease in white blood cell count) with an acute reduction in the level of circulating CD4 T-lymphocytes. The blood levels of virus and viral proteins are high (Levi, 1993). After 2–4 weeks, the immune responses to HIV results in a phenomenal increase in lymphocyte count mainly due to increase in CD8 T-cells (Cooper *et al.*, 1988). The CD4 cell counts also reach pre-infection levels. Specific anti-HIV antibodies are present in the blood, about 2–3 weeks after infection, but this may be delayed in some patients. During the phase of acute infection, most patients manifest clinical symptoms of viral infection.

After the acute phase of primary infection, the levels of both intracellular and circulating HIV decline, probably due to specific lysis of HIV-infected cells by CD8 cytotoxic T-cells (Koup *et al.*, 1994). In vitro studies have shown that activated CD8 cells from HIV-infected patients produce soluble cytokines that inhibit viral replication in CD4 T-cells, without causing their lysis (Walker *et al.*, 1986). An increase in CD8 cell count is observed before seroconversion, in response to acute rise in HIV load and this probably plays a role in controlling virus production (Mackewicz *et al.*, 1994). Later the raised levels of CD8 cells decline but remain much higher than normal, throughout the course of the disease.

5.2 – ASYMPTOMATIC PHASE

After the acute phase resolves, many HIV-infected individuals remain asymptomatic even when the damage to their immune system continues. The damage involves the decline in *number* and *function* of CD4 T-cells. The *annual decline* in CD4 cell count averages 65 cells per μL. This works out to a *daily decline* of about 2×10^8 CD4 cells. This daily decline is compensated by the daily turnover of CD4 T-cells, which is about 2×10^9 in the asymptomatic phase and indicates considerable activity of both the virus and the immune system (Levi, 1993). There is a step-by-step *impairment of function* in CD4 cells. Initially, T-cells lose the ability to respond to recall the antigen, followed by loss of response to foreign cells and finally, loss of response to non-specific mitogens (Clerici *et al.*, 1989).

Mechanisms for Reduction in Existing Cells: Direct HIV-mediated cytopathic effects, such as rupture of T-cell membrane by replicating HIV, and cell fusion and syncytium formation; derangement of normal signalling mechanisms within T-cells leading to premature programmed cell death (called "apoptosis"); HIV-specific immune responses, such as cytotoxic CD8 T-cells, ADCC and NK cells; Binding of free gp 120 to the CD4 co-receptor of uninfected cells, making them the targets for attack by humoral and cellular immune mechanisms; and other mechanisms – autoimmune mechanisms, anergy, super antigens (super infecting organisms).

Mechanisms For Impaired Replacement: Infection of stem cells that produce precursors of T-cells; damage to thymic epithelial cells, which help maturation of T-cells; and prevention of T-cell proliferation in response to contact with many antigens by action of HIV proteins.

5.3 – ADVANCED HIV INFECTION AND AIDS

The decline in *number* and *function* of CD4 cells results in lowered production of cytokines like IL-2. This reduces the anti-HIV activity of CD8 cells. When the CD4 count falls below 200 cells per μL, monocytes, and dendritic cells (that bear the CD8 co-receptor) also get infected by HIV (Livingstone *et al.*, 1996). The HIV-infected CD8 cells contribute to the increase in viral load. The mechanism of infection of monocytes and dendritic cells is not clear. They probably get infected in the thymus when they bear both CD4 and CD8 co-receptors on their surfaces. The disappearance of enlarged lymph nodes indicates poor prognosis. It implies that the immune system is losing the battle against HIV in the lymph nodes.

5.4 – OPPORTUNISTIC INFECTIONS

Though the cellular immunity is mainly affected, HIV destroys all aspects of the immune system. Organisms (that do not cause infection in immunocompetent individuals) cause disease by taking advantage of the lowered host immunity. The usual opportunistic pathogens are commensals (normal bacterial flora), environmental organisms (that do not affect immunocompetent individuals), and endogenous reactivation of dormant infection acquired at an earlier age. Other organisms that cause opportunistic infections include:

1. Bacteria: Mycobacterium tuberculosis, Mycobacterium avium complex, Salmonella
2. Viruses: Cytomegalovirus, Herpes simplex, Varicella-zoster, Epstein-Barr virus (EBV)
3. Fungi: Candida, Aspergillus, Histoplasma capsulatum, Coccidioides immitis, Cryptococcus neoformans, Pneumocystis
4. Protozoa: Leishmania, Toxoplasma gondii, Microsporidia, Cryptosporidia

AIDS-defining opportunistic infections usually appear when the CD4 cell count declines below 200 cells per μL. Since *Mycobacterium tuberculosis* is more virulent, reactivation of this disease can occur even with a relatively higher CD4 cell count (about 400 cells per μL).

Selection: Some of the opportunistic pathogens like *Listeria* and *Nocardia* that are commonly seen in other immune deficient patients, such as cancer patients on chemotherapy and recipients of transplants, are rare in HIV-infected individuals. Conversely, infections like disseminated candidiasis, commonly seen in cancer patients receiving chemotherapy, are rare in HIV-infected individuals. The mechanism for this type of selection is not known.

5.4.1 – Herpes zoster (Shingles)

Shingles is caused by Varicella-zoster virus, which has an infectious stage (chicken pox) and a dormant stage. The virus lives in the nerve tissue and may get reactivated when the immune system gets weakened, as in the elderly or in persons with HIV disease. About 20 per cent of persons who have had chicken pox may develop herpes zoster. The disease starts on one side of the body with pruritus, numbness, tingling, and severe pain in a belt-like pattern, along the dermatomes. Lesions are commonly seen on the trunk, but rarely, they may be present around the mouth, on the face, neck, scalp, in and around the ear, or at the tip of the nose. Few days later, a rash appears on the skin overlying the affected nerve. A vesicular eruption follows. The fluid in the vesicles is highly infectious to others. The vesicles break open and form crusty scabs. Secondary infection of vesicles may require treatment with antibiotics. In most cases, the lesions disappear in a few weeks. In some cases, severe pain (called "post-herpetic neuralgia") can last for months or years. The antiviral drug acyclovir is given orally five times a day. The drug is given intravenously for severe cases. Newly approved drugs famciclovir and valacyclovir are given orally three times a day. Some of the drugs used to treat depression (nortriptyline) or epilepsy (pregabalin) are used to treat the severe pain of herpes zoster. Anaesthetics and/or steroids are being studied as nerve blockers. In 1999, the US Food and Drug Administration (FDA) approved the dermal patch form of lidocaine, an anaesthetic. Zostavax, a shingles vaccine developed by Merck, has been approved by the US FDA but this vaccine has not been studied in persons with immune deficiency, including HIV-infected persons (Fact Sheet 509, 2006).

5.4.2 – *Pneumocystis* pneumonia (PCP)

Pneumocystis is an opportunistic fungal pathogen that causes pneumonia (pneumocystosis) in the immunocompromised host. Organisms derived from humans and rats have been termed *Pneumocystis jiroveci* and *Pneumocystis carinii*, respectively. New nomenclature is still evolving. The taxonomical classification of *Pneumocystis* as a fungus is based on: (a) analysis of gene sequences for ribosomal RNA, mitochondrial proteins, and major enzymes, (b) presence of beta-1,3-glucan in the cell wall, and (c) efficacy of anti-fungal drugs that inhibit beta-glucan synthesis in animal models. In contrast to most fungi, *Pneumocystis* lacks ergosterol and is not susceptible to anti-fungal drugs that inhibit ergosterol synthesis (Walzer, 2005). PCP is among the common opportunistic infections in HIV-infected individuals with decreasing CD4 cell counts. The earliest manifestations of PCP are fever, dyspnoea, and dry cough. Cotrimoxazole is the drug of choice for preventing and treating *Pneumocystis* infection. Pentamidine is usually inhaled as an aerosol to prevent PCP, while it is used intravenously for treatment. Alternative drugs for treatment include atovaquone, clindamycin-primaquine combination, and trimetrexate-leucovorin combination (Walzer, 2005). In comparison with cotrimoxazole, dapsone seems almost equally effective as a therapeutic agent, while pentamidine aerosol is less effective as a prophylactic agent (Fact Sheet 515, 2006). Pentamidine aerosol may cause change in taste, nausea, vomiting, dyspnoea, dizziness, chest pain or tightness, cough, headache, or weakness. When given intravenously, the drug can cause serious adverse effects such as pancytopenia, hyper- or hypo-glycaemia, hyperkalaemia, cardiac arrhythmias, or damage to pancreas, liver, or kidneys (Fact Sheet 537, 2006). Atovaquone is indicated in patients with mild to moderate PCP, who cannot take cotrimoxazole or pentamidine. However, dapsone and cotrimoxazole belong to the sulpha group of drugs and may cause allergic skin rash, accompanied by fever. Allergic reactions can be overcome by using a desensitising procedure: patients are initially started on very small doses of the drug and the dosage is progressively increased till they can tolerate the full dose (Fact Sheet 515, 2006).

5.4.3 – Candidiasis

Candidiasis is a common opportunistic infection in HIV-infected persons and commonly affects the mouth (called "thrush"), throat, or vagina. The fungus may spread deeper into the throat and cause oesophagitis or spread to heart, brain, joints, and eyes. Pharyngeal infection manifests as sore throat, with pain during swallowing, and loss of appetite. Vaginal infection causes pruritus, burning, and thick whitish discharge. Clinically, candidiasis appears as white patches (much like cottage cheese) or red spots. In healthy persons, the normal immune function and presence of commensal bacteria keep candida in check. Use of broad-spectrum antibiotics kill commensal bacteria and may trigger candidiasis. The first-line treatment comprises use of anti-fungal creams, lozenges (that dissolve in the mouth), and vaginal pessaries. Systemic anti-fungal agents such as amphotericin B (given orally or intravenously) are indicated if the infection has spread to the other parts of the body (Fact Sheet 501, 2006).

5.4.4 – Molluscum Contagiosum

Molluscum contagiosum is an opportunistic skin infection, caused by a virus. The lesions are painless, non-pruritic, and have a hard white core. Lesions are commonly found on the face or groin. Shaving with a razor blade can spread the infection. It can also be spread by fomites (non-living materials such as clothing) that come in contact with a lesion. Several methods are available for treating molluscum and recurrences of lesions may need re-treatment. Cryocautery with liquid nitrogen and electrocautery are the common methods used to burn the lesions. Chemicals such as trichloroacetic acid and podophyllin are effective but cannot be used on sensitive skin or near the eyes. Surgical removal of the lesions can be painful and may leave scars. Topical application of tretinoin or oral administration of isotretinoin (both acne drugs) reduces the oil in the skin and consequently, the upper layer of the skin peels off. Antiviral drugs such as cidofovir or imiquimod may be topically applied (Fact Sheet 513, 2006).

5.4.5 – Abnormalities in Cell-Mediated Immunity (CMI)

CMI is essential for defence against intracellular organisms, which are protected from antibody-mediated destruction. The intracellular organisms include all viruses, some bacteria (*Mycobacteria, Salmonella*), protozoa, and fungi. HIV affects all the following three mechanisms by which CMI functions.

Delayed-Type Hypersensitivity (DTH): The intracellular organisms are killed by macrophages, which have been activated by cytokines produced by activated CD4

T-cells. In the early stage of symptomatic HIV disease, loss of DTH is responsible for predominance of skin and mucosal infections. Candidiasis of skin and mucosa, primarily controlled by CMI, are common in HIV infection.

Cytotoxic T-Lymphocytes: On activation by cytokines produced by activated CD4 T-cells, they kill virus-infected cells and malignant cells. Loss of this ability in HIV infection leads to reactivation of viral infections and development of malignancies.

Natural Killer (NK) Cells: These specifically kill virus-infected or tumour cells. Defective activity of NK cells is commonly seen in advanced HIV disease.

5.4.6 – Abnormalities in Humoral Immunity

Antibodies enhance phagocytosis of microorganisms by neutrophils and macrophages by a process called "opsonisation". HIV causes a non-specific stimulation of B-cells leading to overproduction of Ig molecules, particularly IgA and IgG, resulting in hyper-gammaglobulinaemia. B-cells fail to respond though more Ig molecules are produced. In late stages of HIV disease, reactivation of infections, such as cytomegalovirus infection, does not produce IgM response. The IgA and IgG antibody responses may be poor, after vaccination. HIV adversely affects all the functions of macrophages viz. chemotaxis (migration to the site of infection), phagocytosis, intracellular killing, and presenting of antigens to T-cells. Impairment of function of neutrophils leads to increased vulnerability to *Staphylooccus aureus* infection of the skin and pneumonia. ARV drugs (zidovudine (ZDV), ganciclovir) and infection with *Mycobacterium avium* complex (MAC) can cause neutropenia. Autoimmune antibodies may be produced, against platelets, lymphocytes, neutrophils, and myelin. Antibody-mediated drug allergies are more frequent in HIV-infected individuals.

5.5 – HYPERSENSITIVITY REACTIONS

In spite of diminished immune response, HIV-infected individuals exhibit an increased frequency of hypersensitivity reactions, as compared to that in HIV-negative immunodeficient patients and immunocompetent individuals. These reactions may or may not be drug-related. Hypersensitivity reactions related to drug use are anaphylaxis and anaphylactoid reactions (rare), angioedema and urticaria, erythematous skin rash, hypersensitivity in viscera, maculopapular skin rash, and systemic illnesses. Hypersensitivity reactions are caused by numerous drugs such as atavaquone, carbamazepine, cephalosporins, clindamycin, cotrimoxazole, sulfadiazine, dapsone, delaviridine, fluconazole, isoniazid, rifampicin, thiacetazone, neviraprine, penicillins (including synthetic penicillins like amoxicillin), and phenytoin. Dermatitis, eosinophilic folliculitis, localised maculopapular skin rash, morbilliform maculopapular rash, pruritus, rhinitis and sinusitis, and severe local reaction to bites and stings are reactions that are not drug-related.

5.5.1 – Clinical Manifestations

The drug-related hypersensitivity reactions begin about 7–12 days after ingestion of the first dose and are characterized by fever followed by skin rashes. Patients with prior exposure may develop reactions within few hours of administration of the first dose. These reactions may also be delayed and may develop up to 2 weeks after the drug is withdrawn. The skin rash is usually pruritic, erythematous, maculopapular, and most prominent on the trunk and upper extremities. Patients who are on intermittent therapy such as twice-weekly prophylaxis for prevention of PCP may manifest symptoms that wax and wane with each dose. Drug-related hypersensitivity reactions are commonly caused by drugs containing *sulpha* group, such as cotrimoxazole, sulphonamides, and dapsone. Hypersensitivity to multiple drugs has been reported. Cotrimoxazole is known to exhibit cross-hypersensitivity with sulfadiazine, amoxycillin, and dapsone.

5.5.2 – Pathogenesis

The aetiology of hypersensitive reactions in HIV-infected persons is not clear, but it is believed to be immune-mediated. However, these reactions do *not* represent the four main types of immune-mediated hypersensitivity, as described by Gell and Coombs (Carr *et al.*, 1991). The drug hypersensitivity reactions in HIV-infected individuals are *not* mediated by IgE because: (a) most of the IgE-mediated reactions occur within 1 hour of administration of the drug, while in HIV-infected persons, the onset of drug reaction is delayed, (b) other features of IgE-mediated reactions (angioedema, bronchospasm, hypotension) do not occur, and (c) fever, which is seen in drug hypersensitivity reactions in HIV-infected persons, is not a feature of IgE-mediated reactions.

5.5.3 – Mechanisms for Hypersensitivity Reactions

Immune Dysregulation: Irrespective of the causative drug, skin biopsies show epidermal and dermal infiltration by activated CD8 T-cells and macrophages. Cytokines (IL-6 and IL-1-beta) and tumour necrosis factor-alpha may have a role in pathogenesis. HIV infection enhances the sensitivity of CD8 T-cells to drugs, allowing hypersensitivity to develop.

Immune Activation by HIV: HIV-induced immune activation could be another cause.

Dose and Duration of Therapy: Treatment of PCP with high doses of cotrimoxazole is associated with high frequency of hypersensitivity reactions, as compared to the frequency seen with low doses of the same drug used in prophylaxis of PCP. Longer duration of treatment also increases the frequency of drug hypersensitivity reactions.

Structural Similarities of Drugs: Cotrimoxazole, sulphonamides and amoxycillin have a similar benzene-associated para-amino group. *Slow acetylators* (individuals who slowly metabolise the drugs by acetylation) may be at a greater risk for developing hypersensitivity to sulphonamides.

5.6 – NEUROLOGICAL DISORDERS

HIV infection is frequently associated with disorders of the peripheral and central nervous systems, which may be caused by HIV itself, cellular interactions, and neurotoxins produced by HIV-infected macrophages. The degree of neurological deficit (which is due to HIV infection of microglial cells) correlates with the extent of immune dysregulation. HIV does not directly enter the brain in the early stages of the disease, but probably enters the CSF at the time of seroconversion. In late stages of the disease, HIV enters the brain by any of the following routes – chronic infection of the meninges, blood-borne infection via infected T-cells and monocytes (seems more likely), and infection by cell-free viruses, since the blood-brain barrier breaks down in many patients with AIDS (Petito & Cash, 1992).

Cytokines and Neurotoxins: Increasing impairment of the immune system activates compensatory mechanisms, which produce various cytokines (Wesselingh *et al.*, 1994) and toxins. Probably HIV initiates neurological damage, which is enhanced by toxins. These cytokines and toxins damage neural tissues that may be secondarily infected by HIV from circulating infected cells. This can happen only if the circulating cells have macrophage-tropic (or syncytium-inducing) subtype of HIV, since the microglial cells are of macrophage lineage. Neurotoxic components of HIV are probably gp120 and the regulatory genes *tat* and *nef*. Macrophage lineage cells facilitate damage of neurons and seem to release neural toxins that affect astrocytes and oligodendrocytes.

HIV-Related Neurological Disorders: Patients with advanced HIV infection may develop neurological disorder, such as AIDS dementia complex, vascular myelopathy, and peripheral neuropathy. Less common neurological complications such as seizures, transient neurological deficits, and aseptic meningitis occur when the CD4 count drops below 200 cells per µL. These complications may be associated with AIDS dementia complex.

5.7 – MALIGNANCIES

Malignancies are more frequent in immunodeficient disorders, including drug-induced immunodeficiency (Penn, 1990). The malignancies seen in HIV-infected patients are similar in cellular origin and clinical behaviour to those seen in other forms of immunodeficiency. Kaposi's sarcoma, non-Hodgkin's lymphoma, and cervical carcinoma are currently recognised as "AIDS-defining malignancies". Though Hodgkin's disease and cancers of rectum and anus also have an increased incidence in HIV infection, they are currently not included in surveillance criteria for AIDS (Reynolds *et al.*, 1993). However in HIV-infected individuals, there is no increase in frequency of cancers of lung, colon, breast, and prostate, as compared to that in the general population.

Malignancies may occur in HIV-infected persons due to immunodeficiency, aberrant production of cytokines and activation of oncogenic viruses. The process of cell division and differentiation causes genetic aberrations. Many of the genetically altered cells die, but some may survive with the potential for further change, which may lead to the development of malignancies. In immune competent individuals, the immune system prevents malignancies by removing pre-malignant cells, particularly those induced by oncogenic viruses, and clearing foreign antigens that may chronically stimulate the immune system and promote excessive proliferation of cells.

In HIV-induced immunodeficiency, these immune mechanisms are impaired. Latent viruses such as EBV and human papilloma virus (HPV) may also be activated in the immunodeficient state and may contribute to neoplastic changes. EBV and HPV may be involved in the pathogenesis of non-Hodgkin's lymphoma and cervical cancer, respectively. Human herpes virus type 8 is probably linked to development of Kaposi's sarcoma (Chang *et al.*, 1994; Lennette *et al.*, 1996).

5.7.1 – Non-Hodgkin's lymphoma

Lymphoma is cancer of B-lymphocytes. Development of lymphomas is distinctive of HIV-infection. EBV is found in all lymphomas of central nervous system, Hodgkin's disease, and in a majority of non-Hodgkin's lymphomas. This suggests that EBV may be linked to the development of malignancies either directly, as oncogenes, or indirectly, by inducing the production of oncogenes within the host (Gregory *et al.*, 1991). *Cytokines* such as IL-5, IL-6, and IL-10 are detected in reactive lymph nodes of HIV-infected patients. The increased production of these cytokines probably contributes to cellular proliferation leading to malignancy. Lesions of non-Hodgkin's lymphoma may occur in the bone, abdomen, liver, brain, and in many other parts of the body. Lymphomas are treated by chemotherapy. In addition, radiotherapy may be used. Lymphomas of the central nervous system are difficult to treat. Genetically engineered monoclonal antibodies are being studied as possible treatment for non-Hodgkin's lymphoma (Fact Sheet 512, 2006).

5.7.2 – Kaposi's sarcoma

This AIDS-defining malignancy is the most frequent neoplasm in HIV infection but its incidence has dropped dramatically after introduction of ARV therapy. Men outnumber women by 8:1. Before the advent of the HIV epidemic, the disease affected elderly men of Eastern European or Mediterranean background. Kaposi's sarcoma usually affects the skin or mucous membranes of mouth, nose, or eye. It is one of the most visible signs of AIDS because the lesions (purple or red spots on white skin; bluish, brownish, or black spots on dark skin) appear frequently on the face, arms, and legs. It can spread to the liver, lungs, gastrointestinal tract, and lymph nodes (Fact Sheet 511, 2006).

The disease is characterised by development of tiny blood vessels (microangiogenesis). It is a unique malignancy that initially requires a growth factor, but later probably produces its own growth factors to maintain its proliferative capacity. Human herpes virus type 8 has been found in Kaposi's sarcoma, which may or may not be associated with HIV infection. This type of herpes virus probably spreads both vertically and sexually and is present in up to 20 per cent of HIV-infected individuals (Chang *et al.*, 1994; Lennette *et al.*, 1996). *Tat*, the regulatory gene in HIV that accelerates viral replication, is probably responsible for tumour promotion (Ensoli *et al.*, 1994). Cytokines such as basic fibroblast growth factor, oncostatin M, tumour necrosis factor, IL-6, and IL-1-beta (produced by both Kaposi's sarcoma cells and activated endothelium) may stimulate migration, invasion, and proliferation of endothelial cells. This results in angio-proliferation seen in Kaposi's sarcoma (Ensoli *et al.*, 1994; Corbiel *et al.*, 1991; Miles *et al.*, 1992). Beta-human chorionic gonadotrophin, the pregnancy hormone, has an anti-tumour effect (Lunardi *et al.*, 1995). This hormone may confer protection against this disease in women and is probably the reason for predominance of Kaposi's sarcoma in men.

5.7.3 – Carcinoma of Uterine Cervix

Carcinoma of the uterine cervix and its precursor lesion (called "cervical intraepithelial neoplasia" that predisposes to invasive cancer) are found frequently in HIV-infected women. Squamous cell carcinoma of anus, which occurs in HIV-infected men, is probably caused by the same mechanisms as that for cervical carcinoma – infection with HPV and immune dysregulation.

HPV Infection: This stimulates proliferation of epithelial cells of cervix and anus and also promotes genetic instability (which leads to further genetic changes within these cells, resulting in malignancy). Specific oncogenic types of HPV (usually 16, 18, 31, 33, and 35) that are associated with neoplastic changes possess E6 and E7 genes. These genes interact with regulatory proteins. The tat protein of HIV can directly activate HPV.

Immune Dysregulation: In an immunocompetent individual, the CMI response to HPV infection is regulated by cytokines like tumour necrosis factor-alpha, interferon-gamma, and tissue growth factor-beta (Zur Hausen & de Villiers, 1994). But in HIV-related immunodeficiency, the cytokine pathways are disrupted and host defence mechanisms are impaired. In HIV-infected women, the increase in frequency and severity of cervical dysplasia and shedding of HPV is inversely related to decrease in the CD4 cell count (Petry *et al.*, 1994)

5.8 – AUTOIMMUNE DISORDERS

These may manifest clinically in the early stage of HIV disease since the immune system is relatively effective. The genesis of many of these disorders shows important differences, when compared with that of similar conditions in HIV-negative individuals. Autoimmune syndromes may be of the following types:

Antibody-Mediated Syndromes: These include production of auto-antibodies against platelets, red blood cells, cardiolipin, and parietal cells of gastric mucosa. The possible mechanisms for antibody-mediated syndromes are: (a) increased production of IL-1 and IL-6 by HIV-infected macrophages and monocytes, which cause non-specific stimulation of B-cells by bypassing their usual pathway for stimulation, (b) activation of cytomegalovirus or HIV by "anti-self" B-cells, (c) increased production of lymphotoxins and tumour necrosis factor by macrophages and monocytes that may destroy CD8 T-cells, thus reducing suppression of B-cell activity by CD8 cells, and (d) "molecular mimicry": production of cytotoxic anti-lymphocyte antibodies by dysregulated B-cells and when antibodies directed against HIV glycoprotein gp-41 react with MHC Class II molecules (Hoffmann *et al.*, 1991).

CD8 T-Cell Mediated Syndromes: These comprise conditions like polymyositis, lymphocytic interstitial pneumonitis, cardiac myositis, and chronic active hepatitis. Auto-reactive CD8 T-cells infiltrate the affected tissues or organs. In HIV-negative individuals with similar disorders, the CD4 T-cells infiltrate. There is a marked increase in CD8 lymphocytes in peripheral blood. Estimation of level of CD8 lymphocytosis may help in identifying individuals at high-risk of developing CD8 T-cell mediated autoimmune disorders.

Demyelinating Syndromes: Guillian-Barré syndrome, multiple sclerosis-like illness, and demyelinating polyneuropathy may occur early in HIV infection. Their pathogenesis is not well understood.

Immune-Complex Mediated Syndromes: Circulating immune complexes containing detectable HIV antigens are commonly found in HIV-infected persons. These immune complexes may be deposited on vessel walls at any stage of HIV infection and may cause an arteritis, which is similar to polyarteritis nodosa. Primary vasculitis is rare in HIV infection, and is usually secondary to drug reactions or opportunistic infections.

REFERENCES

Carr A., Cooper D.A., and Penny R., 1991, Allergic manifestations of human immunodeficiency virus (HIV) infection. Jour Clin Immunol 11: 55–64.
Chang Y., Cesarman E., Pessin M.S., *et al.*, 1994, Identification of herpes virus-like DNA sequences in AIDS-associated Kaposi's sarcoma. Science 226: 1865–1869.
Clerici M., Stocks M.J., Zajac R.A., *et al.*, 1989, Detection of three distinct patterns of T-helper cell dysfunction in symptomatic human immuno-deficiency virus positive patients. Independence of CD4 + cell numbers and staging. Jour Clin Invest 84: 1892–1899.
Cooper D.A., Tindall B., Wilson E.J., *et al.*, 1988, Characteristics of T-lymphocyte response during primary infection with human immuno-deficiency virus. Jour Inf Dis 157: 889–896.
Corbiel J., Evans L.A., Vasak E., *et al.*, 1991, Culture and properties of cells derived from Kaposi sarcoma. Jour Immunol 146: 2972–2976.
Ensoli B., Gendelman R., Markham P., *et al.*, 1994, Synergy between basic fibroblast growth factor and HIV-1 tat protein in induction of Kaposi's sarcoma. Nature 371: 674–680.

Gregory C.D., Dive C., Henclerson S., *et al.*, 1991, Activation of Epstein-Barr virus latent genes protects human B-cells from death by apoptosis. Nature 349: 712–614.

Hoffmann G.W., Kion T.A., and Grant M.D., 1991, An idiotypic network model of AIDS immunopathogenesis. Proc Nat Acad Sc USA 88: 3060–3064.

Koup R.A., Safrit J.T., Cao Y., *et al.*, 1994, Temporal association of cellular immune responses with the initial control of viraemia in primary human immunodeficiency virus type-1 syndrome. Jour Virol 68: 4650–4662.

Lennette E.T., Blackbourn D.H., and Levy J.A., 1996, Antibodies to human herpes virus type 8 in the general population and in Kaposi's sarcoma patients. Lancet 348: 858–861.

Levi J.A., 1993, Pathogenesis of human immunodeficiency virus infection. Micro Rev 57: 183–189.

Livingstone W.J., Moore M., Innes D., *et al.*, 1996, Frequent infection of peripheral blood CD8 positive T-lymphocytes with HIV-1. Lancet 348: 649–654.

Lunardi J., Skandor Y., Bryant J.L., *et al.*, 1995, Tumorigenesis and metastasis of neoplastic Kaposi's sarcoma cell line in immunodeficient mice blocked by human pregnancy hormone. Nature 357: 64–68.

Mackewicz C.E., Yang L.C., Lilson J.D., and Levy J.A., 1994, Non-cytolytic CD8 T-cell anti-HIV responses in primary HIV-1 infection. Lancet 344: 1671–1673.

Miles S.A., Martinez-Maza O., Razai A., *et al.*, 1992, Oncostatin M as a potent mitogen for AIDS Kaposi's sarcoma-derived cells. Science 255: 1430–1432.

New Mexico AIDS Education and Training Center, 2006, Fact Sheet 501. Candidiasis. University of New Mexico Health Sciences Center. www.aidsinfonet.org. Revised 16 March.

New Mexico AIDS Education and Training Center, 2006, Fact Sheet 509. Herpes zoster (Shingles). University of New Mexico Health Sciences Center. www.aidsinfonet.org. Revised 1 September.

New Mexico AIDS Education and Training Center, 2006, Fact Sheet 511. Kaposi's Sarcoma. University of New Mexico Health Sciences Center. www.aidsinfonet.org. Revised 23 May.

New Mexico AIDS Education and Training Center, 2006, Fact Sheet 512. Lymphoma. University of New Mexico Health Sciences Center. www.aidsinfonet.org. Revised 19 July.

New Mexico AIDS Education and Training Center, 2006, Fact Sheet 513. Molluscum. University of New Mexico Health Sciences Center. www.aidsinfonet.org.Revised 23 May.

New Mexico AIDS Education and Training Center, 2006, Fact Sheet 515. *Pneumocystis* pneumonia (PCP). University of New Mexico Health Sciences Center. www.aidsinfonet.org. Revised 15 February.

New Mexico AIDS Education and Training Center, 2006, Fact Sheet 537. Pentamidine. University of New Mexico Health Sciences Center. www.aidsinfonet.org. Revised 13 July.

Penn I., 1990, Cancers complicating organ transplantation. N Engl J Med 323: 1767–1769.

Petito C.K. and Cash K.S., 1992, Blood-brain abnormalities in acquired immunodeficiency syndrome: immunohistochemical localization of serum proteins in post-mortem brain. Ann Neurol 32: 658–665.

Petry K.U., Scheffel D., Bode U., *et al.*, 1994, Cellular immunodeficiency enhances the progression of human papilloma virus-associated cervical lesions. Int J Cancer 57: 836–840.

Reynolds P., Saunders L.D., Layefsky M.E., and Lemp G.F., 1993, The spectrum of acquired immunodeficiency syndrome (AIDS)-associated malignancies in San Francisco 1980–1987. Am J Epid 137: 19–30.

Walker C.M., Moody D.J., Stiles D.P., and Levy J.A., 1986, CD8 + lymphocytes can control HIV infection in vitro by suppressing virus replication. Science 234: 1563–1566.

Walzer P.D., 2005, Pneumocystis infection. In: Harrison's principles of internal medicine (D.L. Kasper, E. Braunwald, Fauci A.S., *et al.*, eds.). Volume 1, 16th edn. New York: McGraw-Hill, pp 1194–1196.

Wesselingh S.L., Glass J., McArthur J.C., *et al.*, 1994, Cytokine dysregulation in HIV-associated neurological disease. Adv Neuroimmunol 4: 199–206.

Zur Hausen H. and de Villiers E.M., 1994, Human papilloma viruses. Ann Rev Microbiol 48: 427–447.

CHAPTER 6

NATURAL HISTORY OF HIV INFECTION

Abstract

In the first stage of infection, individuals are highly infectious. Antibodies are not detectable by serological tests during the "window period". The infection is clinically silent for 10 years or more. In about 70 per cent of cases, seroconversion illness is seen. The second stage lasts 3–5 years or longer. Antibodies are detectable by serological tests. The patient is less infective to others, as compared to stage 1, and is prone to small range of clinical problems and diseases. In stage 3, which lasts about 3–5 years, the patient becomes vulnerable to a select group of common infections as a result of immune suppression. The CD4 count is 200–500 cells per μL. There may be common symptoms like chronic diarrhoea, loss of body weight, pyrexia of unknown origin, and manifestations of AIDS-related complex (ARC). The fourth or terminal stage is characterised by signs and symptoms of AIDS, including opportunistic infections; the CD4 count falls below 200 cells per μL; and the patients are highly infectious. Wasting (or "slim disease") and tuberculosis are mainly seen in this stage in developing countries.

Key Words

Long-term non-progressors, Opportunistic infections, Oral hairy leukoplakia, Persistent generalised lymphadenopathy, Rapid progressors, Slim disease, Typical progressors, Window period.

6.1 – PROGRESSION OF HIV INFECTION

The range of incubation period is 1–14 years, with an average of 6 years. There are three patterns in the progression of HIV infection. *Typical progressors* comprise about 50–70 per cent of the HIV positive persons. The disease may progress over 8–10 years. *Rapid progressors* constitute about 5–10 per cent and may develop AIDS in 2–3 years. A small number of persons first infected with HIV 10 or more years ago have not developed symptoms of AIDS (NIAID, 2005). Such persons (called "long-term non-progressors") have stable CD4 counts for long periods and comprise about 5 per cent of the HIV positive persons. In *typical progressors*, there are four stages in progression of the disease.

6.2 – WHO CLINICAL STAGING CLASSIFICATION

Stage 1: Only a small number of infected persons have a clinically apparent illness. The symptoms of seroconversion illness are often mistaken for those of other viral infections. Most of the infected individuals are *highly infectious* to others in this stage and may not be aware that they have become infected. HIV is present in large quantities in their blood and genital secretions (NIAID, 2005). Antibodies are not detectable in the *window period*. The infection is clinically silent for 10 years or more. Though the CD4 count may be near normal (the normal CD4 cell count is 950–1700 cells per µL) and the infection may be clinically silent or asymptomatic, and the HIV replication and CD4 cell reversal goes on. There is a selective damage to the immune system and the patient gets progressively immunocompromised (Panteleo *et al.*, 1993).

Stage 2: This stage lasts 3–5 years or longer. The antibodies are detectable by serological tests. The CD4 cell count falls (but is more than 500 cells per µL) and the viral load increases with simultaneous development of immune suppression. Usually, there are no symptoms, but persistent generalised lymphadenopathy (PGL) may be present (NACO, Training Manual for Doctors).

The patient is less infective to others, as compared to stage 1, and is prone to small range of clinical problems and diseases, that include weight loss, minor skin and oral problems, recurrent sinusitis, and herpes zoster (or "shingles"). These clinical manifestations do not affect the normal activities of an individual and require minimal clinical intervention (NACO, Training Manual for Doctors; Panteleo *et al.*, 1993; WHO, 1990).

Stage 3: With further immune suppression, the patient becomes vulnerable to a select group of common and more virulent infections like bacterial pneumonia. These pathogens also cause diseases in individuals with normal and intact immune systems. But in case of HIV positive persons, they occur at much higher rates and with higher mortality. This stage lasts about 3–5 years. The CD4 count is 200–500 cells per µL. This stage is characterised by common symptoms such as chronic diarrhoea, loss of body weight, and pyrexia of unknown origin. Patients may need hospitalisation, specific treatment, and extra follow-up visits to the health centre (NACO, Training Manual for Doctors; Panteleo *et al.*, 1993; WHO, 1990). The signs and symptoms of immune deficiency, seen in ARC include opportunistic infections such as oral thrush, PGL, enlarged spleen, fatigue, acute weight loss (more than 10 per cent of body weight lost in one month), fever, night sweats, and malaise. Oral hairy leukoplakia seems to be unique to HIV-infected individuals. The margins of the tongue show white ridges of fronds on the epithelium. An association with EBV and papilloma viruses has been proposed (Simmonds & Peutherer, 2006).

Stage 4: With more profound immune suppression, the CD4 cell count falls below 200 cells per µL and the individual becomes an easy victim for various opportunistic infections. The individual is considered to have advanced disease – AIDS. Patients are *highly infectious* in this stage. In the absence of specific therapy, AIDS patients present with multiple clinical problems and death may occur. This is the terminal stage, characterised by signs and symptoms of AIDS. In the developing countries, wasting due to HIV infection (also called "slim disease"), and tuberculosis are the most important health problems at the fourth stage of the disease. As a result of human exposure to environmental pathogens, the clinical spectrum of the disease that is seen in poor countries is different and diseases like tuberculosis, pneumonia, and salmonellosis are significantly more common as compared to other opportunistic infections seen in the developed world (Morgan *et al.*, 1998).

The opportunistic infections in this stage are tuberculosis, herpes zoster, fungal, and parasitic infections. Kaposi's sarcoma, non-Hodgkin's lymphoma, and carcinoma of cervix are considered "AIDS-defining malignancies". These diseases are disseminated throughout the body. AIDS encephalopathy and AIDS dementia are rare. AIDS dementia is probably due to direct action of HIV on the central nervous system, since HIV can cross the blood-brain barrier (NACO, Training Manual for Doctors; Panteleo *et al.*, 1993; WHO, 1990).

6.3 – CDC CLASSIFICATION

6.3.1 – Group I: Acute HIV Infection

Seroconversion illness resembles glandular fever with lymphadenopathy and symptoms such as acute onset of fever, malaise, sore throat, myalgia, arthralgia, and skin rash. Only 5–10 per cent of individuals may experience this stage in its entirety though few individuals may experience few symptoms. Encephalitic presentations are rare (Simmonds & Peutherer, 2006). Peripheral blood shows lymphocytosis. Serological test for antibodies are usually negative at the onset of acute stage, but may become positive during its course. The virus itself, viral nucleic acid, or viral p24 antigen may be detected.

6.3.2 – Group II: Asymptomatic Infection

HIV-infected persons are symptomatic but test positive for HIV antibody tests and are *infectious* to others.

6.3.3 – Group III: Persistent Generalised Lymphadenopathy (PGL)

The lymph nodes are symmetrical, painless, and enlarged (Simmonds & Peutherer, 2006). They are more than 1 cm in size and present at *two or more extra-genital sites* for at least 3 months. Other causes of lymph node enlargement such as lymphomas need to be ruled out. PGL is present in 25–30 per cent of HIV-infected individuals, who may be otherwise asymptomatic (Simmonds & Peutherer, 2006).

6.3.4 – Group IV: Symptomatic HIV Infection

SUB-GROUP A = Constitutional disease (ARC)
SUB-GROUP B = Neurological disease
SUB-GROUPS C_1 & C_2 = Secondary infectious diseases
SUB-GROUP D = Secondary cancers
SUB-GROUP E = Other conditions
The infections are based on CD4 counts. When the CD4 count is *less than* 400 cells per mm³, signs and symptoms of immune deficiency are manifested. When the CD4 count drops below 200 cells per mm³, titre of virus increases markedly with an irreversible breakdown of immune mechanism. Most patients die of opportunistic infections and malignancies. AIDS is the terminal stage of HIV infection.

6.4 – DISEASE PROGRESSION

At the individual level, the progression of the disease and survival are both variable. Either the disease rapidly progresses over about 2 years or hardly progresses at all, for a longer period of 10 or 15 years (Morgan *et al.*, 1997). Once the life-threatening severe diseases of clinical AIDS have developed, the life span of the patients is reduced due to scarcity of resources in the developing countries. The *mean survival* may be in the range of 6–7 years especially among patients belonging to the poorer communities (Morgan *et al.*, 1998). The available data suggest that the progression of the disease is independent of race, ethnicity, or gender. Old age and low socio-economic status adversely affect the survival (Chaisson *et al.*, 1995). Very few natural history studies have been conducted in developing countries. Individuals progress through various stages of HIV infection at a variable rate. In rich communities, deaths usually occur after clinical AIDS has developed. But, in poorer communities, HIV positive persons may die in early stages of the natural history of the disease due to higher exposure to virulent and opportunistic infections, and inadequate resources for clinical care at stages 1 and 2 (Morgan *et al.*, 1997; Morgan *et al.*, 1998; Chaisson *et al.*, 1995; Gilks *et al.*, 1996).

6.5 – LONG-TERM NON-PROGRESSORS

Since the onset of the HIV/AIDS epidemic, it has been observed that some HIV-infected individuals (called *long-term non-progressors*) took a long time to progress to AIDS, while some individuals did not get infected at all, in spite of repeated exposures to HIV. These individuals have very low levels of detectable virus in their blood or in the peripheral blood mononuclear cells. On the other hand, they have high levels of active HIV-specific cytotoxic T-cells and CAF, the antiviral factor produced by CD8 T-cells (Barker *et al.*, 1995). It is not known how some HIV-affected individuals have remained asymptomatic for long periods of time. Factors responsible for their non-progression to AIDS could include particular features of their immune system or possible past infection with a less virulent strain of the virus. It is possible that these individuals have received a low dose of infection (equivalent to low-dose vaccination) leading to protective immunity. This has been seen in primates infected with SIV. National Institute for Allergy and Infectious Diseases (NIAID) supported researchers continue to trace how the disease progresses in different people (NIAID, 2005). Genetic factors may have a role since the presence of human leukocyte antigen (HLA) types B-8 and DR-3 are associated with faster progression of HIV infection. If a blood donor has slowly progressing HIV infection, then the recipient also gets the same type of infection (Learmont *et al.*, 1992; Ashton *et al.*, 1994).

REFERENCES

Ashton L.J., Learmont J., Luo K., *et al.*, 1994, HIV infection in recipients of blood transfusion from donors with known duration of infection. Lancet 344: 718–720.

Barker E., Mackewicz C.E., and Levy J.A., 1995, Effect of TH1 and TH2 cytokines on CD8 + T-cell response against human immuno deficiency virus: implications for long-term survival. Proc Natl Acad Sci USA 92: 11135–11139.

Chaisson R.E., Keruly J.C., and Moore R.D., 1995, Race, sex, drug use and progression of HIV disease. N Engl J Med 333: 751–756.

Gilks C.F., *et al.*, 1996, Invasive pneumococcal disease in a cohort of predominantly HIV-1 infected female sex workers in Nairobi, Kenya. Lancet 547: 718–724.

Learmont J., Tindall H., Evans L., *et al.*, 1992, Long-term symptomless HIV-1 infection in recipients of blood products from a single donor. Lancet 340: 863–867.

Morgan D., *et al.*, 1998, Early manifestations (pre-AIDS) of HIV-1 infection in Uganda. AIDS 12: 591–596.

Morgan D., *et al.*, 1997, HIV-1 disease progression and AIDS defining disorders in a rural Ugandan cohort. Lancet 550: 245–250.

NACO. Training manual for doctors. New Delhi: Government of India.

National Institute of Allergy and Infectious Diseases (NIAID), 2005, HIV infection and AIDS: an overview. NIAID Fact Sheet. Bethesda: National Institutes of Health. www.niaid.nih.gov/. March.

Panteleo G., Graziosi C., and Fauci A.S., 1993, The immuno-pathogenesis of HIV. N Engl J Med 528: 327–333.

Simmonds P. and Peutherer J.F., 2006, Retroviruses. In: Medical microbiology (D. Greenwood, R.C.B. Slack, and J.F. Peutherer, eds.), 16th edn. New Delhi: Elsevier, pp 527–538.

WHO, 1990, AIDS – Interim proposal for WHO staging system for HIV infection and disease. Wkly Epidemiol Rec 65: 221–228.

CHAPTER 7

INFECTION CONTROL METHODS

Abstract

Infection control involves good housekeeping (sanitation and dust control), hand washing, using personal protective equipment (PPE) such as gloves, and using natural, physical, or chemical methods to make the environmental conditions detrimental for pathogens. The type of likely pathogens should be considered while choosing the type of disinfection. Nearly all the chemical disinfectants are toxic or harmful to the eyes, skin, and lungs. Sterilisation is recommended for *critical items* that are directly introduced into the blood stream or into the normally sterile areas of the body. *Semi-critical items* come in contact with mucous membranes, do not ordinarily penetrate body surfaces, and require high level chemical disinfection. *Non-critical items* that do not come in contact with the patients or touch their intact skin only, require general housekeeping measures like washing with detergents and water. The basic principle of *universal biosafety precautions* is that blood and body fluids from all patients ought to be considered as potentially infected, irrespective of their serological status. These precautions should be followed during patient care and handling of dead bodies in health care settings.

Key Words

Antisepsis, Barrier nursing, Biomedical waste management, Cohort nursing, D value, Decontamination of spills, Disinfection, Hand washing, Handling dead bodies, Infection control, Needle stick injuries, Personal protective equipment, Sterilisation, Survivor curve, Task nursing, Universal biosafety precautions

7.1 – DEFINITIONS

1. Sterilisation (Latin: *sterilis* = barren): This is a "process by which, an article, surface, or medium is freed of all living entities (including vegetative microorganisms and spores)". An article may be regarded as *sterile* if it can be demonstrated that the probability of viable microorganisms on it is less than one in a million as per pharmacopoeia definition (Simpson & Slack, 2006).
2. Antisepsis (Greek: *anti* = against; *sèpsis* = putrefaction): This is a "process by which, *living tissues* are freed of pathogens". This is usually done by destruction of pathogens or by growth inhibition.
3. Disinfection (Latin: *dis* = reversal of): It is defined as a "process by which, *inanimate* objects, or surfaces are freed of all pathogens". Usually, disinfection does

85

not affect spores. A disinfectant in higher dilution can act as an antiseptic. But, the reverse is not always true. *Prophylactic disinfection* is defined as "measures applied *before* the onset of disease" and includes chlorination of drinking water, pasteurisation of milk, and washing of hands before clinical procedures. *Concurrent disinfection* refers to "measures applied *during* illness, to prevent further spread of the disease" and includes disinfection of patient's excretions, secretions, linen, and materials used in treating the patient. *Terminal disinfection* is defined as "measures applied *after* the patient has ceased to be a source of infection after cure, discharge, or death". This technique is obsolete. Terminal disinfection is now replaced by terminal cleaning of rooms, including ventilation. Rarely, bedding is fumigated (Ananthanarayan & Paniker, 2000; Collins & Grange, 1990; Sathe & Sathe, 1991).

4. Incineration (Latin: *cineris* = ashes): Incineration is "total combustion of all living and organic matter, by dry heat at not less than 800°C"

5. Decontamination (Latin: *dis-* or *de-* = reversal of; *contamunätum* = pollutant): This is a general term that indicates procedures put into practice to make equipment safe to handle. The word "contamination" may refer to chemical, microbiological or radioactive contamination (Simpson & Slack, 2006).

6. Sanitation (Latin: *sanitas* = health): This refers to reduction in the number of pathogens (Ananthanarayan & Paniker, 2000). Sanitation includes cleaning, wet mopping, dust control, environmental hygiene, and safe disposal of waste.

7.2 – KINETICS OF STERILISATION AND DISINFECTION

7.2.1 – Survivor Curve

When microorganisms are subjected to a lethal process, the number of viable survivors decreases exponentially in relation to the extent of exposure to the lethal process. If a logarithm (to the base 10) of the number of surviving organisms is plotted against the lethal dose received such as duration of exposure to a particular temperature, the resulting curve is called the "survivor curve". This survivor curve is independent of the original population of microorganisms. Ideally, the survival curve should be linear. Extrapolation on the survivor curve helps in determining the lethal dose required to give 10^{-6} survivors to meet the pharmacopoeia definition of "sterile" (Simpson & Slack, 2006).

7.2.2 – D Value

While manufacturing sterile products, a figure known as "D value" is used. It is the abbreviation for "death rate value" (Collins & Grange, 1990) and is also called "decimal reduction value" (Simpson & Slack, 2006). The "D value" is the time and dose of exposure, as determined in the laboratory, to reduce the viable count by one log, i.e. one order of magnitude = 1/10 (Collins & Grange, 1990). "D value" is the time and dose of exposure required to inactivate 90 per cent of organisms in the initial population (Simpson & Slack, 2006). The D value remains constant over the full range of the survivor curve. This means that the time and dose required to reduce the population of organisms from 10^6 to 10^5 is the same as that required to reduce the population of organisms from 10^5 to 10^4 (Simpson & Slack, 2006). In order to ensure effectiveness of sterilisation, the magnitude of exposures used is many times more than the "D value", which is calculated according to the known "bio-burden" (Collins & Grange, 1990).

7.2.3 – Components of Infection Control

Infection control involves: (a) good housekeeping (cleaning, wet mopping, and dust control), (b) using PPE (gloves, masks, etc.), and (c) using physical or chemical methods to make the environmental conditions detrimental for pathogens. Many physical methods act by chemical mechanisms. For example, heat kills the pathogens by denaturing cellular proteins (Sathe & Sathe, 1991). The process of disinfection should be technically correct. Many commonly used methods of disinfection are mentioned below, but the type of likely pathogens should be considered while choosing the type of disinfection.

7.2.4 – Classification of Medical Equipment

Medical equipment or items can be divided into three categories (Chitnis, 1997; Simpson & Slack, 2006).
1. Critical Items: All equipment or items that are directly introduced into the blood stream or into the normally sterile areas of the body, e.g. surgical instruments, cardiac catheters, needles, arthroscopes, parenteral fluids, and implants. These articles have to be *sterile* at the time of use.
2. Semi-Critical Items: Articles that come in contact with mucous membranes and do not ordinarily penetrate body surfaces, e.g. non-invasive flexible and rigid fibrooptic endoscopes, endobronchial tubes and ventilation equipment, cystoscopes, aspirators, and gastroscopes. *High-level chemical disinfection* is sufficient for items belonging to this category.
3. Non-Critical Items: Items that do not touch the patients or touch the intact skin only, e.g. blood pressure cuffs, crutches, bed pans, urine pots, and furniture. *General housekeeping measures* like washing with detergents and water are adequate.

7.2.5 – Prior Cleansing

Before subjecting any article or equipment to sterilisation or disinfection, it is essential that the lowest possible bioburden is present at the start of the process. Any used article or instrument is to be soaked in a chemical disinfectant, cleaned with a detergent, followed by thorough rinsing (Mitchell *et al.*, 1997). This is necessary before subjecting the material or equipment to the sterilising process. Cleaning, per se, is also a valuable method of low-level disinfection. Ultrasonic baths are useful in removing dried debris on instruments that are ordinarily difficult to clean. This partially reduces the bioburden. Detergents have surface tension reducing property – they wash away many organisms. The dilution effect of thorough rinsing further reduces the burden and thus increases the probability of successful sterilisation (Simpson & Slack, 2006). Lipid membrane envelope of HIV is highly susceptible to surface tension reducing action of detergents. Hence clothes and utensils may be decontaminated by washing with detergents.

7.2.6 – Factors Affecting Sterilisation and Disinfection

1. Species or Strain of Microorganism: In general, vegetative organisms are more vulnerable, while spores are resistant to action of sterilising and disinfecting agents. There is an interspecies variation in the D value at 60°C – *Escherichia coli* (few minutes) to *Salmonella enterica* subtype Senftenberg (one hour); D value at 70°C – *Staph aureus* (less than 1 minute) and *Staph epidermidis* (3 minutes). Prions (organisms that cause scrapie, bovine spongiform encephalopathy, and Creutzfeldt-Jakob disease) are killed at 134°C for 18 minutes. Hence it is desirable to use gamma-sterilised disposable instruments for operating on nervous tissue including retina because risk of exposure to prions is high (Simpson & Slack, 2006).
2. Growth Conditions: Organisms that grow under nutrient-rich conditions are more resistant to sterilising and disinfecting agents. Resistance usually increases through the late logarithmic phase of microbial growth and declines erratically during the stationary phase.
3. Spore Formation: Bacterial spores are more resistant, as compared to fungal spores. In general, disinfection processes have little or no action against bacterial spores.
4. Micro-Environment: The presence of organic matter (blood, body fluids, pus, faeces, urine) reduces the effectiveness of chlorine-releasing agents (Simpson & Slack, 2006). Presence of salt reduces effectiveness of ethylene oxide (Simpson & Slack, 2006). Chemical disinfectants will inactivate at least 10^5 viruses within few minutes. With the exception of phenols, many chemical disinfectants are inactivated in the presence of organic matter. Hence thorough cleaning is necessary before disinfection (Simpson & Slack, 2006).
5. Bioburden: Higher the initial bioburden (number of microorganisms) the lethal process must be more stringent and extensive to achieve high quality of sterility.
6. Time Factor: All microorganisms do not get killed instantly when exposed to physical agents or to chemical disinfectants because in any population of microorganisms, some will be more resistant than others (Collins & Grange, 1990). Higher the bioburden, longer will be the time taken to destroy all of them (Simpson & Slack, 2006).

7.2.7 – Factors Affecting Action of Chemical Disinfectants

1. Concentration, stability of disinfectant, temperature, and pH during use. "In-use concentration" is the optimal concentration required to produce a standardised disinfecting effect (Simpson & Slack, 2006).
2. Number, type, and accessibility of the microorganisms – Gram-positive bacteria are more sensitive, as compared to Gram-negative bacteria, mycobacteria, and bacterial spores; lipophilic and enveloped viruses are more sensitive, as compared to hydrophilic viruses, e.g. poliovirus. Hepatitis B virus (HBV) is also relatively resistant to action of disinfectants.
3. Presence of inactivators of disinfectants – organic (especially protein) substances, hard water, cork, plastics, organic matter, soaps, detergents, or another disinfectant. Users should refer instructions of manufacturers regarding such inactivators (Simpson & Slack, 2006).

7.3 – PHYSICAL METHODS

7.3.1 – Dehydration

Dehydration (also called "dessication") is lethal to most pathogens, since the organisms lose moisture. Drying can be achieved by exposing the object or article to strong sunlight, or by keeping the object/article in desiccators (Ananthanarayan & Paniker, 2000). *Vacuum drying* is used to preserve the potency of vaccines and the nutritive value of foods. *Adequate ventilation* acts by drying and diluting the number of suspended organisms in the air. Delicate organisms such as meningococci are vulnerable to drying by air (Sathe & Sathe, 1991). Dehydration is unreliable because many viruses and spores are not destroyed. Drying reduces the infectivity of the HIV. Hence, dried serum and blood are not highly infectious. Like other enveloped viruses, HIV must remain in *moist state* (or in solution) in order to be infectious. It is also susceptible to inactivation by physical and chemical agents in the *moist* state (Cunningham, 1997).

7.3.2 – Dry Heat

Dry heat acts by denaturing proteins of the microorganism.
Flaming: Exposure of scalpels or necks of flasks to a flame for a few seconds is of uncertain efficiency. Inoculating loops and needles are sometimes immersed in methylated spirit or alcohol and burnt off. But, this method does not produce sufficiently high temperature. In addition, there is the flammable risk of alcohol. The following items can be sterilised using the blue (oxidising) flame of a Bunsen burner – spatulas, inoculating loops, glass slides. Use disposable inoculating loops when dealing with highly pathogenic organisms. This is because flaming may cause "spluttering" of unburnt material, which is dangerous (Simpson & Slack, 2006).

Incineration: Unfortunately, many incinerators are inefficient for burning hospital waste. The waste may be merely scorched and the infected waste may escape with the smoke and pollute the atmosphere. The basic requirements in design of incinerators for hospital waste include: (a) easy attainability of high temperatures (at least 800°C) with the "load", and (b) presence of an "after-burner", i.e. a chamber, where smoke and other gaseous effluents are heated to similar or even higher temperatures (Collins & Grange, 1990). It is essential to train incinerator operators on what type of materials can (or cannot) be burned and how to mix loads in order to ensure adequate combustion, with the minimum of toxic effluent. Even an intrinsically efficient incinerator may fail if it is improperly used (Collins & Grange, 1990). During the process of incineration, the pathogens are destroyed along with the contaminated article/object. This method is recommended for disposal of low-value non-reusable articles like soiled dressings, swabs, dry waste, etc., and for incinerating animal carcasses and biomedical waste (Collins & Grange, 1990; Sathe & Sathe, 1991).

Hot Air Oven: The articles are wrapped in heat-resistant paper before they are placed in the hot air oven. The recommended temperature and duration is 160°C for 1 hour. This method is used for sterilising sharp instruments, glassware (syringes), dusting powders (French chalk, antibiotic powders), vaseline, and paraffin.

7.3.3 – Methods using Moist Heat

Moist heat is more lethal than dry heat. The cell wall of the microorganisms encloses protein particles in colloidal suspension. Heat coagulates cellular protein, which results in death of microorganisms. Coagulation of protein is instantly lethal and takes place at a lower temperature in the *presence of moisture*. Hence, moist heat is more lethal than dry heat. The vegetative forms contain more moisture and therefore, their cellular proteins coagulate faster. Spores contain less moisture and are consequently more resistant to action of heat (Sathe & Sathe, 1991).

Pasteurisation: Rapid heating, followed by sudden cooling destroys or inactivates most of the pathogenic organisms. Pasteurisation is chiefly used for milk and milk products. The methods of pasteurisation are:
(a) *Holder method* – Milk is heated to 63°C for half an hour and is rapidly cooled
(b) *Flash process* – Also known as "high temperature, short time (HTST) method". Milk is rapidly heated to 72°C within 15 seconds and is quickly cooled
(c) *Ultra-high temperature (UHT) method* – Milk is superheated to 125°C in 15 seconds and is rapidly cooled. Milk and milk products pasteurised by UHT method have a longer shelf life.

Water Baths: Used for disinfecting sera and other products that are destroyed or denatured at high temperatures. The recommended temperature and duration is 60°C for 1 hour (Ananthanarayan & Paniker, 2000).

Boiling: Boiling destroys all vegetative organisms within 5 minutes. But, spores may remain viable (Ananthanarayan & Paniker, 2000). Moist heat is not suitable for woolens and may cause shrinkage. Boiling is useful for disinfecting linen, crockery, utensils, bottles, and glassware. The articles should be thoroughly washed with soap or detergent before they are boiled. A bundle of clothes should be boiled at least for half an hour so that the moist heat can penetrate the bulky mass. Sputum, collected in a metal container, should be boiled for 1 minute after adding some water (Sathe & Sathe, 1991). HIV in solution is inactivated by heat at 56°C within 10–20 minutes. In lyophilised protein preparations, such as Factor VIII, HIV is killed at 68°C within 2 hours.

7.3.4 – Steaming at Atmospheric Pressure

Types of Steam: *Dry steam* does not contain suspended droplets of water. *Wet steam* contains suspended droplets of water at the same temperature and is less efficient as a sterilising agent. *Saturated steam* holds all the water it can, in the form of transparent vapour. For effective sterilisation, steam should be both *dry* and *saturated* (Simpson & Slack, 2006). *Superheated steam* is at a higher temperature than the corresponding pressure would allow. This type of steam behaves in a manner similar to hot air and is less penetrative. Mixture of steam at low temperature and formaldehyde gas combines the thermal effect of steam generated at subatmospheric pressure and chemical effect of formaldehyde gas to give effective sporicidal action. This method is useful for reprocessing heat-sensitive instruments. However, safety requirements make the process unsuitable for routine hospital use (Simpson & Slack, 2006).

7.3.5 – Autoclaving

The autoclave (Greek: *auto* = self; *clavis* = key) is essentially a pressure cooker. The autoclave has a cylindrical body made of strong alloy. The lid, made of gun-metal, can be sealed with "butterfly screws". The autoclave has an outlet for steam, a safety valve, and a pressure gauge. In modern autoclaves, the process, temperature, and time are controlled automatically. Articles to be sterilised are kept on a perforated stage inside the autoclave cylinder. The water level is to be checked every time the autoclave is used and should be below this perforated stage. Gas burner or electricity may be used for heating. Autoclaving is a reliable method of sterilisation, which destroys all living entities (pathogenic as well as non-pathogenic microorganisms). Vegetative organisms are killed instantly and most spores within 2 minutes. Trained personnel are required for handling and maintenance. Articles are moist soon after they are removed from the autoclave. Sharp instruments lose their sharpness and hence cannot be autoclaved (Collins & Grange, 1990; Sathe & Sathe, 1991).

The autoclave is used in laboratories for sterilising all biochemical and bacteriological media, except those containing heat labile substances like blood, serum, or eggs (Sathe & Sathe, 1991). In health care settings, autoclaves are used for sterilising linen, gloves, gowns, and surgical instruments (other than sharps). Needles and syringes may also be autoclaved in certain situations though use of disposable, gamma-sterilised needles and syringes is recommended (Collins & Grange, 1990; Sathe & Sathe, 1991).

7.3.6 – Working Principles

Charles' Law: The higher the pressure, the higher is the temperature, when the volume is constant. At normal atmospheric pressure (at sea level), water boils at 100°C. A pressure of 15 pounds per square inch (PSI) is equivalent to *one atmosphere* pressure. Metric (SI) units are not used in autoclaving. When water is subjected to a pressure of 15 PSI above the atmospheric pressure, water will boil at a temperature of 121°C if air is completely expelled from the closed container. If air is not expelled, water will boil at a lower temperature. Pressure, *per se*, does not ensure effective sterilisation since microorganisms can withstand high pressures. Pressure acts by raising the temperature at which water boils, and increasing penetration of steam. The raised temperature is instrumental in ensuring sterilisation (Collins & Grange, 1990).

Latent Heat of Steam: For converting water into steam at the same temperature, an additional heat of 540 calories per gram has to be supplied. Conversely, when 1 g of steam condenses back to form water at the same temperature, this heat is instantly released without any change in temperature. Hence, this is called "latent heat" (latent = hidden). The latent heat is instantly delivered to the article on which the steam condenses resulting in instantaneous death of the microorganisms that may be present (Collins & Grange, 1990).

Sudden Reduction in Volume During Condensation: About 1,700 mL of steam at 100°C condenses on a relatively cooler surface to form 1 mL of water at 100°C and releases latent heat. Condensation causes a local drop in pressure, which draws in more steam. This movement of steam continues till the article reaches temperature equilibrium (Collins & Grange, 1990).

Action of Moist Heat: Once the surface layer of the article/object has reached temperature equilibrium the steam does not condense on the surface layer since the temperature is the same. The steam penetrates into the next cooler layer and condenses. Thus, moist heat under pressure is more penetrative. This action continues till the entire article or object is penetrated by steam (Collins & Grange, 1990).

7.3.7 – Prior Cleansing

Before subjecting any article or instrument to autoclaving, it is essential that the lowest possible bioburden is present at the start of the process. Any used article or instrument is to be soaked in a chemical disinfectant, cleaned with a detergent, followed by thorough rinsing. Cleaning, per se, is also a valuable method of low-level disinfection. Ultrasonic baths are useful in removing dried debris on instruments that are ordinarily difficult to clean. This partially reduces the bioburden. Detergents have surface tension reducing property and wash away many organisms. The dilution effect of thorough rinsing further reduces the burden and thus increases the probability of successful sterilisation (Simpson & Slack, 2006).

7.3.8 – Elimination of Air

Since air is a bad conductor of heat, its presence inside the autoclave chamber reduces the maximum temperature that can be achieved, and thus diminishes the penetrating power of steam. Steam under a pressure of 15 PSI reaches a temperature of 121°C only when the air is completely removed. Air can be removed from the autoclave chamber by a vacuum pump. Alternatively, air can be removed by downward displacement. Since cooler air (being heavier) tends to settle at the bottom of the chamber, steam is let in from the top to displace air downwards. When steam starts coming out of the discharge outlet, it indicates that the air is completely removed from the chamber. The steam outlet valve is then closed and the pressure inside the chamber is allowed to rise (Collins & Grange, 1990).

7.3.9 – Sterilisation Time

Sterilisation time has the following components:
(a) *Heating or penetration time*: Time taken to increase the temperature of the article to that of steam.
(b) *Holding time*: Time during which the contents of the chamber are maintained at the selected temperature (usually 121°C at 15 PSI). *Prions*, the causative agents of scrapie, bovine spongiform encephalopathy, and Creutzfeldt-Jakob disease, are killed at 134°C for 18 minutes (Simpson & Slack, 2006). Hence, to ensure *absolute sterility*, a temperature of 135°C is to be attained. The holding time is determined by the *thermal death point* of heat-resistant bacterial spores.
(c) *Safety period*: This is usually 50 per cent of the holding time for dressing drums and is equal to the holding time for fabric bundles. It is essential to follow the instructions mentioned in the autoclave manufacturer's operating manual (Collins & Grange, 1990).

7.3.10 – Tests to Ensure Completeness of Sterilisation

Commonly, the pressure gauge is relied upon to check for the completeness of sterilisation. However, it is essential to have a thermometer fitted and to record the actual temperature attained, and the duration for which the temperature was maintained. Equipment with vacuum-assisted air removal cycle are fitted with air detectors. Temperature-sensitive probes (thermocouples) may be inserted into standard test packs (Simpson & Slack, 2006). The *Bowie-Dick test* monitors the penetration of steam by a bubble of residual air in the pack (Simpson & Slack, 2006). In the original test, an adhesive indicator tape is pasted on the surface of the articles or objects to be sterilised, before they are placed in the autoclave chamber. This indicator tape, usually green in colour, changes to black if the article has been exposed to the recommended temperature (Collins & Grange, 1990; Sathe & Sathe, 1991). The change in colour should be uniform along the entire length of the indicator tape (Simpson & Slack, 2006).

Biological indicators comprise dried spore suspensions of a reference heat-resistant thermophilic spore-bearing bacterium such as *Bacillus stearother-mophilus, Bacillus thuringensis*, or *Bacillus subtilis*. Spores of one of these organisms are kept in a sealed glass ampoule and placed inside the autoclave chamber. After the process of autoclaving, the ampoule is sent to the laboratory where the spores are incubated to check for bacterial growth. Presence of growth indicates that the spores have remained viable. This procedure requires laboratory support, is expensive, and the results of the laboratory tests are not available immediately (Collins & Grange, 1990; Sathe & Sathe, 1991). These biological indicators are no longer considered for routine testing. Spore indicators are essential in low-temperature gaseous processes like ethylene oxide, in which physical measurements are not reliable (Simpson & Slack, 2006).

7.3.11 – Tips for Successful Autoclaving

The manufacturer's operating manual should be carefully followed and autoclave operators should be trained. Materials such as talcum powder cannot be penetrated by steam and should not be autoclaved. Before using the autoclave, water should be maintained at a level recommended by the manufacturer. Steam outlet and safety valve should be checked and cleaned if necessary. All articles should be wrapped in kraft paper or cloth and then placed in individual trays. The barrel and plunger of the syringes should be disassembled and wrapped in cloth or kraft paper. Linen should be wrapped in loose flat bundles since larger bundles would need longer time for sterilisation. The *thermo-chemical indicator tape* is stuck on the cloth covering each tray. It is important to ensure that all the air is expelled, since a mixture of air and steam will not have the same penetrative effect as saturated steam alone. The steam outlet is closed only after all the air in the cylinder is expelled and excess steam starts coming out from the steam outlet. The temperature, pressure, and the sterilisation time are recorded. The "holding time" is counted after the steam outlet is closed. It is safer to rely on the temperature attained and the time period for which this temperature is maintained. To ensure absolute sterility, a temperature of 135°C is to be attained. The autoclave should cool on its own. Autoclaved articles should be prevented from contamination during drying, transportation, and storage. For sterilising small quantities, a domestic pressure cooker may be used. The "holding time" is counted after the first "whistle" (expulsion of steam) from the steam outlet of the pressure cooker (Collins & Grange, 1990; Sathe & Sathe, 1991).

7.3.12 – Radiation

Ionising Radiation: The articles/objects exposed to ionising radiation do not get heated. Hence, this method, also called "cold sterilisation", is the most cost-effective and safe method of sterilisation. Sterilisation is achieved by using high-speed electrons (beta-rays) from a linear accelerator or gamma-rays from an isotope such as Cobalt-60. A dose of 255 kilo Gray is adequate for large-scale sterilisation (Simpson & Slack, 2006). The articles travel through the facility on a conveyor belt. Ionising radiation is used for large-scale sterilisation of pre-packed single-use syringes, needles, catheters, antibiotics, ophthalmic medicines, microbiological media, heat-sensitive plastics, and other heat-sensitive instruments (Ananthanarayan & Paniker, 2000).

Non-Ionising Radiation: Substances exposed to non-ionising radiation get heated. *Short wave* radiation (such as ultraviolet rays) is more bactericidal than *long wave* radiation (such as infrared rays). Infrared radiation is used for sterilising glassware. Ultraviolet (UV) radiation is a low-energy, non-ionising type of radiation, with poor penetrating power. Ultraviolet rays are lethal to microorganisms under optimal conditions. UV lamps produce effective UV radiation with wavelength of 240–280 nm. The UV rays with shortest wavelength that reach the earth's surface have a wavelength of 290 nm. UV lamps are used for sterilising closed areas, such as operation theatres, wards, neonatal intensive care units, and laboratory safety cabinets (Simpson & Slack, 2006). Commercially available household water purifiers use UV radiation for sterilising small quantities of water for laboratory or household use. Its disinfecting action on water is impaired if water is turbid. UV light is also used for therapeutic purpose, to accelerate the conjugation of bilirubin in neonates with jaundice. It should be noted that eye protection is essential because exposure to UV light may lead to premature cataract.

7.4 – CHEMICAL METHODS

Lipophilic viruses such as HIV, HBV, and cytomegalovirus are *highly sensitive* to chemical disinfectants (Gangakhedkar, 1999). Disinfectants rapidly inactivate HIV in suspension but are less effective against HIV in dried body fluids (Cunningham *et al.*, 1997). HIV is inactivated by 70 per cent isopropanol (3–5 minutes), 70 per cent ethanol (3–5 minutes), 2 per cent povidone iodine (15 minutes), 4 per cent formalin (30 minutes), 2 per cent glutaraldehyde (30 minutes), household bleach (diluted) containing 1 per cent available chlorine (30 minutes), and 6 per cent hydrogen peroxide (30 minutes). For decontaminating used medical equipment, 2 per cent glutaraldehyde may be used.

Criteria For Selecting Disinfectants:
(a) Area of health care facility, where the disinfectant will be used
(b) Spectrum of action – bacteria, lipophilic and hydrophilic viruses, mycobacteria, fungi, and spores

(c) Rapidity of action and residual effect – antimicrobial action is to be sustained for prolonged periods
(d) Should not cause allergy or irritate the skin or mucous membrane
(e) Microorganisms should not develop resistance on repeated use
(f) Odour and colour should be acceptable
(g) Should not stain skin or clothing

Factors Affecting Disinfection: Before disinfection, the articles or surfaces must first be thoroughly cleaned with detergent and thoroughly rinsed with water. Effective chemical disinfection depends on multiple variables such as concentration of the disinfectant, temperature and pH, presence of organic substances, and contact period (period of exposure to the disinfectant). Since it is difficult to consider all the variables that can affect chemical disinfection, it would be easier to follow the manufacturer's guidelines. A minimum "contact period" of 30 minutes is recommended.

7.4.1 – Standard Operating Procedures

The hospital infection control committee (HICC) should agree on a sterilisation and disinfection policy and procedures involved. Once the policy is finalised, all concerned staff members should be made aware of the policy and trained in the Standard Operating Procedure (SOP). The sterilisation and disinfection policy should mention choice of sterilising or disinfecting process is required for equipment, instrument, skin, mucous membrane, furniture, floors, and biomedical waste. The available choices or options are to be restricted to avoid unnecessary costs, confusion, and chemical hazards. The processes are categorised and the SOPs for items in the hospital that are to be disinfected or sterilised are described. Copies of the SOP should be circulated to all concerned staff members. The policy and SOPs may be updated periodically by the HICC (Simpson & Slack, 2006).

7.4.1.1 – Components of standard operating procedure

1. Details of methodology
2. Site where the procedure is to be done
3. Time schedule for the procedure
4. Persons responsible for carrying out the various steps in the entire procedure
5. Safety precautions and type of protective equipment to be worn during each procedure
6. Supervision of entire procedure including safety considerations

7.4.2 – Limitations of Disinfectants

Under ideal conditions, chemical disinfectants destroy most of the vegetative microorganisms. Few kill bacterial spores, fungi, and viruses that have lipid capsids (Collins & Grange, 1990). Presence of organic materials, protein, hard water, rubber, and plastics may impair the action of some disinfectants. The recommendations of the manufacturer are to be followed for correct dilution, storage after dilution, and use. Nearly all the chemical disinfectants are toxic or harmful to the eyes, skin, and lungs. Therefore, they should be cautiously selected and used. Disinfectants are not a substitute for efficient cleaning with detergent and water. If pre-diluted to working concentrations, chemical disinfectants may rapidly lose their strength on storage. In health care settings, sterilisation by autoclaving is the most reliable method. When neither autoclaving, nor boiling is possible, chemical disinfectants are to be used.

7.4.3 – Quality Control of Disinfectants

Unfortunately, chemical disinfectants are overused and abused, and then they are most inefficient and may give a false sense of security. Disinfectants that are used in health care settings should be monitored regularly (Chitnis, 1997). A wide range of testing methods has been developed for medical and veterinary use. In Europe, standardisation of test methods include: (a) simple screening tests of rate of kill, (b) laboratory tests simulating in-use conditions (skin antisepsis and inanimate surface disinfection tests), and (c) in-use tests on equipment and samples of disinfectants (in "in-use dilutions"). These tests determine survival and multiplication of contaminating pathogens. In-use tests are used for monitoring effectiveness of the disinfectant or antiseptic and monitoring method of use. The *Rideal Walker coefficient* (RWC), also known as carbolic coefficient or phenol coefficient, compares the bactericidal power of a given disinfectant with that of phenol. A major limitation of this test is that it compares disinfectants, without the presence of organic matter such as pus, faecal matter, blood, or body fluids (Chitnis, 1997). *Chick Martin test* compares the disinfecting action of two disinfectants in the *presence* of organic matter (Chitnis, 1997). *Capacity test of Kelsey and Sykes* gives guidelines for the dilution of the disinfectant to be used (Kelsey & Sykes, 1969). After a particular disinfectant is selected at the desired dilution using the capacity test of Kelsey and Sykes, routine monitoring should be done using the *"in use" test of Maurer* (Maurer, 1969).

7.4.4 – Coal-Tar Derivatives

Phenols are relatively cheap, stable, and not readily inactivated by organic matter with the exception of chlorxylenol. Adding ethylene diamine tetraacetic acid (EDTA), as a chelating agent, can improve the activity against Gram-negative organisms. *Black* and *white* phenols (insoluble in water) leave stains on surfaces and *clear soluble* phenols are replacing these. *Corrosive phenols* are used for disinfection of floors, in discarding jars in laboratories, disinfection of excreta, etc. *Non-corrosive phenols*, such as chlorxylenol are less irritant and are used for topical antisepsis (Chitnis, 1997). With the exception of Chlorhexidine, phenols are incompatible with cationic detergents. Contact with rubber and plastic is to be avoided since they may get absorbed. Due to slow release of phenol fumes in closed environments and corrosive action on skin, phenols are being replaced in hospitals by detergents for cleaning, and by hypochlorites for disinfection (Simpson & Slack, 2006).

Carbolic Acid: Joseph Lister first used carbolic acid (pure phenol) for antisepsis. Crystals of carbolic acid are colourless, but when exposed to air, they turn pinkish and then dark red. Pure phenol is not a good disinfectant. It is toxic to tissues and can penetrate intact (unbroken) skin. Carbolic acid is used as a standard to compare the disinfecting power of disinfectants. For phenol, Rideal Walker coefficient is one. It is added to fuchsin dye to prepare carbol fuchsin, which is used for staining acid-fast bacilli. Formerly, carbolic acid was used as 0.5 per cent solution in glycerine as mouthwash, ear drops, and topical antipruritic; as analgesic in dentistry; and as 5 per cent solution in almond oil for sclerosing haemorrhoids.

Crude Phenol (Household Phenyl): This is a mixture of phenol and cresol and is available as a dark, oily liquid, used as general-purpose household disinfectant in a concentration of 1–2 per cent. Phenyl is effective, even in the presence of organic matter, against most Gram-positive and Gram-negative bacteria. However, it is an irritant to living tissues and is *not* effective against spores and acid-fast bacilli.

Cresol: This is a mixture of ortho-, meta-, and para-isomers of methyl phenol. It is effective against both Gram-positive and Gram-negative bacteria and is safer than pure phenol. But, it is an irritant to living tissues and only mildly effective against acid fast bacilli. It is used as an all-purpose disinfectant in the following concentrations: for faeces or sputum (10 per cent); for floors in wards and operation theatres (5 per cent).

Saponified Cresols: Lysol, izal, and cyllin are emulsions of cresol, prepared by mixing with surface-active agents. These are very powerful disinfectants. Being irritants to living tissues, they are used as all-purpose disinfectants for faeces, sputum, and floors in wards and operation theatres, and for the sterilisation of sharp instruments like scalpels.

Chlorhexidine (Hibitane): It is a chlorinated phenol containing cationic biguanide (Chitnis, 1997). It is effective against *vegetative forms* of Gram-positive, Gram-negative bacteria and fungi, and has moderate action on *Mycobacterium tuberculosis*. It is fast acting and has residual action for 5–6 hours. Hibitane is effective over a wide range of pH (5–8) and can be safely used on living tissues (skin and mucous membrane). However, this disinfectant is not effective against spores and is inactivated by soap (anionic detergents), organic matter, hard water, and natural materials like cork liners of bottle closures. It is combined with compatible (cationic) detergent and used for hand wash or combined with 70–90 per cent alcohol for use as "hand rub". Hibitane is a component of Savlon (Cetrimide + Hibitane) and is used as a skin antiseptic for wounds and burns. Hibitane is also available as lotion or cream (Ananthnarayan & Paniker, 2000; Sathe & Sathe, 1991).

Hexachlorophane: It is a chlorinated diphenyl (bis-phenol) and has a restricted role as a disinfectant due to its limitations (Chitnis, 1997). It is very effective against Gram-positive organisms and is compatible with soap, i.e. anionic detergents

(Ananthnarayan & Paniker, 2000; Sathe & Sathe, 1991). In infants, hexachlorophane may get absorbed through the skin and cause neurotoxicity. Hence, a concentration of more than 3 per cent should not be used in neonatal units. Triclosan is used for controlling outbreaks of methicillin-resistant *Staphylococcus aureus* (MRSA) in nurseries (Simpson & Slack, 2006). It is poorly effective against Gram-negative organisms, mycobacteria, fungi, and viruses. Hexachlorophane is used in germicidal soaps and in deodorants to prevent bacterial decomposition of apocrine sweat. As a topical ointment, it is employed in the treatment of seborrhoeic dermatitis and impetigo (Ananthanarayan & Paniker, 2000; Chitnis, 1997).

Chloroxylenol: This is chlorinated derivative of phenol is available as concentrated solution, liquid soap, and soap cake under the brand name Dettol. Being relatively safe, it is used as a household skin antiseptic and for disinfecting plastic equipment. It requires a minimum contact period of 15 minutes for action on Gram-positive bacteria and more time is required in case of Gram-negative bacteria (Ananthanarayan & Paniker, 2000; Chitnis, 1997) Chloroxylenol is easily inactivated by organic matter and hard water and is not recommended for hospital use (Simpson & Slack, 2006).

Triclosan: It is chemically phenyl ether and is highly effective against Gram-positive bacteria. It has moderate activity against Gram-negative bacteria, fungi, and viruses (Chitnis, 1997). Triclosan is used for controlling outbreaks of MRSA in nurseries (Simpson & Slack, 2006).

7.4.5 – Alcohols

Ethyl, isopropyl, and methyl alcohols have rapid and high levels of initial activity, but no residual activity. Optimal disinfecting activity is at 70–90 per cent concentration; antimicrobial activity decreases at both higher (more than 90 per cent), and lower (less than 70 per cent) concentrations (Chitnis, 1997; Simpson & Slack, 2006). Alcohols have good action on viruses, limited action on mycobacteria, and no action on spores, with reduced activity in presence of organic matter. Alcohol-based formulations of chlorhexidine, Savlon, and povidone iodine are used for hand-washing (Simpson & Slack, 2006).

Ethanol (synonym: ethyl alcohol): 100 per cent ethyl alcohol (called "absolute alcohol") has poor antiseptic and disinfectant properties, while 60–70 per cent ethyl alcohol is a good general purpose skin antiseptic and can also be used to dilute other antiseptics, such as tincture iodine, chlorhexidine, and Savlon. It inactivates vegetative bacteria in a few seconds. Being a fat solvent, it can dissolve sebaceous secretions on the skin. However, its action on spores, viruses, and fungi is not reliable. It is volatile in warm climates. Methyl alcohol is added to ethyl alcohol to prepare *methylated spirit* to prevent consumption.

Isopropanol (synonym: isopropyl alcohol): It is more bactericidal, more of a fat solvent, and less volatile than ethanol. However, it is more toxic than ethanol, and its action on spores, viruses, and fungi is *not* reliable. Isopropanol is used at 70 per cent for skin antisepsis and for disinfecting clinical thermometers, incubators, and cabinets.

Methylated Spirit (synonyms: rectified spirit, denatured spirit): Methyl alcohol is added to ethyl alcohol to prepare methylated spirit to prevent consumption. A 70 per cent solution has bactericidal, fungicidal, and virucidal action. However, it is a volatile compound. It is used for decontaminating surfaces, such as metals and table tops, where use of household bleach and hypochlorites is contraindicated. A mixture of 70 per cent methyl alcohol and 1 per cent glycerine is used as a hand-washing antiseptic (called "alcohol hand wash"), in all clinical settings. Glycerine is used in a concentration of 1 per cent as emollient to counter the drying effect of alcohol on the skin (Simpson & Slack, 2006).

7.4.6 – Aldehydes

Most aldehyde disinfectants are based on formulations of glutaraldehyde or formaldehyde, alone or in combination. Since they are an irritant to the eyes, skin, and respiratory mucosa, health and safety authorities in many countries control their use. Aldehydes are to be used only with adequate protection (protective clothing, safety hoods) of staff and ventilation of the working environment (air scavenging equipment). The treated equipment is to be thoroughly rinsed with sterile water to remove toxic residues. All chlorine-releasing agents should be removed from areas where formaldehyde fumigation is to be done, in order to prevent release of carcinogenic products (Simpson & Slack, 2006).

Glutaraldehyde: When buffered with sodium bicarbonate to pH 7.5–8.5, this has potent bactericidal, mycobactericidal, sporicidal, fungicidal, and virucidal action. The buffered solutions should be used within 2 weeks of preparation (Chitnis, 1997). Alkaline-buffered solution of glutaraldehyde is claimed to have a residual effect for several days but this will depend on the amount of contaminating organic material (Simpson & Slack, 2006). After disinfection with glutaraldehyde, the equipment should be thoroughly washed with sterile water. It is available under the brand name Cidex. A 2 per cent solution provides a high level of disinfection, which approximates sterilisation. It is the only disinfectant that can be reused. It is less toxic than formaldehyde and does not damage optical lenses and cementing material of endoscopes. However, glutaraldehyde has a strong odour and is expensive, as compared to formaldehyde. At alkaline pH (more than 8.5), glutaraldehyde gets polymerised, resulting in loss of antimicrobial activity. It is used for disinfecting sharp instruments, surfaces/instruments (which are destroyed by household bleach and sodium hypochlorite), face masks, catheters, endoscopes, endotracheal tubes, and respirators. An aqueous solution of 2 per cent is used topically for treatment of idiopathic hyperhidrosis of the palms and soles. Flexible endoscopes are to be disinfected by special closed washer–disinfector systems. These systems use oxidising agents, such as peracetic acid, chlorine dioxide, and super-oxidised water (Simpson & Slack, 2006).

Formaldehyde: Commercially available formaldehyde (formalin) contains 35–40 per cent formaldehyde. The vapours are toxic and irritant. A solution used in 1:10 dilution (containing 3.5–4 per cent formaldehyde) inactivates vegetative forms of bacteria, viruses, and fungi in less than 30 minutes. Any equipment that is disinfected with formaldeyde should be thoroughly rinsed with sterile distilled water before reuse (Chitnis, 1997). It is a cheap disinfectant, which kills all vegetative bacteria, most viruses, and fungal spores, while mycobacteria and bacterial spores are killed slowly. It does not damage metals or fabric, but is an irritant to the eyes and respiratory tract. It can damage optical lenses and cementing material used in endoscopes. Formaldehyde tablets (used for disinfection of equipment and nursery incubators) are not reliable and need to be subjected to quality control after use (Chitnis, 1997). As *aqueous solution* (formalin), it is used for preserving biological specimens and cadavers, and for destroying anthrax spores in animal hair and woollen products. Woollen products are soaked in a 2 per cent solution of formalin at 40°C temperature for 2 minutes. Hair and bristles are soaked in a 2 per cent solution at 60°C temperature for 6 hours. This process is called "duckering". A mixture containing one part formalin + one part glycerine + 20 parts water is used as a liquid spray for disinfecting walls and furniture in operation theatres, prior to fumigation. A 3 per cent solution may be used for removal of warts on palms and soles. Formalin in a concentration of 4 per cent is recommended for decontamination of spills of blood and body fluids. Formaldehyde vapour is used for fumigating operation theatres, wards, beds, books, and mattresses. Formaldehyde vapour is generated in special fumigators or by pouring 280 mL of 40 per cent formalin on 45 g of potassium permanganate ($KMnO_4$) crystals per 1,000 cubic feet of space. The exposure period for fumigation is 36–48 hours (Chitnis, 1997). The excess gas is neutralised using ammonia vapour.

7.4.7 – Halogens

Halogens (compounds of chlorine and iodine) are relatively inexpensive and have a broad spectrum of action. Hence, they are commonly used for decontamination of surfaces (Chitnis, 1997). However, they are required in higher concentrations in the presence of organic substances such as blood, body fluids, excretions, or secretions.

Compounds of chlorine kill vegetative organisms and inactivate most viruses by strong oxidising effect of nascent chlorine. The disinfecting power of all chlorine compounds is expressed as "percentage of available chlorine" for solid compounds and as "percentage" or "parts per million" (PPM) for solutions. All chlorine-releasing agents should be removed from areas where formaldehyde fumigation is to be done, in order to prevent release of carcinogenic products (Simpson & Slack, 2006). Chlorine-releasing agents are corrosive and most compounds deteriorate rapidly (Ananthanarayan & Paniker, 2000; Chitnis, 1997).

Sodium Hypochlorite: This compound is available as solution containing 5 per cent available chlorine, powder containing 60 per cent available chlorine, or tablets containing 1.5 g of available chlorine per tablet. The solution is available under various brand names such as Milton's solution, Saaf, Chlorwat, Kloroklin. It is bactericidal, virucidal, cheap, and more effective than bleaching powder ($CaOCl_2$) solution. However, it deteriorates rapidly, corrodes nickel and chromium plating of metallic instruments, and has an offensive odour. Sodium hypochlorite is most commonly used for bleaching and disinfecting linen and clothing: decontaminating spills of blood and body fluids; sterilising infant feeding bottles; and emergency disinfection of drinking water, fruits, and raw vegetables during epidemics.

Household Bleach: Also commercially available as fabric stain remover under various brand names, the solution contains 5 per cent available chlorine. It is bactericidal, virucidal, relatively cheap, and more effective than the solution of bleaching powder (calcium hypochlorite). For disinfecting materials and surfaces contaminated by blood or body fluids, household bleach should be made available in plastic recyclable bottles in hospitals and clinics. Household bleach deteriorates rapidly, corrodes galvanised buckets, nickel and chromium plating of metallic instruments, and has an offensive odour. The bleaching action may affect fabric and carpets while disinfecting spillage of blood or body fluids (Simpson & Slack, 2006).

Bleaching Powder: Chemically, this is calcium hypochlorite ($CaOCl$), also called "chlorinated lime". Bleaching powder is an unstable compound and should contain 33 per cent of available chlorine. When stabilised by mixing with lime (calcium oxide) it is called "stabilised bleach" or "tropicalised chloride of lime" (TCL). Bleaching powder is used in powder form or as a *freshly* prepared solution. A 1 per cent solution of bleaching powder can be prepared by mixing one-fourth teaspoonful (1 g) of bleaching powder in 1 L of water. This gives a chlorine concentration of 1,000 PPM. The minimum contact period is half an hour. The OCl^- ion is responsible for disinfection. *Uses of bleaching powder*: (a) decontaminating spills of blood and body fluids, (b) as a deodorant and disinfectant for toilets and bathrooms, (c) chlorination of water in wells (2.5 g of bleaching powder per 1,000 L of well water; contact period = 1 hour), (d) disinfection of faeces and urine (400 g of bleaching powder per litre; contact period = one hour), and (e) preparing Eusol, which was formerly used in dressing wounds.

Povidone Iodine: An "iodophor" is a loose complex of elemental iodine or tri-iodide with an anionic detergent, which increases solubility of iodine and functions as sustained-release iodine reservoir. "Povidone iodine" is a water-soluble complex of iodine and polyvinyl pyrrolidone (Simpson & Slack, 2006). Povidone iodine, which is available under various brand names, such as Betadine and Microshield, is an intermediate level disinfectant with bactericidal, fungicidal, and virucidal action. It does not stain or irritate the skin and is water miscible. However, it has poor residual effect. A *freshly* prepared dilution of povidone iodine is to be used everyday. It is relatively expensive and should not be used on copper and aluminium (Chitnis, 1997). It is used as a 0.75–1 per cent solution for hand wash; for disinfecting instrument trays, head-rests, and equipment; and for decontaminating spills of blood and body fluids. Aqueous or alcohol-based povidone iodine is used for skin antisepsis (preoperative preparation), peritoneal wash, and treating superficial mycoses, such as *Tinea circinata* (ringworm) and oral/vaginal moniliasis. One per cent solution used as gargles for pharyngitis. Due to its virucidal action, it is used for local treatment of wound in case of dog bite.

Sodium Dichloroisocyanurate (NaDCC): Also known as "Troclosene", this compound is available as white powder, granules, and tablets. NaDCC powder containing 60 per cent available chlorine is recommended for decontamination of blood spills. The chemical is readily soluble in water. In solution, combined forms of chlorine (di- and mono-chloroisocyanurate) exist in a 50:50 equilibrium with hypochlorous acid (HOCl) and nascent chlorine (Cl) – combined forms of chlorine. As the chlorine is used up, the free chlorine is released from combined forms so as to maintain the 50:50 equilibrium. The process continues till all the combined chlorine is depleted. NaDCC is used for disinfection of drinking water, decontamination of floors, blood spills, laboratory glassware, bedpans, and urine bottles. NaDCC kills bacteria, mycobacteria, fungi, viruses, and spores. Its disinfecting action is not generally affected by presence of organic matter such as faeces, algae, blood, pus, and serum. It has a long shelf life (up to 2 years) and is effective over a wide pH range of 4–9. It is relatively less corrosive to metals and relatively safe. The antidote for accidental ingestion is drinking plenty of milk.

7.4.8 – Surface-Active Agents

Anionic Detergents: Common soap is an anionic detergent. Hard soap contains saturated fatty acids and hydroxide of sodium, or potassium; while soft soap contains unsaturated fatty acids and hydroxide of sodium, or potassium. Anionic detergents reduce surface tension and wash away microorganisms, sebaceous secretions from the skin, and dirt. Soaps are moderately bacteriostatic against Gram-positive bacteria. They are used for washing hands and body parts, and for soap water enema (evacuation enema). However, soaps are incompatible with cationic detergents and are precipitated by hard water.

Cationic Detergents: These quaternary ammonium compounds are water-soluble and act mainly on Gram-positive vegetative bacteria. They reduce surface tension and wash away microorganisms, sebaceous secretions from the skin, and dirt. These compounds are heat-stable, even on autoclaving and are most active at neutral pH and slightly alkaline pH. However, they have no effect on most Gram-negative organisms and mycobacteria, and are inactivated by anionic detergents (common soap) and acidic pH. Cationic detergents combine with protein in organic matter such as pus, sputum, blood, and body fluids, and thus get inactivated. They are adsorbed by cotton, rubber, and porous material and potentially hazardous to neural tissue. Cationic detergents are used as general-purpose skin antiseptics in 1:1,000 dilution. For burns and wounds, 1:2,000 dilution is recommended. They are also used as disinfectants in 1:2,000 dilution for nylon (*not* rubber) tubings and catheters; and in 1:20,000 dilution for irrigation of bladder and urethra in catheterised patients. For preventing nappy rash, nappies may be washed in 1:8,000 dilution. A 5 per cent w/v (weight by volume) solution is recommended for topical use for treatment of dandruff (*Pityriasis capitis*). *Benzalkonium chloride* is a quaternary ammonium compound with strong action against Gram-positive bacteria. When EDTA is added as a chelating agent, it acts on Gram-negative organisms also (Chitnis, 1997).

Savlon: This is a combination of cetavlon (or Cetrimide, a cationic detergent) and chlorhexidine (Hibitane). Though it is inactivated by anionic detergents (common soap) and acidic pH, its advantages are its solubility in water and alcohol. When diluted with alcohol, it is a better disinfectant than when diluted with water. It reduces surface tension and washes away microorganisms and dirt. It is effective against Gram-positive, Gram-negative bacteria, and fungi. It is heat stable even on autoclaving and acts through a wide pH range of 5–8, but most effective at neutral and slightly alkaline pH. Savlon is effective even in the presence of organic matter. It is chiefly used as a skin antiseptic for general use. A 1 per cent solution of Savlon is used for disinfecting plastic appliances, clinical thermometers, and Cheatle's forceps. The solution is to be changed daily.

7.4.9 – Miscellaneous Agents

7.4.9.1 – Hydrogen peroxide (H_2O_2)

Hydrogen peroxide (H_2O_2) is an unstable compound. A solution containing 6 per cent w/v (weight by volume) releases 20 times its volume of nascent oxygen (O). Such a solution is called 20 volumes H_2O_2. It releases nascent oxygen (O), when applied to tissues. Nascent oxygen (O) prevents multiplication of anaerobic bacteria and inactivates HIV in 30 minutes. Effervescence mechanically removes tissue debris from inaccessible regions. It has good microbicidal and deodorant action (Simpson & Slack, 2006). However, presence of proteinaceous organic matter reduces its activity (Chitnis, 1997). Release of nascent oxygen (O) and effervescence are of short duration. H_2O_2 should not be used in closed cavities, such as urinary bladder, due to its effervescence. It is to be stored in amber colour bottles and kept in a cold place. Being highly reactive and corrosive to skin and metals, it should not be used on copper, aluminium, brass, or zinc (Simpson & Slack, 2006). H_2O_2 is used for irrigating wounds, abscesses, and septic pockets to remove anaerobic conditions. In 1:8 dilution it is used as a deodorant gargle and mouthwash. It is also used for removing earwax and for disinfecting ventilators (Chitnis, 1997).

7.4.9.2 – Other oxidising agents

Peracetic acid, chlorine dioxide, and superoxidised water are used as oxidising agents for disinfecting flexible endoscopes in special closed washer-disinfector systems (Simpson & Slack, 2006). Endoscopes are immersed in 0.35 per cent per-acetic acid or 1100 PPM chlorine dioxide (ClO_2) for 5 minutes. This alternative method is used since personnel handling glutaraldehyde may develop adverse reactions (Gangakhedkar, 1999).

7.4.9.3 – Ethylene oxide gas

Ethylene oxide is a highly penetrative, non-corrosive, and microbicidal gas. It is used in industry for sterilising heat-sensitive medical devices, such as prosthetic heart valves and plastic catheters. The devices to be sterilised are exposed to ethylene oxide gas at a concentration of 700–1,000 mg per litre for about 2 hours. When using ethylene oxide gas, the process of disinfection is complex, and requires appropriate temperature (45–60° Celsius) and high relative humidity (more than 70 per cent) for action, followed by post-treatment aeration (Simpson & Slack, 2006). *Before sterilisation* the devices to be sterilised should be cleaned thoroughly and wrapped in a material that allows the gas to permeate (Chitnis, 1997). Presence of salt reduces effectiveness of ethylene oxide (Simpson & Slack, 2006). *During sterilisation* adequate precautions should be taken because the gas is inflammable and is potentially explosive. It is also toxic to personnel. *After sterilisation* the product is aerated to remove residual gas (Simpson & Slack, 2006).

7.4.10 – Recommended Chemical Disinfection

For preventing transmission of blood-borne pathogens including HIV and HBV, the recommended concentrations of disinfectants are given below. "Clean" condition indicates that the item or surface has been already cleaned and is free from organic contamination. "Dirty" condition refers to contamination of item, article, or surface with organic substances such as blood, body fluids, and excretions.

Chlorine Releasing Compounds
1. Household bleach containing 5 per cent available chlorine: for "clean" condition – 1:50 dilution (20 mL/litre), and for "dirty" condition – 1:5 dilution (200 mL/litre)
2. Sodium hypochlorite solution containing 5 per cent available chlorine (0.1 per cent = 1,000 PPM = 1 g/L; 1.0 per cent = 10,000 PPM = 10 g/L): for "clean" condition – 0.1 per cent or 1:50 dilution (20 mL/litre), and for "dirty" condition 1.0 per cent or 1:5 dilution (200 mL/litre)
3. Calcium hypochlorite (bleaching powder) containing 70 per cent available chlorine: for "clean" condition – 1.4 g per litre, and for "dirty" condition -14 g per litre
4. Sodium dichloroisocyanurate (NaDCC) powder containing 60 per cent available chlorine: for "clean" condition – 1.7 g per litre, and for "dirty" condition –17 g per litre.

5. Sodium hypochlorite-based tablets containing 1.5 g of available chlorine per tablet: for "clean condition – 1 tablet per litre, and for "dirty" condition – 4 tablets per litre.

6. Chloramine containing 25 per cent available chlorine: for "clean" condition – 20 g per litre, and for "dirty" condition – 20 or 40 g per litre. Chloramine releases chlorine slowly. The solution is *freshly* prepared every day, in non-metallic containers. Acids should not be used concomitantly since they displace chlorine gas.

Iodine Compounds
1. Tincture of Iodine (iodine 0.5 per cent + alcohol 70 per cent): for both "clean" and "dirty" conditions – 2.5 per cent.

2. Povidone iodine usually 10 per cent solution containing 1 per cent available iodine: for both "clean" and "dirty" conditions – 2.5 per cent. *Freshly* prepared dilution is to be used everyday. Povidone iodine should not be used on copper and aluminium.

Aldehydes
1. Glutaraldehyde (buffered): for both "clean" and "dirty" conditions – 2.0 per cent.

2. Formalin (40 per cent formaldehyde + 10 per cent methanol in water): for both "clean" and "dirty" conditions – 2.5 per cent.

Alcohols
1. Ethyl alcohol (ethanol): for both "clean" and "dirty" conditions – 70 per cent.

2. Isopropyl alcohol (isopropanol): for both "clean" and "dirty" conditions – 70 per cent.

3. Methylated spirit (also known as denatured or rectified spirit): for both "clean" and "dirty" conditions – 70 per cent.

Phenol Derivatives
1. Cresol: for "clean" conditions – 2.5 per cent, and for "dirty" conditions – 5.0 per cent.

2. Lysol (a saponified cresol): for both "clean" and "dirty" conditions – 2.5 per cent.

3. Chloroxylenols (4.8 per cent w/v is marketed as "Dettol"): for "clean" conditions – 4.0 per cent, and for "dirty" conditions – 10 per cent.

4. Chloroxylenols (4.8 per cent w/v) + EDTA: for "clean" conditions – 3.0 per cent, and for "dirty" conditions – 6.0 per cent.

Miscellaneous Disinfectants
1. Savlon (Cetrimide + Hibitane): for "clean" conditions – 5.0 per cent, and for "dirty" conditions – 10 per cent.

2. Ethylene oxide gas: 450–800 mg per litre is used for "clean" conditions. This compound is not recommended for use under "dirty" conditions.

3. Hydrogen peroxide (30 per cent stabilised solution): 6 per cent weight by volume (w/v) *freshly* prepared solution is used for "clean" conditions. This compound is not recommended for use under "dirty" conditions. H_2O_2 is stored in amber colour bottles in a cold place. It should not be used on copper, aluminium, brass, or zinc.

7.5 – BARRIER NURSING

Barrier nursing refers to use of physical or chemical barriers by all categories of hospital personnel and is not restricted to nursing staff. The objective of barrier nursing is to prevent the spread of pathogens through the intermediary of hospital staff. Health care providers with cuts, injuries, or infectious diseases should not be involved in patient care. While imparting mouth-to-mouth resuscitation, the risk of transmission of HIV is low. However, it is *safer* to use a barrier. Placing at least a gauze piece on the patient's mouth is to be advocated (Gangakhedkar, 1999).

Pre-requisites for barrier nursing are:
1. Induction and periodic in-service training to all staff in asepsis
2. Disinfection and periodic health check-up of personnel to rule out carrier state
3. Fully functional and efficient Central Sterile Supplies Department (CSSD)
4. Availability of stand-by autoclave
5. Availability of separate isolation ward or special infectious disease hospital. During epidemics, makeshift arrangement may be made in school buildings and community centres, or home isolations can be organised

The procedures for effective barrier nursing include: (a) repeated hand washing after attending to each patient, (b) concurrent and terminal disinfection, (c) using PPE, (d) establishing multidisciplinary HICC, (e) periodic supervision of disinfection, and (f) microbiological surveillance.

Barrier nursing is indicated in high-risk areas such as infectious disease wards and hospitals, neonatal wards, premature baby units, intensive care units, post-operative wards, and burns wards.

There are three types of barrier nursing techniques. In *cubicle nursing*, patients suffering from the same type of disease are kept in a cubicle and a separate set of PPE is kept for each cubicle. *Cohort nursing* is practised in premature baby and neonatal units. A group of babies born on the same day are cared for by a nurse. In *task nursing*, each nurse performs a given task such as giving tablets or making beds, and washes her hands after attending to each patient.

7.6 – UNIVERSAL BIOSAFETY PRECAUTIONS

Synonym: Universal work precautions. The basic principle is that blood and body fluids from *all patients* ought to be considered *potentially infected*, irrespective of their serological status. In health care facilities, hepatitis B is much more transmissible than HIV, while hepatitis C is considered to be of intermediate risk. However, HIV/AIDS has caused fear in the minds of lay persons and also health care personnel due to its: (a) association with sexual minorities, socially marginalised groups, and heterosexual promiscuity, (b) lack of preventive vaccine or curative therapy, and (c) fear of impending death if infected with HIV (Gangakhedkar, 1999).

The rapidly increasing prevalence rates of HIV infection in the general population increases the likelihood of occupational exposure to HIV. Transmission of HIV may be direct (contact with blood or body fluids) or through infected instruments. The major risk of transmission is through percutaneous exposure to infected blood or body fluids. Though infective doses of HIV are present in CSF, semen, blood, and cervico-vaginal secretions, HIV is present in different concentrations in almost all *body fluids* – amniotic, synovial, pleural, peritoneal, pericardial, sweat, faeces, nasal secretions, sputum, tears, urine, and breast milk of infected persons. Though CSF has the highest concentration of HIV, the likelihood of accidental occupational exposure to CSF is extremely low (CDC, 1988). The risk of transmission of HIV through exchange of fluids (other than sexual fluids and blood) tends to be extremely low or insignificant unless there is visible contamination with blood (Gandakhedkar, 1999). Lipophilic viruses such as HIV, HBV, and cytomegalovirus are *highly sensitive* to chemical disinfectants. In comparison, *Mycobacteria, Pseudomonas, Staphy-lococci*, spore-forming bacteria, fungi (*Candida, Cryptococci*), and hydrophilic viruses (polio virus, rhino virus) are not very susceptible to chemical disinfectants (Gangakhedkar, 1999).

7.6.1 – Controversies about Universal Precautions

It has been argued that universal precautions are too costly and time-consuming to be applied in case of *every* patient in *every* health care facility. The *suggested alternative* is to screen all patients for HIV antibodies before routine invasive procedures or surgery so that specific extra safety precautions can be taken. Extra safety precautions include selection of trained and experienced staff, restricting staff in the operating room to a bare minimum, and use of "surgical armour" for HIV-positive patients (2–3 pairs of gloves, impervious head gear, full-length waterproof apron, goggles, and rubber boots). Some surgeons are uncomfortable in such gear and feel their dexterity is compromised.

The pros and cons (Mitchell *et al.*, 1997; Gangakhedkar, 1999) are given below.

Argument 1: Mandatory HIV screening will uncover previously undiagnosed HIV infection and special protective gear can be used for HIV-infected patients. From the community point of view, routine screening along with suitable counselling may limit HIV transmission by detecting more HIV-positive persons who would have otherwise remained undetected. Initiation of ARV therapy and follow up may prolong survival (Rhame & Mahi, 1988).

Counterpoint 1: Mandatory screening will not identify HIV-infected individuals during *window period* when there is *high viraemia*. A non-reactive HIV test during the window period may lead to complacency among staff and it is possible that procedures may be undertaken without adequate precautions. On the other hand, false positive results may lead to unnecessary panic among staff. It is possible that HIV-infected patients may be refused health care causing mental agony for the patient. Often, patients may require emergency surgery before

HIV test result becomes available. Mandatory screening for HIV will not protect against hepatitis B/C and other unknown blood-borne pathogens. Mandatory screening cannot replace universal precautions since most cases of accidental occupational transmission have occurred from patients with *known* HIV infection. A study of 1,307 consecutive surgical procedures at San Francisco General Hospital found that knowledge of patient's HIV sero-status did not decrease the risk of occupational exposure (Gerberding *et al.*, 1990).

Argument 2: Use of protective gear and disposables is expensive.
Counterpoint 2: Mandatory screening for all patients before surgery is more expensive than the cost of universal precautions. Universal precautions are meant to protect against exposure to a range of blood-borne pathogens, besides HIV.

Argument 3: Mandatory screening of patients before surgery is similar to mandatory screening of donated blood, organs, and tissues.
Counterpoint 3: Before HIV testing, it is obligatory to counsel the patient and obtain informed consent. In case some patients refuse HIV testing their treatment cannot be compromised. Mandatory screening of donated blood, organs, and tissues does not require the donor's consent. Mandatory screening of patients raises issues regarding human rights and confidentiality and may hamper doctor–patient relationship. These issues do not arise while screening donated blood, organs, and tissues. Screening cannot prevent patient-to-patient cross infection. Hence, universal precautions are necessary.

Argument 4: Using universal precautions in all procedures is time-consuming and will delay emergency procedures.
Counterpoint 4: In most countries including India, it is mandatory to impart pre-test counselling and obtain informed consent before blood is collected for HIV testing. These procedures are also time-consuming and will delay emergency procedures. Moreover, result of the HIV test is also not available immediately, or may be inconclusive. Practising universal precautions is less time-consuming than waiting for result of HIV test.

7.6.2 – Components of Universal Precautions

1. Hand washing
2. Careful handling and disposal of sharps
3. Safe decontamination of spills of blood and body fluids
4. Safe techniques and safe method of transporting biological material
5. Using single-use injection vials.
6. Decontaminating, pre-cleaning and sterilising or disinfecting all multiuse equipment before reuse
7. Disposal of disposable and reusable items/materials as appropriate
8. Compliance with hospital protocol and norms on sterilisation and disinfection
9. Using PPE and provision for their regular and adequate supplies
10. Covering wounds and weeping skin lesions with waterproof dressing

11. Immunising health care providers with hepatitis B vaccine
12. Segregation and safe disposal of biomedical waste
13. Periodic training of personnel to prevent occupational exposure
14. Establishing *needle-stick audit* and *spills audit* to prevent recurrence

7.6.3 – Organisms on Skin

Organisms present on the skin are categorised as:
(a) Resident Organisms: These survive and multiply in the superficial layers of the skin and include coagulase-negative *Staphylococci*, diphtheroids, and *Candida*;
(b) Transient Organisms: These pathogens are acquired from infected or colonised patients or from the hospital environment. They survive for a limited time period on the skin of health care providers. Examples include *Escherichia coli* and *Staphylococcus aureus*.

7.6.4 – Hand Washing

The transfer of microorganisms through the hands of health care providers is the most important mode of transmission of hospital-associated infections (Sathe & Sathe, 1991). Hand washing is the most important measure to prevent the spread of infections. There are *three* types of hand washing as described below.

Social Hand Washing: This involves the use of *plain soap* and *water*. Most of the transient organisms are removed from moderately soiled hands. This type of hand washing is indicated whenever hands are soiled, before handling food, before eating or feeding patients, after using the toilet, and *before* and *after* nursing procedures, such as bed-making. Surfaces of both hands are to be cleaned using plain soap and water. The cleaning of hands should be carried out for at least 10–15 seconds. Hands should be rinsed under a stream of running water and dried with a disposable paper towel. *In the absence of running water*, stored water from an elevated drum with a spout or tap may be used as running water. Alternatively, a clean bowl containing *pathogen-free water* may be used. The bowl is to be cleaned and water changed after each use. *In the absence of paper towel*, a clean cloth may be used for drying hands and this cloth is to be discarded in a laundry bag after each use.

Hygienic Hand Washing: In this method, hands are washed using *antiseptics*. Four per cent Chlorhexidine (Hibitane) or povidone-iodine containing 0.75 per cent available iodine diluted in water is used. Alternatively, 0.5 per cent Chlorhexidine (Hibitane) or povidone-iodine containing 0.75 per cent available iodine in 70 per cent isopropanol or 70 per cent ethanol is used. Other antiseptics are diluted as per manufacturer's guidelines. Hygienic hand washing is indicated in any situation where microbial contamination or contact with blood or body fluids of patients is likely to occur; *before* handling immunocompromised patients; and *before* and *after* using gloves. Hands are to be washed thoroughly with the antiseptic, at least for 10–15 seconds. Subsequently, hands are rinsed and dried, as mentioned above. Alternatively, *alcohol hand wash* or *alcohol rub* is used, wherein the hands get automatically dry. Rinsing and drying of hands is not required. *Alcohol hand wash* solution contains 70 per cent methyl alcohol with 1 per cent glycerine (used as an emollient to prevent the skin-drying effect of alcohol).

Surgical Hand Washing: In this method, antiseptics are used for destroying transient organisms and decreasing the resident organisms. This method is indicated while carrying out invasive procedures to prevent wound contamination, when there is a risk of possible damage to gloves. During the washing procedure, it is *mandatory* to scrub hands, fingernails, and forearms (up to the elbows) at least *twice*. While washing, the hands are kept in an upright position with forearms flexed at the elbow. This ensures that water from the unwashed areas does *not* drip down to the washed (disinfected) areas. After washing, the tap is closed using the elbow. Alternatively, a foot-operated tap may be provided. The washed area is dried using a sterile towel or a disposable sterile paper towel. After drying, surgical gown and gloves are worn.

7.6.5 – Immunization against Hepatitis B

If the exposed person has not been previously vaccinated against hepatitis B, hepatitis B immunoglobulin in a dose of 0.06 mL per kg body weight is given intramuscularly. Additionally, complete primary course of hepatitis B vaccine is given. If the exposed person has previously received a complete course of hepatitis B vaccine, booster dose of the vaccine is given (NACO, Training Manual for Doctors).

7.6.6 – Preventing Needle Stick Injuries

Double gloves are to be worn (outer pair should be half size larger) during surgical procedures or where prolonged contact with blood or body fluids is likely. This prevents perforation of both the gloves at the same site and reduces risk of contact with blood or body fluids. Generally, needle stick injuries occur on the index finger and thumb of the non-dominant hand during procedures such as suturing. Injuries are usually caused by rash approach or lack of caution due to fatigue at the end of a surgical procedure. Exercising caution reduces risk of needle stick injury. Sharp instruments should *never* be handed over during a surgical procedure but should be placed on a tray so that the surgeon picks up the instrument by its blunt end. Needles should not be recapped after use. Used sharps (needles and other sharp instruments) are disposed off in a puncture-resistant container filled with freshly prepared 1 per cent hypochlorite solution. The needles should remain in the solution for at least half an hour. Used syringes are disposed of after heat-sealing their nozzles to prevent their reuse. Alternatively, used syringes are sent for incineration. Workers involved in disposal of sharps or infectious hospital waste must wear thick India rubber gloves that are reusable (Gangakhedkar, 1999).

7.6.7 – Transporting Biological Material

All body fluids and tissues from each and every patient should be considered potentially infectious, irrespective of the patient's HIV status. Body fluids and tissues should be sealed in a tightly closed inner container, which should be placed in a leak-proof transparent bag with a clearly visible "biohazard" label.

7.6.8 – Safe Decontamination of Spills of Blood and Body Fluids

All spills of blood or body fluids are potentially infectious. Health care personnel should adhere to the following procedure: (a) wear heavy duty India rubber gloves throughout the procedure, (b) cover the spill with absorbent material (paper napkin, thick blotting material, old newspaper), (c) pour *freshly prepared* hypochlorite solution containing 1 per cent available chlorine (10 gm per litre or 10,000 PPM) around the spill and over the absorbent material, (d) cover with absorbent material and place it in a waste container for contaminated waste, (e) wipe the surface again with disinfectant, (f) sweep the broken glass or fractured plastic containers, collect in a plastic scoop and dispose of in contaminated waste container, and (g) report all spills to hospital authorities. Hospital authorities should maintain a written record of all such incidents. After critical analysis they should issue recommendations to prevent recurrences.

7.6.9 – Handling Dead Bodies

Universal biosafety precautions are to be strictly followed while handling and cleaning the dead body and the trolley. HIV-positive cadaver is to be labelled on the right arm with a red patch, which is *suggestive* of HIV seropositivity, but maintains confidentiality. In order to maintain confidentiality, HIV status of the deceased should not be mentioned in the "Cause of Death" certificate. The secondary or opportunistic infection is to be mentioned as the cause of death. The details of the deceased are to be mentioned in a routine mortuary register. All natural orifices are plugged with cotton wool soaked in 1 per cent sodium hypochlorite, 2 per cent glutaraldehyde, or 10 per cent formalin to prevent spillage or seepage of body fluids from the cadaver. The body should be double-bagged in heavy plastic, irrespective of the HIV status of the deceased (Gangakhedkar, 1999).

If the deceased was known to be HIV-positive, the seropositive status of the deceased should be confided to the next of kin before handing over the body so that the relatives can take due precautions. The next of kin are to be advised that the body should be *preferably* be cremated. However, religious sentiments should be respected. After handing over the dead body to the next of kin, the following decontamination procedure must be adhered to, irrespective of the HIV status of the deceased: (a) first clean the trolley with soap and water, (b) *repeat* the cleaning process using 1 per cent sodium hypochlorite, 2 per cent glutaraldehyde, or 10 per cent formalin, (c) leave some of the disinfectant solution on the trolley for 1 hour, (d) subsequently, clean again with soap and water. Where possible, all materials that have been used in the diagnosis, nursing care, or treatment of the deceased should be incinerated.

Post-Mortem Examination: The number of persons in the autopsy room should be kept to a *minimum*. All persons should wear PPE (gown, impermeable apron, head cover, mask, gloves, and boots). All collected specimens (*except* specimens for culture or where fresh specimens are required) must be placed in 10 per cent formalin. After autopsy, the body should be double-bagged in heavy plastic, irrespective of the HIV status of the deceased. While handing over the body to relatives, they should be instructed not to disturb the plastic double bag before disposing of the body. If the deceased was known to be HIV-positive, the next of kin should be advised as mentioned above. After handing over the dead body to the next of kin, the decontamination procedure mentioned above is to be followed, irrespective of the HIV status of the deceased (Gangakhedkar, 1999).

Embalming: Strict adherence to universal precautions is essential while embalming. Body fluids and faecal matter are decontaminated using 0.5 per cent hypochlorite solution (Gangakhedkar, 1999).

Cadavers For Anatomical Dissection: All cadavers are injected intra-arterially with 10 per cent formalin before they are provided for anatomical dissection. HIV does not survive under these conditions. Experimentally, it has been shown that HIV gets inactivated within 48 hours even with one-tenth of this concentration of formalin (Gangakhedkar, 1999).

7.7 – PERSONAL PROTECTIVE EQUIPMENT (PPE)

PPE for blood-borne pathogens include – disposable solid front gowns (tied at the back) with cuffed sleeves, impermeable (waterproof) aprons, impervious head gear or head cover ("bouffant") with face mask, disposable gloves (surgical gloves and latex examination gloves with cuff), alcohol ("waterless") hand wash solution, absorbent laboratory mat, disposable bags with "biohazard" labels, rubber boots or shoe (foot) covers, and non-fog goggles or UVEX glasses (CDC, 2003a).

Face Masks: Health care personnel should discard face masks after 4–6 hours of use. Before performing procedures or surgeries where splashing of blood or body fluids is likely, it is better to wear safety hoods. Surgical masks are not splash-proof (CDC, 2003b).

Goggles (Eye Wear): Protective eye wear is to be worn before performing invasive procedures. Alcohol-based solutions are used to disinfect goggles prior to reuse. UVEX goggles may be worn with glasses (CDC, 2003a).

Latex Gloves: Latex is the milky, viscous sap of certain rubber-yielding trees that coagulates on exposure to air. Being a natural product, the availability of latex gloves is limited. Indiscriminate use of latex gloves would be wasteful. Heavy-duty

India rubber gloves should be used by workers involved in disposal of hospital waste. They should not wear latex gloves. The indications for using latex gloves are: (a) major surgical and invasive procedures, (b) procedures involving contact with blood (collecting blood by venepuncture, starting and removing intravenous lines, (c) procedures involving contact with body fluids (internal body examinations: per vaginal, per rectal, oral, dental), and (d) examination of infectious lesions – STIs and contagious infections of skin and mucosa.

Gloves should not be regarded as a substitute for hand washing. Hand washing is mandatory before sterile gloves are worn and after removing gloves. Gloves are not complete impermeable barriers because there is a risk of puncture or tear during surgery. Gloves should be changed after a procedure on a patient and even between two procedures on the same patient. If the gloves are torn during a procedure resulting in splash of blood or body fluids on the skin, the method for decontamination is as follows: (a) remove the torn gloves under *running* tap water, (b) wash hands thoroughly with soap and water for 2 minutes, (c) dip hands for 15 seconds in undiluted Savlon, and (d) wear a fresh pair of gloves before resuming the procedure (Gangakhedkar, 1999).

Gloves are sterilised by gamma radiation or autoclaving. Gamma-radiation sterilised disposable items should be used where possible. Ideally, surgical and examination gloves are to be used only once. While dealing with highly infectious diseases, gloves should never be washed or reused (CDC, 2003a). However, in resource-poor settings, reuse of gloves may be considered if not used in highly infectious situations. In such cases, gloves are to be reprocessed using the following procedure: (a) rinse gloved hands thoroughly in hypochlorite solution; then rinse gloved hands thoroughly under running tap water to remove residues of hypochlorite, which may cause deterioration of gloves; (b) wash gloved hands with soap or detergent and water; then rinse gloved hands thoroughly under running tap water to remove residues of soap or detergent, since these agents may enhance penetration of liquids through undetected holes in the gloves; (c) remove gloves and hang them up by their cuffs to dry; (d) wash hands thoroughly again with soap and water, (e) before reusing gloves, test for holes in the gloves by filling each glove with 325 mL water and 25 mL air at room temperature; twist them through 360° and place them in a rack for 2 minutes; detect leakage by visual and tactile means; and (f) dust gloves with French chalk powder or talcum powder before sending them for autoclaving (Gangakhedkar, 1999).

7.7.1 – General Guidelines for using PPE

Arrange all necessary equipment, material before starting the procedure. Wear and remove PPE in the order mentioned below. Decontaminate all used PPE, seal them in disposal bags and send for incineration. Do not reuse PPE.

Procedure for Wearing PPE: Wear shoe covers or rubber boots with trousers tucked inside. Wear face mask, head cover, and goggles. Rubber boots are preferable where the floor is likely to be wet or heavily contaminated. After surgical hand wash, wear impermeable (waterproof) apron. Wear gloves with gown-sleeve cuff tucked into the glove.

Procedure for Removing PPE: Wash gloved hands in hand-wash solution (containing more than 60 per cent alcohol) such as Sterillium. Using gloved hands, remove boots and place in receptacle containing 1 per cent bleach. Using gloved hands, remove the waterproof apron, gown, and shoe covers, without contaminating the clothing underneath. Touch only the outside of apron, gown, and shoe covers. Place the waterproof apron, gown, and shoe covers in a disposal bag with "biohazard" label. Remove outer gloves in such a way that fingers are under the cuff of the second glove. This will avoid contact between the skin and the outside of the inner glove. Wash hands in hand-wash solution (mentioned above). Remove goggles and place in receptacle for cleaning with alcohol-based disinfectant. The person cleaning the goggles should use the same PPE procedures. Remove head cover and mask. Place it in a disposal bag with "biohazard" label. Wash hands thoroughly up to elbow in hand-wash solution (mentioned above). Then, wash hands up to elbow, thoroughly with soapy water. Change into street clothing and wash hands once again in soapy water.

7.8 – MANAGEMENT OF BIOMEDICAL WASTE

In 1998, the Indian Government notified that all establishments (hospitals, nursing homes, animal houses, blood banks, research institutions) that generate biomedical waste should get registered and should install a suitable biomedical waste treatment or disposal facility in their premises, or set up a common facility. As per the government notification, biomedical waste is any waste, which is generated during the diagnosis, treatment, or immunisation of human beings or animals or in research activities pertaining thereto or in the production or testing of biological products, and other products mentioned in Schedule-I of the Biomedical Waste (Management and Handling) Rules, 1998 (MOEF, 1998).

7.8.1 – Salient Features of the Rules

Segregation of Waste: All waste must be segregated into the containers/bags at the point of generation itself and properly labelled as provided in the rules. The biomedical waste shall not be mixed with other waste.

Transportation of Waste: Requisite information such as "category of waste", sender's name and address, receiver's name and address, shall be provided while transporting the waste.

Storage and Disposal of Waste: The waste must be disposed of within a period of 48 hours and in case of storage beyond this period, specific permission of the prescribed authority shall be obtained. An annual report shall be submitted by 31 January every year to the prescribed authority in the prescribed format.

Record Keeping: Records shall be maintained pertaining to the generation, collection, reception, storage, transportation, treatment, disposal and/or any other form of handling of biomedical waste. The records should be kept ready for inspection any time. Any accident involving biomedical waste should be reported to the prescribed authority.

Local Authorities: Municipal Corporations and Councils have been made responsible for providing suitable site for common disposal or incineration in the area under their jurisdiction and taking an initiative for providing common waste treatment facility (CWTF) so that the biomedical waste generated in various institutions can be handled, treated, and disposed of in a scientific manner. Municipal Corporations and Councils must continue lifting non-biomedical waste, as well as treated biomedical waste for disposal in dumping grounds.

7.8.2 – Segregation at Point of Generation

Different types of waste are collected in category-specific colour-coded containers or bags at the site of generation of waste (Table 1). At this stage, wastes are segregated into different streams. Incorrect classification of wastes at this stage may create many problems later. If the infectious waste (which forms a small part of total hospital waste) is mixed with other non-infectious hospital waste, the entire waste will need to be treated as "infectious" waste (an expensive option). Otherwise, the entire hospital waste would have the potential to cause infection during handling and disposal. Segregation helps in reducing the bulk of waste, preventing the spread of infection to general waste, and reducing treatment cost and overall cost of waste handling and disposal in health care settings.

7.8.3 – Safe Storage and Transport

The category-specific colour codes for the containers or bags have been notified by the Government of India (Table 1). The bags are tied tightly after they are three-fourths full. Waste should *not* be stored at the place of generation for more than 2 days. Date of collection and other details of biomedical waste should be clearly mentioned on *red labels*.

The biomedical waste containers should be provided with a well fitting lid; protected from insects, birds, animals, and rain; and should *not* be accessible to rag pickers and scavengers. All waste should be transported *without spillage* in vehicles specifically authorised for this purpose.

7.9 – INFECTION CONTROL CHECKLIST

This checklist contains six indicators and may be used to assess the status of infection control measures in health care settings.

Table 1. Specifications for waste containers and disposal options (MOEF, 1998)

Container and colour code	Category of waste	Disposal options
Yellow plastic bags	Category 1: Human anatomical wastes (organs, tissues, blood, and body fluids) Category 2: Animal/slaughter house waste Category 3: Microbiology and biotechnology waste Category 6: Soiled wastes (linen, bedding, dressing contaminated with body fluids)	Incineration or deep burial
Red disinfected container or red plastic bag	Category 3: Microbiology and biotechnology waste Category 6: Soiled wastes (linen, bedding, dressing contaminated with body fluids) Category 7: Disposable items (other than sharps)	Autoclaving, microwaving, or chemical treatment
Blue or white transparent plastic bag or puncture-proof container	Category 4: Waste sharps (that can cause cuts and punctures) Category 7: Disposable items (other than sharps)	Autoclaving or microwaving or chemical treatment destruction and shredding
Black plastic bag	Category 5: Discarded and cytotoxic drugs Category 8: Incineration ash Category 9: Chemical wastes (solids)*	Chemical treatment followed by disposal in secured land fills

* For chemical wastes (liquids), chemical treatment is followed by discharge into drains.

Hand Washing:
• Soap, adequate quantity of clean water, and clean towels available (observed)
• Washing of hands and drying done every time after contact with body fluid, removal of gloves, or contact with patient (observed)
• Observe hand washing technique (whether correct?)

Use of PPE:
Disposable gloves, face masks, head gear, protective eye wear, plastic (water-proof) apron, foot or shoe covers or rubber boots (observed)

Waste Disposal:
• Evidence of segregation and safe storage of waste
• Correct colour coding of waste containers
• Whether waste treatment is done on premises or waste is sent to common waste treatment facility

Instruments:
* Whether steriliser is in working condition
* Whether instruments are cleaned thoroughly after use
* Whether clean instruments are stored in cupboards

Prevention of Accidental Occupational Injuries:
* Correct puncture-proof container for waste sharp
* Container less than three-quarters full at the time of observation
* Whether sharps are protruding from the container
* Observe recapping of needle and syringe
* Whether hospital staff members know whom to report incidents of accidental injuries

Infection Monitoring Mechanisms:

* Evidence of microbiological monitoring of operation theatres, wards, and labour room
* Whether disinfectants or antiseptics are periodically assessed for efficacy on locally existing organisms
* Whether written documentation of SOP exists for disinfecting or sterilising each item
* Whether all concerned staff are aware of such a document

REFERENCES

Ananthnarayan R. and Paniker C.K.J., 2000, Textbook of microbiology. 6th edn. Hyderabad: Orient Longman.

Centers for Disease Control (CDC), 2003a, www.cdc.gov/ncidod/sars.

Centers for Disease Control (CDC), 2003b, Biosafety in microbiological and biomedical laboratories manual. www.cdc.gov/od/biosfty/bmbl4/

Centers for Disease Control (CDC), 1988, Update – Universal precautions for prevention of transmission of HIV, hepatitis B and other blood borne pathogens in health care settings. Morb Mort Wkly Rep Suppl 37: 377–382, 387–388.

Chitnis D.S., 1997, Choice of disinfectants in hospitals. Bombay Hosp Jour 39(1): 42–48.

Collins C.H. and Grange J.M., 1990, The microbiological hazards of occupations. Leeds: Science Reviews.

Cunningham A.L., Dwyer D.E., Mills J., and Montagnier L., 1997, Structure and function of HIV. In: Managing HIV (G.J. Stewart, ed.). North Sydney: Australasia Medical Publishing.

Gangakhedkar R.R., 1999, Universal precautions in hospital setting. AIDS Research and Review 2(3): 146–150.

Gerberding J.L., Littel M.P.H.C., *et al.*, 1990, Risk of exposure of surgical personnel to patients' blood during surgery at San Francisco general hospital. N Engl J Med 322: 1788.

Kelsey J.C. and Sykes G.A., 1969, A new test for the assessment of disinfectants with particular reference to their use in hospitals. Pharmaceutical J 202: 607–609.

Maurer I.M., 1969, A test for stability and long-term effectiveness in disinfectants. Pharmaceutical J 203: 529–534.

Ministry of Environment and Forests (MOEF), 1998, Biomedical Waste (Management and Handling) Rules (1998). New Delhi: Government of India, 27 July. Amended on 6th March and 2nd June 2000.

Mitchell D.H., Sorrell T.C., and McDonald P.J., 1997, HIV Control in Medical Practice. In: Managing HIV (G.J. Stewart, ed.). North Sydney: Australasian Medical Publishing, pp 149–153.

National AIDS Control Organisation (NACO). Training manual for doctors. New Delhi: Government of India.

Rhame F. and Mahi D., 1988, The case for wider use of testing HIV infection. N Engl J Med 320: 1248–1254.

Sathe P.V. and Sathe A.P., 1991, Epidemiology and management for health for all. Mumbai: Popular Prakashan.

Simpson R.A. and Slack R.C.B., 2006, Sterilisation and disinfection. In: Medical microbiology (D. Greenwood, R.C.B. Slack, and J.F. Peutherer, eds.). 16th edn. New Delhi: Elsevier, pp 73–82.

SECTION TWO

DIAGNOSTIC ASPECTS

CHAPTER 8

CASE DEFINITIONS

Abstract

Several surveillance case definitions are used for HIV sentinel surveillance and surveillance reports should indicate which definition has been used. An appropriate national AIDS surveillance definition is recommended for systematic reporting of AIDS cases. Surveillance case definitions should not be used as a guide for clinical diagnosis or for other purposes and should not be confused with clinical staging systems for HIV infection, which are meant for clinical management of patients and clinical research purposes. Clinical case definitions are available for children and adults. A range of AIDS-defining clinical conditions have been described.

Key Words

AIDS-defining conditions, Bangui definition, Candidiasis, Caracas definition, Clinical case definitions, Cryptococcal meningitis, Cytomegalovirus retinitis, E/R/S tests, Herpes zoster, HIV sentinel surveillance, Kaposi's sarcoma, *Penicillium marneffei* infection, Persistent generalised lymphadenopathy, *Pneumocystis* pneumonia, Surveillance case definitions, Toxoplasmosis.

8.1 – INTRODUCTION

Surveillance of HIV infection and AIDS cases is necessary for monitoring the course of the HIV pandemic and planning suitable public health interventions. The WHO recommends HIV sentinel surveillance for HIV infection and systematic reporting of AIDS cases (using an appropriate national AIDS surveillance definition) for their surveillance. HIV sentinel surveillance is a method for measuring the prevalence of HIV infection in specific populations. In this method, blood obtained in the health care setting for other purposes such as screening for syphilis in antenatal clinics is tested for antibody to HIV after all patient identifiers have been removed. Surveillance case definitions should not be used as a guide for clinical diagnosis or for other purposes and should not be confused with clinical staging systems of HIV infection, which are meant for clinical management of patients and clinical research purposes (WHO, 1994).

123

8.1.1 – Various Case Definitions

WHO Surveillance Case Definition: is simple to use and inexpensive since it does not involve HIV serological testing. This is a modification of the provisional WHO clinical case definition (or "Bangui definition"). Its relatively low sensitivity and specificity is especially so because HIV-negative tuberculosis patients have similar clinical presentations.

Expanded WHO Surveillance Case Definition: has a higher specificity than the WHO case definition and is also simple to use. However, it requires HIV serological testing which may be expensive and logistically difficult.

1999 CDC Revised Case Definition for AIDS Surveillance: is available on the website of the Centers for Disease Control and Prevention: www.cdc.gov.

Pan American Health Organization (PAHO): also called "Caracas definition" is a case definition for AIDS surveillance.

European Case Definition for AIDS Surveillance: does not use immunological criteria. It includes recurrent pneumonia, pulmonary tuberculosis, or invasive cervical cancer as AIDS indicator conditions in persons with confirmed HIV infection (WHO, 1994).

One or more of the above-mentioned case definitions should adequately meet the AIDS surveillance requirements for most countries of the world. Case definitions that require HIV serology are suitable for use in settings with intermediate to high levels of diagnostic capabilities. The surveillance reports should indicate which definition has been used (WHO, 1994).

HIV infection and/or disease is suspected on clinical symptomatology and confirmed by serological tests. The case history should include past and present history of injecting drugs, multiple sexual contacts, history of genital ulcer or discharge, transfusion of blood and/or blood products, and scarification or tattooing (NACO). The difficulties in diagnosis are that the clinical signs and symptoms are not pathognomonic and may mimic those of many common illnesses.

The CDC case definition (CDC, 1993) is more rigorous and involves sophisticated diagnostic procedures. The surveillance case definition of the WHO (1986) is simplified and correlates well with CDC case definition (CDC, 1993).

8.2 – NACO CASE DEFINITION FOR CHILDREN (UP TO 12 YEARS OF AGE)

This case definition is given by NACO, Government of India. Two positive serological HIV tests in children older than 18 months, or confirmed maternal HIV infection in children younger than 18 months, and presence of at least two major and two minor signs in absence of known cause of immune suppression. *Major signs*: loss of body weight or failure to thrive, diarrhoea (intermittent or continuous) for more than 1 month, and fever (intermittent or continuous) for more than 1 month. *Minor signs*: repeated attacks of common infections such as pneumonitis, otitis and pharyngitis; PGL, oropharyngeal candidiasis, persistent cough for more than 1 month, and disseminated maculo-papular dermatosis.

8.3 – NACO CASE DEFINITION FOR ADULTS

Two positive serological tests for HIV by ELISA, rapid, or simple (E/R/S) tests and any *one* of the following criteria:

1. Weight loss (more than 10 per cent of body weight lost in 1 month), or HIV-related cachexia and diarrhoea (intermittent or continuous) for more than 1 month, or fever (intermittent or continuous) for more than 1 month
2. Tuberculosis (extensive pulmonary, miliary, and extra-pulmonary)
3. HIV-related neurological impairment preventing independent daily activities
4. Candidiasis of oesophagus (diagnosable by oral candidiasis with odynophagia)
5. Clinically diagnosed life-threatening or recurrent episodes of pneumonia, with or without etiological confirmation
6. Kaposi's sarcoma
7. Presence of other conditions – Cryptococcal meningitis, neuro-toxoplasmosis, cytomegalovirus retinitis, *P. marneffei* infection, disseminated molluscum, and recurrent or multidermatome herpes zoster

8.4 – WHO CLINICAL CASE DEFINITIONS

8.4.1 – Adults

AIDS in an adult is suspected by the presence of *at least two* major signs associated with *at least one* minor sign, in the absence of known causes of immune suppression such as cancer and severe malnutrition.

Major signs: weight loss (more than 10 per cent of body weight lost in previous 1 month), diarrhoea for more than 1 month, and prolonged fever for more than 1 month. *Minor signs*: persistent cough for more than 1 month, generalised pruritic dermatitis, recurrent herpes zoster, oropharyngeal candidiasis, chronic progressive and disseminated herpes simplex infection, and generalised lymphadenopathy.

8.4.2 – Children

AIDS is suspected in an infant or child presenting with *at least two* major signs associated with *at least two* minor signs, in the absence of known causes of immune suppression such as cancer and severe malnutrition.

Major signs: weight loss or abnormally slow growth (failure to thrive), diarrhoea for more than 1 month, and prolonged fever for more than 1 month. *Minor signs*: generalised lymphadenopathy, oropharyngeal candidiasis, repeated common infections such as otitis and pharyngitis, persistent cough for more than 1 month, generalised dermatitis, and confirmed maternal HIV infection.

8.5 – AIDS-DEFINING CLINICAL CONDITIONS

The CDC has defined a range of AIDS-defining clinical conditions in adults and adolescents. This includes atypical mycobacterial infection (disseminated or extra-pulmonary), candidiasis of oesophagus and lower respiratory tract, cancer of uterine cervix, coccidioidomycosis, cryptococcosis, cryptosporidiasis (for more than 1 month), cytomegalovirus disease, herpes simplex virus infection (for more than 1 month), histoplasmosis, HIV-related encephalopathy, isosporiasis (for more than 1 month), Kaposi's sarcoma, lymphomas (primary, Burkitt's, immunoblastic), *pneumocystis* pneumonia, recurrent septicemia due to *Salmonella*, toxoplasmosis of brain, tuberculosis at any site, and wasting syndrome due to HIV (also called "slim disease").

8.6 – SYSTEMIC MANIFESTATIONS

In addition to the clinical presentations mentioned below, patients may present with signs and symptoms pertaining to cardiac, renal, endocrine, reproductive, and haematological systems.

8.6.1 – Gastrointestinal System

Dysphagia may be seen in oral and oesophageal candidiasis, cytomegalovirus oesophagitis, oral hairy leukoplakia, gingivitis, ulcer. OHL seems to be unique to HIV-infected individuals. The margins of the tongue show white ridges of fronds on the epithelium. An association with EBV and papillomaviruses has been proposed (Simmonds & Peutherer, 2006). Persistent diarrhoea is a feature of infection with *Salmonella, Shigella, Entamoeba histolytica, Giardia lamblia, Microspora, Cryptosporidia,* and *Isospora*. Colitis is present in Kaposi's sarcoma and cytomegalovirus infection. Individuals with proctitis or ulcers due to herpes virus may experience perianal discomfort.

8.6.2 – Respiratory System

Patients with tuberculosis may present with persistent cough, breathlessness, haemoptysis, and pleural effusion. Rapid breathing and cyanosis are seen in atypical mycobacterial infection, bacterial pneumonia, *Legionella* infection, cytomegalovirus pneumonia, PCP, interstitial pneumonia, candidiasis, and herpes simplex infection.

8.6.3 – Neurological System

When HIV-infected individuals complain of visual changes, it is important to rule out cytomegalovirus retinitis, microsporidiasis, toxoplasmosis, and keratoconjunctivitis. Headache and lethargy may be seen in HIV encephalopathy, though fatigue is commonly experienced as the disease progresses. Meningism or meningitis may be due to infection with crytococcus, *M. tuberculosis*, or other bacteria. Neurological deficits may occur as a consequence of lymphoma and abscesses caused by mycobacteria, cryptococcus, and toxoplasma. Lymphoma and HIV vasculitis are among the causes of peripheral neuropathy in HIV-infected persons. Ataxia, altered personality, convulsions, and incontinence are seen in lymphoma, AIDS dementia complex, cryptococcal meningitis, and herpes virus infection.

8.6.4 – Skin and Mucous Membranes

Infections of skin and mucous membranes include herpes zoster, herpes simplex, *candidiasis*, *cryptococcosis*, *histoplasmosis*, *molluscum contagiosum*, *folliculitis*, and *hairy leukoplakia*. Kaposi's sarcoma, lymphoma, and basal cell carcinoma are cutaneous malignancies seen in HIV sero-positive individuals. Pruritic popular dermatitis, seborrhoeic dermatitis, drug eruptions, vasculitis, and gingivitis are among the other manifestations.

REFERENCES

Centers for Disease Control and Prevention (CDC), 1993, Revised classification system for HIV infection and expanded surveillance case definition for AIDS among adolescents and adults. Morb Mortal Wkly Rep 41(RR-17): 1–19.

National Aids Control Organisation (NACO), Training manual for doctors. New Delhi: NACO.

Simmonds P. and Peutherer J.F., 2006, Retroviruses. In: Medical microbiology (D. Greenwood, R.C.B. Slack, and J.F. Peutherer, eds.), 16th edn. New Delhi: Elsevier, pp 527–538.

WHO, 1986, Acquired immunodeficiency syndrome (AIDS). Wkly Epidemiol Rec 61: 69–73.

WHO, 1994, WHO case definitions for AIDS surveillance in adults and adolescents. Wkly Epidemiol Rec 69(37): 273–275.

CHAPTER 9

LABORATORY DIAGNOSIS OF HIV INFECTION

Abstract

Tests for diagnosis of HIV infection include specific tests for HIV infection, detecting immune deficiency, and diagnosing opportunistic infections and malignancies. The *specific* tests for diagnosis of HIV infection are *screening tests*: Enzyme linked immunosorbent assay (ELISA), rapid assays, and simple agglutination assays, collectively known as "E/R/S"; *confirmatory* or *supplemental assays*: immunoblot or Western blot (WB), immuno-fluorescence assay (IFA), radio-immuno precipitation assay (RIPA), and radio-immuno assay (RIA); and *tests for detecting HIV antigen, viral nucleic acids, viral components, or the virus itself*: detection of p24 viral antigen, RT assay, virus culture, and detection of HIV nucleic acids. HIV-infected women tend to have higher CD4 cell counts as compared to their male counterparts, even when the viral loads are similar. HIV-infected women also tend to develop opportunistic infections at higher CD4 counts and this fact should be considered while initiating ARV treatment.

Key Words

ELISA, HIV testing policy, HIV testing protocols, Polymerase chain reaction, Rapid assays, Simple assays, Strategies for HIV testing, Supplemental assays, Western blot

9.1 – PURPOSE OF HIV TESTING

Mandatory Screening: All donated blood, blood products, semen, cells, tissues, cornea, bone marrow, kidney, and other organs are screened for the presence of HIV antibodies, as required by the law of the country. Antenatal screening may also be carried out. An individual found to be positive for HIV antibody should *never* donate blood or any body parts. Though HIV antibody tests may be negative during the early stage of infection, mandatory screening can eliminate large majority of HIV positive persons among potential donors. An individual who tests positive in early stage of infection is infectious to others.

Sero-Epidemiological Studies: Serological tests are useful in epidemiological studies to study incidence, prevalence, geographical, and demographical distribution of HIV infection.

Diagnosis: HIV antibody tests are used to determine when infection has occurred after accidental occupational exposure or in cases of rape. The antibody tests may be negative in acute illness (CDC Group I) and also in late stages when immune suppression has occurred. The test kits should be able to detect anti-bodies to both HIV-1 and HIV-2.

Prognosis: Absence of detectable anti-p24 antibodies indicates clinical deterioration. This is associated with increase in HIV antigens and viral load in peripheral blood. The following tests are used for monitoring the course of HIV infection: (a) CD4 cell count: less than 500 cells per mm^3 of peripheral blood indicates progression of disease and the need for ARV treatment; less than 200 cells per mm^3 indicates risk of severe opportunistic infections; (b) viral load: measurement of HIV RNA by reverse transcriptase (RT)-PCR; and (c) measurement of beta-2-microglobin and neopterin in serum or urine: increasing levels indicate progression of HIV disease

9.2 – TYPES OF TESTS

9.2.1 – Laboratory tests for HIV infection and related conditions

1. *Specific tests* for the diagnosis of HIV infection are grouped as follows: (a) screening (E/R/S) tests, (b) confirmatory or supplemental assays, and (c) tests for detecting HIV antigen, viral nucleic acids, viral components, or the virus itself – detection of p24 viral antigen, RT assay, virus culture, and detection of HIV nucleic acids.
2. *Tests for detecting immune deficiency* are not specific for detecting HIV infection alone. Abnormal results may also be obtained in other diseases and/or infections. They provide extra information about the health status of the patient. These comprise: (a) surrogate markers: serum prolactin (increased), dehydro epiandrosterone (decreased), anti-infectivity factor protein to HIV-1 (present), serum IgE (increased); and (b) indirect predictors: total leukocyte count (less than 400/mm^3), CD4: CD8 ratio (normal = 2:1; AIDS = 0.5:1), CD4 cell count (decreased), thrombocyte (platelet) count (decreased), beta-2-microglobulin (increased), serum neopterin (increased), alpha-1-thymosin (increased), IgG and IgA levels (both raised), tuberculin and other skin tests (indicate reduced CMI).
3. *Tests for diagnosing opportunistic infections and malignancies*: most opportunistic infections are diagnosed by microscope and culture. Serology is unreliable due to immune suppression.

In most cases, the duration of viraemia is approximately less than 1 week before the appearance of anti-HIV antibodies. Anti-HIV antibodies are detectable by commonly employed tests within 4–6 weeks of infection and in virtually all infected individuals within 6 months and these antibodies persist for life. The level of viraemia is a predictive marker for progression of HIV disease (WHO, 2003).

9.2.2 – CD4 Cell Counts

The normal absolute count of CD4 cells is 950–1,700 per mm^3 of peripheral blood. The percentage of CD4 and CD8 cells is 20–45 per cent and 30–60 per cent, respectively. These normal counts are based on Western literature. Studies conducted in Lucknow and Vellore show variations in these counts for the Indian population. Hence, a nationwide baseline study is necessary. 3 mL of blood is collected aseptically by venepuncture in a vaccutainer containing EDTA (an anticoagulant). The blood is collected after 10–12 hours of overnight fasting. Ideally, the sample should be processed *immediately* to prevent erroneous results due to disintegration of T-lymphocytes. CD4 and CD8 counts are usually determined by flow cytometry. This method measures the wavelength of light emitted by cells as they flow in a stream, to determine the type of cell. The percentage of cells of each type is multiplied by the total lymphocyte count per micro litre to calculate the absolute number of CD4 or CD8 cells per mL of blood. Manual count kits, enzyme immunoassays, and microsphere assays are also used for CD4 and CD8 cell counts. Though easier to perform and less expensive, these methods are less accurate, compared with flow cytometry. CD4 counts show considerable interassay variation.

As compared with adults, all counts of T-lymphocytes are higher in children. In women, the CD4 counts are influenced by the phase of the menstrual cycle. The CD4 cell counts are higher in the morning and day-to-day variations are also seen. It varies during stress, exercise, acute illness, and drug therapy. Therefore, any large unexpected change in the CD4 cell counts should be confirmed by repeat testing, few days later. The *percentage* of CD4 cells is less variable than the absolute count and may be more reliable in evaluating response to ARV therapy or progression of HIV disease. HIV-infected women tend to have higher CD4 cell counts as compared with their male counterparts, even when the viral loads are similar. HIV-infected women also tend to develop opportunistic infections at higher CD4 counts and this fact should be considered while initiating antiretroviral treatment.

9.2.3 – HIV Diagnosis in the New Born

Conventional antibody tests cannot establish the presence of HIV infection in babies born to HIV-infected mothers because presence of anti-HIV antibodies in newborns may be due to either primary infection or passive transfer of anti-HIV antibody from the mother to the uninfected baby. These anti-HIV antibodies may persist up to 18 months. Appropriate methods for diagnosing HIV infection in babies less than 18 months are by detection of viral DNA or RNA, viral culture, and detection of IgA antibodies to HIV. IgA does not cross the placenta (WHO, 2003).

9.3 – SPECIFIC TESTS

This is the simplest and most frequently used diagnostic technique. Antibodies to both *core* (p24) or *envelope* (gp120, gp41) proteins are detected. There are two types of serological tests: screening (E/R/S) tests and supplemental or confirmatory tests. Among the tests for detecting HIV infection, antibody detection remains the method of choice (Joshi & Chipkar, 1997). Antibodies can not be detected during the "window period" when initial viral replication occurs (Table 1). IgM antibodies appear in circulation only about 3–4 weeks after infection. Subsequently, IgG antibodies appear. IgM antibodies disappear after 8–10 weeks, but IgG antibodies remain throughout. With the onset of immune suppression, some anti-HIV antibodies such as anti-p24 may disappear.

9.3.1 – Screening (E/R/S) Tests

The following three tests are collectively known as "E/R/S" and includes (a)enzyme linked immunosorbent assay (ELISA); (b) rapid assays – dot blot assay, particle agglutination (latex, gelatine), HIV spot, and comb tests; and (c) simple tests based on ELISA principle.

CLINICAL SPECIMENS: The specimen of choice for HIV testing is serum or plasma. Assays for detecting anti-HIV antibodies in whole blood, saliva, urine, and dried blood spot have also been developed (WHO, 2003).

Saliva: In injecting drug users, blood vessels are collapsed and serum samples can not be easily obtained. Saliva is an acceptable alternative clinical specimen for antibody testing by ELISA. Use of saliva yields comparable results. Collection of saliva is a non-invasive and painless procedure with no risk of needle-stick injuries to health care workers (Frerichs *et al.*, 1992). However, tests that use saliva are not very sensitive in relation to recent sero-conversion.

Urine: HIV antibody levels in urine are much lower than that in plasma. This could lead to false negative reaction, especially in early seroconverters. It is also difficult to achieve sterility of urine to prevent microbial contamination and enzymatic activity (Joshi & Chipkar, 1997).

9.3.1.1 – ELISA

ELISA is recommended as a *first-line test* by NACO. ELISA techniques require an ELISA reader and are suitable for use in laboratories where more than 30 samples are tested at a time (WHO, 2003). Direct solid phase ELISA is most

Table 1. Relation of tests to clinical stage

Stage of infection	Antigen detection			Antibody detection	
	p24[a]	RT[a]	Virus isolation	ELISA	Western blot
Window period	+ (< 50%)	+ (< 50%)	++	–	–
Acute HIV infection	+	+	±	+	+ (Partial p24 and/or gp120)
Asymptomatic infection	–[b]	–	–	+	+ (Full pattern)
Symptomatic disease	+	+	+	+	+ (p24 absent)

[a] *p 24 = p24 antigen; RT = Reverse Transcriptase assay.*
[b] *Decline of p24 antigen signals progression of disease and indicates need for ARV therapy.*

commonly used. Most commercially available ELISA kits detect both HIV-1 and HIV-2.

Procedure: The antigen is prepared from viral lysate or recombinant protein and/or synthetic peptides (WHO, 2003). The viral antigen is coated on surface of microtitre wells and the test serum is added. If antibody is present, it binds to the viral antigen; unbound serum is washed away. Anti-human goat Ig linked to a suitable enzyme and a colour-forming substrate are added. Development of colour (detected by photometer and read by ELISA reader) indicates *positive* test.

Advantages: Easy to perform, high sensitivity and specificity, can be automated for testing a large number of samples, reagents have a shelf life of 6–12 months, and economically advantageous over rapid test.

Disadvantages: Takes 3 hours to yield results. In general, higher the prevalence of HIV positive, greater the probability of true positives, greater is the positive predictive value (PPV). Causes of *false positive* ELISA test include blood malignancies, any infection due to DNA-virus, autoimmune disorders, multiple myeloma, primary biliary cirrhosis, alcohol-induced hepatitis, chronic renal failure, positive RPR test for syphilis, vaccination against influenza and hepatitis B, treatment with antisera-containing antibodies, antibodies to class II lymphocytes, and Steven-Johnson syndrome. *False negative* ELISA test results may be obtained when the test is performed during the "window period". Other causes include immunosuppressive therapy, replacement transfusion, malignant disorders, B-cell dysfunction, bone marrow transplantation, use of kits that primarily detect p24 antibodies, and presence of laboratory glove starch powder (NACO Training Manual).

Modifications of ELISA: Various modifications of ELISA are possible, depending on solid phase, source of antigen, enzyme conjugate-substrate detection system, and sequence in which the reagents are used in ELISA (Joshi & Chipkar, 1997). The solid phases used are microtiter plate (common), polystyrene beads (common), nitrocellulose membrane, nylon membrane, red blood cells, gelatine particles, latex particles, and microscope slides.

Sources of Antigen: (a) viral lysate: since this is produced from T-cell culture material, the contaminating cellular antigens can increase the false positive reactions. This is used in the first generation ELISA; (b) recombinant antigen: incorporating the genetic material of HIV into "vehicles" such as *Escherichia coli* produces recombinant antigens. This is used in second generation ELISA; (c) synthetic peptides: these are chemically synthesized amino acid residues corresponding to specific epitopes of viral antigens. They are free from contaminating cellular material and are used in third generation ELISA.

Enzyme Conjugate-Substrate Detection System: A substrate is a reagent, which is degraded in the presence of the conjugate due to enzymatic activity. The enzymes used are alkaline phosphatase and horseradish peroxidase. *Horseradish peroxidase* is the preferred enzyme due to its low cost, easy conjugation to protein and the wide variety of substrates that can be used in combination with it.

   Based on the sequence in which the reagents are added, ELISA can be classified as follows:

1. Indirect ELISA: This is the most common type of ELISA used for antibody detection. The HIV antigens are coated on to the solid phase such as wells of microtitre plate or polystyrene beads. Antibodies, if present in the sample will bind to the antigen. This antigen-antibody complex can be detected by enzyme-labelled conjugate. A colour reaction is produced on addition of a suitable substrate. This colour reaction is directly proportional to the concentration of antibodies in the given sample.

2. Competitive ELISA: The sample from the patient and the enzyme-labelled antibody (conjugate) are added simultaneously on to the solid phase. HIV antibodies, if present in the sample, compete with the antibodies in the conjugate and reduce the binding capacity of the labelled antibody on the solid phase. If a colour reaction is produced on adding substrate, it indicates the absence of HIV antibodies in the given sample.

3. Antigen Sandwich ELISA: It is a modification of indirect ELISA, to increase the sensitivity and specificity of the test. This assay detects all classes of HIV antibodies. The components of this test are similar to that of indirect ELISA, except that an enzyme-labelled antigen is used instead of enzyme-conjugated anti-human immunoglobulin.

4. Antigen-Antibody Capture Assay: The solid phase is coated with an antibody agent (usually goat or sheep anti-human immunoglobulin), which "captures" the antibody, if it is present in the sample being tested. The antibody is detected by using either an enzyme-labelled antigen, or antigen followed by an enzyme-labelled antibody. If an antibody is present in the given sample, on addition of the substrate, a colour change is obtained, in presence of the conjugate. This colour change is directly proportional to the amount of antibody present in the specimen.

5. Chemiluminescence: This is a modification of ELISA, involving the use of light enhancing substances with the substrate. A special reader is used, which is able to detect the slightest variation in optical density (Joshi & Chipkar, 1997).

9.3.2 – Rapid Tests ("R")

Rapid tests are visual tests that do not require the ELISA reader. These tests are available in smaller test packs and are therefore suitable for a laboratory processing smaller number of samples (WHO, 2003). Rapid tests include dot-blot assays, particle agglutination (gelatine, erythrocyte, latex, and microbeads), HIV-spot and comb test, and fluorometric microparticle technologies. These tests use HIV-1 or HIV-2 recombinant or synthetic peptide antigens separately. This allows differentiation of HIV-1 and HIV-2 infection.

Advantages: Most rapid assays have a built-in immunological control "dot" to confirm that the test has been performed accurately. Though most rapid tests have sensitivity and specificity comparable to that of ELISA, they are technically simple to perform and results are available in less than 10 minutes. No specific equipment is required for identifying positive reaction.

Disadvantages: These are expensive tests with subjective end-point. Stored, contaminated, or lipaemic samples can give faulty results. Most of the rapid assays use only one synthetic peptide. Hence, early seroconverters or weak positive samples may not be detected. Automation is not possible and thus, only a few tests can be done at a time (Joshi & Chipkar, 1997).

9.3.3 – Simple Tests ("S")

These are based on the ELISA principle. Simple tests take a little longer (more than 30 minutes), as compared with rapid tests which are used to confirm the diagnosis. The WHO recommends that a sample tested positive by one type of ELISA/rapid test should be retested by another ELISA/rapid test, based on a different principle, and using a different antigen preparation.

9.3.4 – Agglutination Assays

These assays have a solid phase (usually gelatin particles, red blood cells, latex particles, or microbeads are used), which acts as the indicator system. The HIV antigens get non-specifically bound to these carrier particles. HIV antibodies, if present in the given sample, form a lattice network with the antigen-coated particles. This lattice network is visualised as "clumping" or "agglutination".

Advantages: These tests are easy to perform, have good sensitivity, require little or no equipment, and are relatively cheap. Their results are comparable to that of ELISA, and are less time consuming.

Disadvantages: These tests cannot be automated for mass screening of samples; false negative reactions or "prozone phenomenon" can occur, especially when gelatin particles are used as the solid phase (false negative reactions are prevented by diluting the samples); interpretation of reactions is subjective; and species-specific agglutinins are a problem when red blood cells are used as solid phase. Red blood cells (erythrocytes) have a tendency for *rouleaux* formation, which could be misinterpreted as a positive reaction (Joshi & Chipkar, 1997).

9.3.5 – Confirmatory (or Supplementary) Tests

The WHO recommends that a sample tested positive by one type of ELISA/rapid test should be re-tested by another ELISA/rapid test, based on a different principle, and using different antigen preparation. When a tested sample is reactive once, by a system of E/R/S, the test is repeated by a different system to confirm the diagnosis. When a tested sample is reactive a second time, a confirmatory/supplemental test is used with the same sample to confirm the diagnosis (Joshi & Chipkar, 1997; NACO).

Rationale: Screening tests are highly sensitive, but may lack specificity. False positive reactions can be caused by cross-reactivity with contaminating host cellular antigens or due to technical errors. Therefore, more specific confirmatory tests are essential for discriminating between a true positive and false positive reaction.

Limitations: Confirmatory assays do not always produce conclusive results ("positive" or "negative"). Additional tests should be performed in case the results are inconclusive or "indeterminate". Individuals who test indeterminate should be re-tested after several weeks or months (Joshi & Chipkar, 1997).

9.3.5.1 – Western Blot

This is the standard confirmatory assay, which should be used only to resolve indeterminate results or diagnose HIV-2 infection (WHO, 2003). Combination kits for detection of HIV-1 and HIV-2 antibodies are available, where HIV-1 viral lysate antigens and HIV-2 envelope-synthetic peptides are incorporated into nitrocellulose strips. Individual viral antibody detection kits are also available. The western blot (WB) usually detects antibodies to p24 (*gag* gene, core protein) and gp 120, gp 41, gp 160 (*env* gene, envelope protein). False positive and indeterminate results are avoided since these assays do not contain contaminating cellular components (Joshi & Chipkar, 1997).

Technique: The HIV proteins (antigens) are separated according to their molecular weight by polyacrylamide gel electophoresis. These separated proteins (antigens) are transblotted on to a nitrocellulose membrane. This blotted membrane is cut into thin strips and are reacted with test serum. If antibodies to HIV are present in the test serum, they combine with different protein fragments of HIV. The strips are then washed and reacted with enzyme conjugated anti-human globulin. A suitable substrate is then added, which produces colour bands. Position of the colour band on the strip indicates fragment of antigen with which antibodies have reacted

Interpretation: If the test is *positive*, bands will show multiple protein fragments. Bands representing the 3 genes (gag, env, pol) provides conclusive evidence of HIV infection. WB test is *positive* if strips show bands representing at least *two* out of four HIV proteins – p24, gp160, gp120, gp41. If bands representing p24 or gp120 are seen only at one site, interpretation of WB test is difficult. This situation occurs in early HIV infection or may be non-specific. A reference strip and criteria for interpretation have been issued by the WHO, the American Red Cross, and the Consortium for Retrovirus Serology Standardization (CRSS). (Joshi & Chipkar, 1997).

Line Immunoassay (LIA): The single strip line immunoassay is a modification of the WB. This technique uses synthetic peptides and recombinant antigens coated as discrete lines on a nylon strip. Combination kits for line immunoassay are available for detecting HIV-1 and HIV-2 (Joshi & Chipkar, 1997).

9.3.5.2 – Immunofluorescence assay (IFA)

HIV-infected cells are fixed onto a glass slide and reacted with test serum. Fluorescein-conjugated anti-human gamma globulin is then added. If the test is positive, *apple-green* fluorescence appears when examined under fluorescent microscope. Immunofluorescence assay is less time consuming as compared with the WB technique, but the performance is comparable to that of the WB technique. However, it requires an expensive fluorescence microscope and trained personnel to interpret the results (Joshi & Chipkar, 1997).

9.3.5.3 – Radioimmuno precipitation assay (RIPA)

This method is a research-type assay, which can be used as an alternative to, or in combination with the WB. Though extremely sensitive, it is expensive and involves radioisotope banding and cultivation of HIV. HIV-infected T-lymphocytes are cultured in presence of radiolabelled amino acids. These T-lymphocytes are then lysed. HIV antibodies, if present, forms antigen-antibody complexes, which are precipitated. This precipitate is subjected to electrophoresis to separate the labelled complexes by their molecular weight. The bands are visualised by autoradiography (Joshi & Chipkar, 1997).

9.3.5.4 – Radioimmunoassay (RIA)

Radioactive iodine[125] I, which is usually used as a *detector*, is covalently bound to an antigen, antibody, or antibody-binding reagents. A gamma-counter is used to measure the gamma radiation produced by the bound[125] I. This assay is highly sensitive and specific. However, it is very expensive and measures response to a single viral protein only. It takes a long time to standardise the assay (Joshi & Chipkar, 1997).

9.3.6 – Tests for Detecting Antigen or Virus

9.3.6.1 – Detection of HIV antigen

When a *single* infective dose is high, p24 viral antigen and RT antigen may be detected in the blood after about 2 weeks. The p24 antigen appears early, disappears from the circulation during the prolonged asymptomatic phase and then reappears with the onset of symptomatic disease. Decline of p24 antigen during the asymptomatic phase signals progression of disease and indicates need for ARV therapy. Recurrence of p24 antigenemia corresponds to disappearance of anti-p24 antibody from the circulation. This antigen is detected by p24 antigen ELISA capture assay that uses anti-p24 antibody as the solid phase. Clinical samples for detecting HIV antigen are serum, plasma, CSF, and cell-culture supernatant. This technique is expensive. A "positive" test confirms HIV infection; while a "negative" test does not rule out HIV infection. On incubating the anti-HIV antibody with the patient's serum, free HIV antigen reacts with the anti-HIV antibody. Addition of rabbit anti-HIV p24 antibody and enzyme conjugate produces a change in colour, which is read at 450 nanometres. The concentration of p24 is determined from a standard curve. Antigen detection tests are used to detect HIV infection in the newborn, when maternal antibodies to HIV-1 confound serological diagnosis and to detect HIV infection during early "window" phase. There is a burst of viral replication, a few weeks after the primary infection. This method is used to resolve equivocal results of WB test. It is also used to: (a) diagnose involvement of the central nervous system or terminal illness. Due to immune collapse during this stage, the HIV antibodies against the viral core antigen may disappear or the HIV antigens may be absent; (b) monitor response to ARV therapy; and (c) monitor progression of HIV disease (Joshi & Chipkar, 1997).

Reverse Transcriptase (RT) Assay: The enzyme RT catalyses the transcription of genetic code of the HIV from its RNA genome to a double stranded DNA, by using radio-labelled thymidine. The activity of RT can be detected as the DNA molecules are synthesised using radio-labelled thymidine, which gets incorporated into them. The amount of radio-labelled thymidine is directly proportional to the concentration of RT, which is directly proportional to the concentration of HIV. RT assay is not used routinely because radio-labelled thymidine, which has short shelf-life, is required to detect RT activity. Handling radioactive material is hazardous. Expensive material and equipment are required, and the assay is not cost-effective.

9.3.6.2 – Isolation of virus

It is a reference method for identifying HIV infection. Once infected with HIV, an individual remains infected for life. HIV is present in blood and body fluids mostly within CD4 lymphocytes. Clinical samples for isolation of virus are peripheral blood, serum, and bone marrow. HIV titres are *high* in the early phase of infection, before antibodies appear. Antibodies do not neutralise the virus but coexist with HIV. During the asymptomatic phase, HIV titre is *low* and may not be detectable. With the onset of clinical symptoms, HIV titre rises again. A 98 per cent positivity is obtained from confirmed HIV positive individuals. Lymphocytes from patients are co-cultured with an indicator cell-line or peripheral blood mononuclear cells from a healthy, adult, sero-negative donor. Every 3–4 days, the culture has to be supplemented with fresh peripheral blood mononuclear cells, or indicator cells. These cultures are maintained for at least 4 weeks. Assays are done weekly for evidence of viral specific markers like p24 or RT. Viral culture can not be used as a routine assay because the laboratory should maintain strict sterile conditions for the viral culture. Since the virus is directly being cultured, the material is hazardous to maintain. Laboratory staff may be exposed to high concentration of HIV.The culture has to be monitored for 2–7 days and it takes several weeks for the final results. The necessary reagents are very expensive and the assay labour-intensive. The success rate is only about 60–70 per cent. Viral culture is used mainly for characterisation of the virus, determining drug sensitivity, and for vaccine studies (Joshi & Chipkar, 1997).

9.3.6.3 – Detection of viral nucleic acid

Viral nucleic acid can be detected by PCR even in the window period. Being expensive, PCR tests are indicated only when the results of other diagnostic tests are inconclusive. In DNA PCR, peripheral lymphocytes are lysed and proviral DNA is amplified. This test has high sensitivity and specificity. RNA PCR is used both for diagnosing HIV infection and for monitoring level of viremia.

Clinical Samples: peripheral blood, plasma, semen, vaginal or cervical secretions, tissues, CSF, or other specimens such as saliva, urine, breast milk, tears, and amniotic fluid.

It is necessary to amplify target HIV sequences or the molecular probes bound to the target sequences, due to the relatively low viral load in early seroconverters, in individuals in the window period and in infants with transplacental transmission of HIV. PCR, the commonly used amplification technique, is a multistep repetitive process. One PCR cycle takes only a few minutes, and the cycle is repeated about 30–40 times. Theoretically 30 amplification cycles can produce up to a million copies from a single target DNA. Being an extremely sensitive technique, PCR is used to: (a) detect HIV infection in the window period and in neonates born to HIV positive mothers; (b) resolve indeterminate WB test and detect genetic variability among HIV isolates; (c) identify mutations of RT; (d) monitor viral loads in patients on ARV therapy; and (e) differentiate between latent HIV infection from active viral transcription, and HIV-1 and HIV-2 infections. However, PCR can not be routinely used as a screening assay since there may be doubts regarding cross-contamination, and problems of carry-over of amplified products leading to false positive results. It is also an expensive assay, requiring costly equipment and highly trained personnel (Joshi & Chipkar, 1997; NACO).

Viral Load Assay: This involves quantitative estimation of HIV RNA in plasma using PCR. Being highly sensitive, viral load assay is used for determining base-line viral load at start of ARV therapy, for monitoring progress of therapy, and for predicting disease progression and clinical outcome (Joshi & Chipkar, 1997).

9.4 – TESTING PROTOCOLS

9.4.1 – Methods of Testing

1. Unlinked Anonymous Testing: This is an epidemiological method for measuring HIV prevalence which involves the use of already collected blood.
2. Voluntary Confidential Testing: It is mandatory to maintain confidentiality in this diagnostic technique.
3. Mandatory Testing: This is done without consent, as required by law in some countries. In India, mandatory testing is restricted to screening of donors of semen, ova (for in vitro fertilization), blood, tissues, and organs (NACO).

9.4.1.1 – Testing protocols

The procedures should be documented so that the staff members know how to carry out various tests. A sample testing protocol is given below. After pre-test counselling and informed consent, collect the desired specimen from the client and label the tube with a code. Verify the specimen to be collected (whole blood, serum, and plasma), the mode of collection (by venepuncture or finger prick), and procedures for specimen collection. Record a code on the clinic form. The code can be linked to personal identifying information, demographic, and medical history of the client. Perform the HIV test based on the testing strategy (ELISA, or other simple, rapid tests). The exact procedures should be documented. Record the HIV test result on the clinic form. Give the result to the client by name, during post-test counselling (NACO).

9.4.1.2 – National HIV testing policy

Every HIV testing should be done after the patient's explicit consent and accompanied by pre-test and post-test counselling. Consent is however, *not* required for mandatory screening of donors of semen, ova, blood, organs, and tissues. No individual should be made to undergo a mandatory HIV test. HIV testing should not be imposed as pre-condition for employment, or for provision of health care. India's armed forces are exempted from this provision (NACO).

9.5 – STRATEGIES FOR HIV TESTING

Positive result obtained in any one screening test (E/R/S) has to be confirmed by a supplemental test. The WB test is expensive and time consuming and therefore, different strategies are followed for confirmation. Two different types of ELISA, or one ELISA test along with any of the rapid tests is performed. The clinical sample is considered "positive" if it is positive in both sets of tests. If inconclusive, the test sample is re-tested after 1–2 weeks. The *first* E/R/S test selected for any of the following three strategies should have high *sensitivity*. The *second* and *third* E/R/S tests should have high *specificity* to eliminate likelihood of false positive results.

1. Strategy-I: The clinical sample is subjected once to E/R/S using a highly sensitive and reliable test kit. If *positive*, the clinical sample is considered HIV infected and if negative, it is taken as HIV free.
2. Strategy-II: The clinical sample is subjected to a *first* E/R/S test. If "negative", the sample is taken as HIV free; if "positive", the sample is re-tested with a *second* E/R/S test based on a different antigen preparation and/or a different test principle. If the *second* E/R/S test is also *positive*, it is reported as "positive". Otherwise it is "negative".
3. Strategy-III: This is similar to Strategy-II, with an additional confirmation by a *third* E/R/S test that is based on a different antigen preparation and/or a different test principle. If the clinical sample tests *positive* on all 3 E/R/S, it is reported "positive". A clinical sample that is negative in the *third* E/R/S test is considered "equivocal" and re-testing advised after 2 weeks. If the

second sample collected after 2 weeks also shows "indeterminate" results, it should be tested by a confirmatory assay such as WB. If the confirmatory assay fails to resolve the serodiagnosis, follow-up testing is repeated at 4 weeks, 3 months, 6 months, and 12 months. After 12 months, such indeterminate results should be considered "negative" (WHO, 2003).

The WHO recommends that the strategy for HIV testing should be based on the objective of testing and the prevalence of HIV infection in the population. For ensuring *transfusion or transplant safety*, for all levels of prevalence of HIV infection in the population, Strategy-I is to be adopted. For *HIV surveillance*, Strategy-I is recommended when prevalence of HIV infection exceeds 10 per cent and Strategy-II when it is less than 10 per cent. For *diagnosing* HIV infection in the *presence of clinical manifestations*, Strategy-I is advocated when the HIV prevalence in the population exceeds 30 per cent and Strategy-II when it is less than 30 per cent. For *diagnosing* HIV infection in *asymptomatic* individuals, Strategy-II is to be used when the HIV prevalence in the population exceeds 10 per cent and Strategy-III when it is less than 10 per cent (WHO, 2003).

9.6 – QUALITY ASSURANCE AND SAFETY

The following checklist would be helpful in ensuring quality and reliability of HIV testing.

Pre-Analysis Phase:
1. Trained staff are available and are capable of performing HIV tests.
2. Laboratory set-up is safe and safety issues have been covered during the training.
3. Procedures for specimen collection and labelling are clear.
4. Conditions for transport of specimens are adequate.
5. Specimens are processed correctly before testing.
6. Expiry dates of test kits have been checked.
7. There is a system of rotation of stock of test kits.
8. Test kit reagents are stored at appropriate temperatures, as specified by the manufacturer.

Analysis Phase:
1. A written SOP manual exists.
2. Specimens are processed and stored as specified in the procedure manual.
3. Equipment is maintained and their performance is checked periodically.
4. Reagents are prepared and used correctly.
5. Internal quality controls are included in the HIV test kits.
6. There is a procedure for monitoring quality of tests.

Post-Analysis Phase:
1. HIV test results are correctly interpreted and recorded and records are properly maintained.
2. Data are entered in appropriate records.
3. Quality control is regularly reviewed.

REFERENCES

National AIDS Control Organisation (NACO). Training manual for doctors. New Delhi: Government of India.

Joshi S.H. and Chipkar S.S., 1997, Laboratory detection of HIV infection. Bombay Hospital J 39(1): 99–108.

Frerichs RR., Htoon M.T., Eskes N., and Lwin S., 1992, Comparison of saliva and serum for HIV surveillance in developing countries. Lancet 340: 1496–1499.

WHO, 2003, Guidelines on Standard Operating Procedures for diagnosis of HIV infection. New Delhi: South-East Asia Regional Office. 23 July.

CHAPTER 10

LABORATORY DIAGNOSIS OF COMMON REPRODUCTIVE TRACT AND SEXUALLY TRANSMITTED INFECTIONS

Abstract

Laboratory tests are essential for confirming the clinical diagnosis in symptomatic individuals and diagnosing infection in asymptomatic individuals. Vaginal discharge and urethral and endocervical infection may be caused by a variety of pathogens. Syphilis is diagnosed by testing a patient's serum for antibodies to *Treponema pallidum*. The *non-specific* tests for syphilis include VDRL, RPR, and Wassermann tests. Specific serological tests detect the *specific* treponemal antibodies in the serum.

Key Words

Bacterial vaginosis, Reproductive tract infections, Sexually transmitted infections, Syphilis, *Trichomonas vaginalis*.

10.1 – BACTERIAL VAGINOSIS

Bacterial vaginosis (BV) was formerly called "non-specific vaginitis" or "anaerobic vaginitis". The new term implies that there is no infection, but a change in microbial flora of the vagina. BV is caused by an imbalance in the microbial flora of the vagina with overgrowth of anaerobic bacteria and lack of normal lactobacilli. This condition may be associated with presence of several pathogenic and non-pathogenic organisms, including anerobic bacteria (Goyal *et al.*, 2004; Beigi *et al.*, 2005; Hillier *et al.*, 1993). The microbial flora of the vagina contains many anaerobic and facultative bacteria such as *Trichomonas vaginalis, Gardnerella vaginalis, Mobiluncus curtissi, Mobiluncus mulieris, Candida* spp., *Bacteroides, Peptostreptococcus, Provabella, Mycoplasma hominis,* and *Ureaplasma urealyticum*.

The estimated prevalence of this condition varies from 15 to 40 per cent in different populations. Indian studies have reported prevalence ranging from 11 to 71 per cent in women with abnormal vaginal discharge. Postulated *risk factors* include age at first sexual intercourse (Amsel *et al.*, 1983), multiple sexual partners, use of vaginal medications, intrauterine contraceptive devices, and use of tobacco (Joesosef *et al.*, 2001). Though sexual association has been

143

reported, the condition is also seen in lesbians (Marrazzo *et al.*, 2002) and in women who have never been sexually active. The aetiopathogenesis of BV is still unclear but the following hypotheses have been put forward: (a) increased oestrogen concentration during the follicular phase of menstrual cycle favours growth of various pathogenic and non-pathogenic bacteria; (b) altered estrogen-progesterone ratio in pregnancy reduces prevalence of BV; (c) increase in mucinase and sialidase enzymes in vaginal discharge favours penetration of microorganisms; (d) endocrine changes causing disappearance of endogenous flora, facilitating growth of other bacteria; and (e) sexual transmission. In 50 per cent of women, BV is asymptomatic. The commonest symptom is whitish grey thin vaginal discharge, which has a characteristic fishy odour. The discharge is adherent to the vaginal wall. Other manifestations include vaginal pruritus and vulvovaginitis.

Diagnosis: New diagnostic tests (affirm VP III DNA probe test, QuickVue advance pH and amine test, QuickVue advance *G. vaginalis* test) have questionable sensitivity and hence are not widely used. According to the *composite clinical criteria* described by Amsel *et al.* BV is present if three out of the following four signs are present: (a) thin, homogeneous vaginal discharge; (b) vaginal pH more than 4.5; (c) positive amine odour with 10 per cent potassium hydroxide ("whiff test"); and (d) presence of "clue cells". Clue cells are denuded vaginal epithelial cells covered by bacteria in which the margins of these cells are not defined. The absence of clue cells indicates absence of BV, irrespective of vaginal pH, or clinical appearance of vaginal discharge. Thus, presence of clue cells is pathognomonic of BV. BV is defined as "cured" if three out of four criteria are absent and as "partially cured" if only two criteria are present (Amsel *et al.*, 1983).

Grading of Gram-Stained Slides: In 1991, Nugent *et al.* described a cheap, simple, and objective scoring method for grading Gram-stained slides. Score for the following is summated to obtain a total score between 1 and 10 – large Gram-positive rods (*Lactobacillus* spp.), small Gram-positive rods (*Bacteroides* spp.), small Gram-variable rods (*G. vaginalis*), curved Gram-variable rods (*Mobiluncus* spp.), and Gram-positive cocci. A total score of more than 7 is diagnostic for BV (0–3 = normal vaginal flora; 4–6 = intermediate).

Treatment and Prognosis: Metronidazole and clindamycin are recommended for the treatment of BV (McDoland *et al.*, 2005). The same treatment regimen is recommended for pregnant women also. The mutagenicity associated with long-term use of metronidazole in animals has not been reported in humans in the first trimester of pregnancy (Burtin *et al.*, 1995; Piper *et al.*, 1993). Both metronidazole and clindamycin are secreted in breast milk. In the recommended doses, these drugs are not harmful to the infant. Metronidazole acts on anaerobes, spares H_2O_2-producing lactobacilli and does not alter the normal vaginal flora. The recommended dose is 400 mg orally, twice daily, for 7 days. Metronidazole cream (0.75 per cent; 5 g *per vaginum* at night) is not recommended since it does not reach the upper genital tract and is very expensive. However, metronidazole cream (1 per cent) may be used *per rectum* as a prophylactic prior to gynecological surgeries. Clindamycin is administered orally (300 mg twice daily for 7 days) or clindamycin cream (2 per cent; 5 g *per vaginum* at night). However, clindamycin destroys the H_2O_2-producing lactobacilli. Though BV is also sexually transmitted, studies on current treatment of the male partner have been inconclusive. In some women, BV may recur with the onset of menstruation and disappear during mid-cycle or it may follow vaginal candida infection. Avoiding vaginal douches and alkaline agents (soap, shower gels) in the genital region can prevent recurrence. Once-a-month treatment has been advocated for recurrent BV, but its efficacy is not proved.

10.2 – INFECTIONS IN NON-PREGNANT WOMEN

The acidic environment of the vagina and hydrogen peroxide produced by lactobacilli protect against many bacterial and viral infections. *G. vaginalis* reportedly acts as an HIV-1 inducing factor (Cohn *et al.*, 2005). Antibodies to *Neisseria gonorrhoeae, Chlamydia trachomatis,* and *M. hominis* have been demonstrated in serum from patients with acute salpingitis (Mardh *et al.*, 1981) while antibodies to *M. hominis* have been demonstrated in serum from patients with genital infections (Mardh & Westrom, 1970b).

Bacterial Vaginosis: The production of CD4 lymphocytes is blocked; while that of interleukin-10 is increased. It has been found to be associated with adnexal tenderness, indicative of pelvic inflammatory disease (Eschenbach *et al.*, 1988). It has also been found to act as a cofactor to HPV in causing carcinoma in situ (Platz-Christensen *et al.*, 1994). Pre- and post-operative metronidazole therapy has been shown to reduce the frequency of vaginal cuff infection from 9.5 to 2 per cent in women with BV who underwent abdominal hysterectomy (Larsson & Carlsson, 2002). *M. hominis*, an organism associated with BV, has been isolated from tubal and cervical cultures (Mardh & Westrom, 1970a). There is an increased risk of post-abortal upper genital tract infection in women who had "clue cells" in their vaginal secretions. Initiation of antibiotic therapy before medical termination of pregnancy (MTP) was found to have a protective effect (Crowley *et al.*, 2001).

10.3 – INFECTIONS IN PREGNANT WOMEN

Pregnant women with BV have a 40 per cent higher risk of preterm birth. Preterm birth accounts for 8–10 per cent of all births and is a major cause of neonatal morbidity and mortality. In patients with BV, peptostreptococci, and *G. vaginalis* are the common causes of endometritis that may occur within 2 days after delivery by caesarean section. This is because infection is introduced directly into the endometrium. However, in case of normal vaginal delivery, endometritis may occur up to 6 weeks after delivery due to ascending infection. In view of the documented adverse pregnancy outcomes, asymptomatic pregnant women with a past history of BV should be ideally screened for this condition (Coli *et al.*, 1996; Silver *et al.*, 1989). Earlier the infection in pregnancy, higher is the risk of preterm birth. Treatment of BV in pregnant women reduced the risk by 27–50 per cent (Coli *et al.*, 1996). Amniotic fluid from patients with chorioamnionitis has been found to contain *G. vaginalis, M. hominis,* and some anerobes (Silver *et al.*, 1989).

10.4 – NEED FOR LABORATORY DIAGNOSIS

Community-based studies reveal that common reproductive tract infections (RTIs) in women are cervicitis, pelvic inflammatory disease, and vaginitis (Aggarwal & Kumar, 1999; Bang *et al.*, 1989). Though preventable and treatable, inadequate treatment or neglect of RTIs may lead to serious complications like infertility, ectopic pregnancy, pregnancy wastage, and low birth weight (Misra *et al.*, 1997).

The frequency of vaginal discharge as reported in several studies (Thakur *et al.*, 2002; Palai *et al.*, 1994; Das *et al.*, 1994; Kambo *et al.*, 2003; Nandan *et al.*, 2001) varies from 4.9 per cent (Kambo *et al.*, 2003) to 71.17 per cent (Ram *et al.*, 2006). A community-based study (Aggarwal & Kumar, 1999) in rural Haryana (on ever-married women of reproductive age), and a syndromic approach-based study (Ranjan *et al.*, 2003) on married women have reported RTIs in 70 per cent and 37 per cent of respondents, respectively.

Most RTIs go unrecognised and are considered as "normal" by women. In HIV-positive individuals, the clinical presentation and course of STIs may be modified. In such cases, laboratory diagnosis is essential. Due to the limitations of syndromic management (Gupta & Mahajan, 2003) laboratory tests are essential for confirming clinical diagnosis and diagnosing infection in asymptomatic individuals.

10.5 – TESTS FOR VAGINAL DISCHARGE

Vaginal discharge, a common complaint, may be due to infection of vagina, cervix, and/or uterus. The possible pathogens include *T. vaginalis, G. vaginalis, M. curtissi, Bacteroides* spp., *M. mulieris, Candida* spp. and *Peptostreptococcus*.

Materials Required: sterile gloves, sterile cotton wool swab sticks, grease-free microscope slides, spirit lamp or bunsen burner, vaginal speculum, freshly pre-pared reagents, and sterile physiological saline.

Collecting Vaginal Specimen: (a) moisten the speculum with sterile warm water and insert it into the vagina (do not lubricate the speculum with a gel that may be bactericidal); (b) cleanse the cervix using a swab moistened with sterile phys-iological saline; (c) pass a sterile cotton wool swab for about 20–30 mm into the endocervical canal and gently rotate the swab against the endocervical wall, to obtain a specimen; and (c) when gonorrhoea is suspected, inoculate the Petri dish containing the culture medium, taking all sterile precautions.

Examination and Interpretation:

10.5.1 – Colour and Odour

1. Yellow-green purulent discharge (*T. vaginalis*)
2. White colourless discharge (*Candida albicans*)
3. Grey, offensive smelling (fishy ammoniacal), thin discharge (*G. vaginalis*)

10.5.2 – pH Test

The normal pH of the vaginal discharge from puberty to menopause is *acidic* and the pH ranges from 3.0 to 3.5. The pH can be measured using Whatman pH paper.

Materials Required: sterile gloves, sterile cotton wool swab sticks, grease-free microscopic slides, sterile physiological saline, spirit lamp or bunsen burner, vaginal speculum, Whatman pH paper.

Steps: Take pH indicator paper strips in the range of ±3.8 to ±6.0. Touch the specimen swab on the pH paper (or) touch the pH paper to the tip of the vaginal speculum after removing it from the vagina, (or) touch the pH paper to the wall of the vagina directly.

Precaution: The pH paper should not come in contact with the cervical secretions.

Interpretation: Normal adult vagina has a pH of 3.0–3.5. pH of more than 5.0 is seen in *Trichomoniasis* and BV; pH of less than 5 may indicate candida infection.

False Positive/Negative Results: Presence of menstrual blood, cervical mucus, and semen may also raise vaginal pH.

10.5.3 – Potassium Hydroxide Wet Mount

Materials Required: sterile gloves, sterile cotton wool swab sticks, grease-free microscopic slides, sterile physiological saline, spirit lamp or bunsen burner, vaginal speculum, freshly prepared reagents.

Steps: (1) Place the vaginal discharge specimen on a clean grease-free microscope slide; (2) add two drops of 10 per cent potassium hydroxide (KOH, 10 g per 100 mL) on the specimen and mix well. KOH being corrosive, should be handled with care; (3) put a clean cover slip over the specimen, ensuring that there is no trapping of air bubble between the specimen and the cover slip; (4) pass the slide gently over a flame of the bunsen burner or spirit lamp for 10–20 seconds, taking care that the specimen does not boil; and (5) observe the slide under high power objective (×40) of light microscope.

Interpretation: Round or oval yeast cells of the size of 5–7 μm in diameter and the presence of mycelia or pseudohyphae are diagnostic of candidiasis (Cheesbrough, 2000).

Significance of this Test: A majority of cases of candidiasis can be diagnosed by this method. Saline wet mount can also be used as a substitute for KOH, but adding KOH is better since the mycelia are separated and clearly visible because it digests other epithelial cells and debris.

10.5.4 – Wet Mount

Materials Required: sterile gloves, sterile cotton wool swab sticks, grease-free microscopic slides, sterile physiological saline, spirit lamp or Bunsen burner, vaginal speculum, freshly prepared reagents.

Steps: (1) use a sterile swab to collect a specimen from the vagina; (2) transfer a sample of the exudate on a clean grease-free microscope slide. Alternatively, a drop of vaginal fluid can be used, if available; (3) add a drop of *sterile* physiological saline and mix; (4) cover with a cover slip; (5) observe the slide first under ×10 magnification. Any field, which shows the suspected organism, or parasite, is seen under ×40 magnification of light microscope.

Precautions: The preparation should not be too thick and should be examined as soon as possible, after the specimen is collected. Only *sterile* saline or saline that is checked daily by the laboratory should be used to exclude contamination by motile organisms that can be mistaken for *T. vaginalis*. In tropical climates, saline solutions get easily contaminated. The condenser and iris diaphragm should be sufficiently closed to give good contrast.

Interpretation: Trophozoites of *T. vaginalis* are larger than pus cells, measuring 10–20 µm in diameter. They are round or oval in shape. There are four anterior flagella, and a fifth flagellum forms an undulating membrane. The parasite moves actively with jerky movements. Incubating the preparation in a Petri dish containing a damp piece of cotton wool, at 35–37°C for a few minutes can revive motility. The presence of "clue cells" suggests the diagnosis of BV. These cells do not have a well-defined edge because of the presence of bacteria and the disintegration of cells (Cheesbrough, 2000).

10.5.5 – Amine Test

Materials Required: sterile gloves, sterile cotton wool swab sticks, grease-free microscopic slides, sterile physiological saline, spirit lamp or Bunsen burner, vaginal speculum, freshly prepared reagents.

Steps: (1) take a drop of vaginal fluid on a clean, grease-free microscope slide; (2) add one drop of 10 per cent KOH on the vaginal fluid; and (3) bring the slide close to the nose and smell immediately.

Precautions: The preparation becomes odourless soon after the test is performed; hence should be read immediately.

Interpretation: An intense putrid fishy odour indicates a positive reaction, which is suggestive of infection with BV organisms such as *G. vaginalis*, or with anaerobes such as *Bacteroides, Peptostreptococcus*, and *Mobiluncus*.

10.5.6 – Hydrogen Peroxide Test

This test is very simple and can be easily performed at peripheral laboratories (Cheesbrough, 2000).

Materials Required: sterile gloves, sterile cotton wool swab sticks, grease-free microscopic slides, sterile physiological saline, spirit lamp or Bunsen burner, vaginal speculum, freshly prepared reagents.

Steps: (1) take a drop of vaginal fluid on a clean grease-free slide; (2) add a drop of 3 per cent hydrogen peroxide on the specimen; and (3) on mixing, "foaming bubbles" (effervescence) is seen on the slide.

Precautions: If the specimen is collected from the endocervix (in a case of gonorrhoea, or chlamydia infection), a *false positive* test for *trichomoniasis* may be obtained.

Interpretation: presence of "foaming bubbles" (effervescence) indicates the presence of white blood cells, which are seen in *trichomoniasis*.

10.6 – TESTS FOR URETHRAL AND ENDOCERVICAL DISCHARGE

Possible pathogens are *Neisseria gonorrhoea, Streptococcus pyogenes, U. urealyticum, C. trachomatis*, and *T. vaginalis* (occasionally). It is essential to collect the specimen from the site of the lesion, without touching the surrounding area, for the correct laboratory confirmation of clinical diagnosis.

Materials Required: sterile gloves, sterile cotton wool swab sticks, grease-free microscopic slides, sterile physiological saline, spirit lamp or Bunsen burner, vaginal speculum, freshly prepared reagents, and glass marking pen.

Steps for Collecting Urethral Discharge: Collect the specimen while wearing sterile gloves. Cleanse around the urethral opening using a swab moistened with sterile physiological saline. Gently massage the urethra from above downward and collect the pus with a sterile cotton wool swab. If there is no discharge, insert a sterile thin cotton wool swab 2–3 cm into the urethra and rotate for 5–10 seconds to scrape the mucosa. Put the swab in Amie's transport medium or sterile test tube and label (Cheesbrough, 2000).

Steps for Collecting Endocervical Discharge: Moisten the vaginal speculum with sterile warm water and insert the speculum into the vagina. Do not lubricate the speculum with a gel that may be bactericidal. Cleanse the cervix with a swab moistened with sterile physiological saline. Inspect the exocervix for lesions and insert another cotton wool swab up to 2 cm into the cervical canal. Rotate the swab for 5–10 seconds and withdraw. Put the swab in Amie's transport medium, or sterile test tube and label. (Cheesbrough, 2000).

Precautions: The patient should not have passed urine, preferably for 1–2 hours before the specimen is collected. Antiseptics should never be used, as very delicate organisms like the *gonococci* are likely to get destroyed. The specimen should be processed *immediately* in the laboratory, as the organisms are highly autolytic and may not be visible if there is a delay in processing the sample.

Microscopic Examination: Take a microscopic slide and place a drop of water on it. If the slide is clean and grease-free, a thin film of this drop can be made on the slide. Otherwise, water collects on the slide in the form of fine droplets and a film can not be made.

10.6.1 – Preparing the Smear

1. Take a clean *grease-free* microscopic slide and wipe it with gauze. The slide should be free from scratches. Pass the slide 2–3 times through the flame of a Bunsen burner or a spirit lamp in order to remove traces of grease from the slide.
2. Mark the central part of the slide with two vertical lines, 2–3 cm apart, with the help of a glass-marking pen.
3. Roll the cotton wool swab with the specimen, on to this marked area on the slide. *Note*: When making a smear, a swab of the discharge should be gently rolled on the slide to avoid damaging the pus cells. This helps in better visualisation of intra-cellular *gonococci*.
4. Make a smear of the size of 2 × 1 cm. Allow the smear to dry in air. *Note*: Heat fixation is contraindicated if gonococcal infection is suspected.
5. Label the smear on the right or left-hand corner of the slide.

10.6.2 – Heat Fixing the Smear

The objective of heat fixing is to: (a) prevent autolytic changes in the smear and preserve the smear and to prevent the wash out of the smear during the staining process; (b) make the organisms more permeable to different stains, thus giving good staining characters; and (c) render the smear non-infectious to some extent.

Steps for Heat Fixing: Hold the slide in such a way that the smear is on the upper side. Pass the slide over the flame of a Bunsen burner or a spirit lamp twice or thrice. Now judge the temperature of the slide by feeling it on the back (dorsum) of the hand. The slide should be hot enough for the heat to be intolerant.

Precautions during Heat Fixing: *Excess heating* will char the smear and nothing can be visualized under the microscope. On the other hand, with *inadequate heating*, the smear may get washed out during the staining process and nothing can be visualized under the microscope.

Contraindication for Heat Fixation: Since the *gonococci* are delicate organisms, heat fixation is contraindicated. Instead of heat fixing, methanol is used for fixing the smear.

10.6.3 – Reagents for Gram's Stain

- Crystal Violet – Solution A: crystal violet powder (20 g) in 200 mL ethyl alcohol. Solution B: 8 g ammonium oxalate in 800 mL distilled water. Mix Solutions A and B, and filter.
- Gram's Iodine – 20 g resublimated iodine is mixed in 100 mL one normal NaOH solution (4 g sodium hydroxide in 100 mL distilled water).
- Acetone-Alcohol – 100 mL acetone in 500 mL ethyl alcohol.
- Safranin – The stock solution contains 10 g safranin in 200 mL ethyl alcohol. The working solution is prepared by diluting 100 mL of the stock solution in 900 mL of distilled water.

10.6.4 – Procedure for Gram's Staining

1. Cover the smear with crystal violet solution (the "dye") and allow it to act for 20–30 seconds. *Note*: Flooding the whole slide will only cause wastage of reagents. Pour the crystal violet solution and wash the slide under slow running water, keeping the slide at an angle, so that only the crystal violet solution is washed out and not the smear.
2. Cover the smear with freshly prepared Gram's iodine solution and allow it to act for 30 seconds. This acts as a mordant, by forming a dye-iodo complex.
3. Wash the slide under slow running water and add acetone-alcohol mixture. This acts, as a decolorizing agent and the end point of decolourisation is that violet colour ceases to come off the slide. This can be confirmed by holding the slide against a white background. (Absolute alcohol, i.e. 100 per cent ethanol can be used as a substitute for acetone-alcohol mixture for decolourisation.) This is a critical stage of Gram's staining, as overdecolourisation can make a Gram-positive organism to appear as Gram negative, and lead to faulty diagnosis.
4. After washing the slide with water, pour safranin on the slide and allow it to act for 1–2 minutes. The purpose of safranin is to counterstain the Gram-negative organism, pus cells and the background.
5. Wash the slide with water and gently blot the slide dry, between two blotting papers. Put a drop of liquid paraffin on the stained smear and observe under oil immersion lens (×100).

10.6.5 – Microscopic Examination

1. Put a drop of liquid paraffin on the stained smear and observe under oil immersion lens (×100).

2. Push the condenser of the microscope upwards and open the iris diaphragm so that maximum light passes through the smear. This helps in a clear visualisation of the organisms, pus cells, and other elements.
3. Examine the smear for epithelial cells, polymorphonuclear leukocytes (pus cells), organisms and their location – whether extracellular or intracellular.

Interpretation: The ideal oil immersion field for microscopic examination is the one that which shows a clear contrast between Gram-positive material (stained *purple*) and Gram-negative material (stained *pink*). The severity of the infection can be judged from the number of organisms and pus cells per oil immersion field. *Gonococci* are Gram negative (stained *pink*), intracellular, bean-shaped, and are usually arranged in pairs ("diplococci"). In asymptomatic *gonococcal endocervicitis, gonococci* may not be visible under Gram's stain. Culture methods are required in such cases. If the whole slide appears *pink*; this indicates excessive decolourisation.

10.7 – LABORATORY TESTS FOR SYPHILIS

Syphilis is primarily transmitted through sexual contact, though it can also be spread by blood transfusion or by percutaneous route, as in occupational exposure. The disease has an *early infectious stage*, which occurs within the first 2 years of infection, this includes: (a) primary syphilis (a characteristic primary lesion called "chancre" appears at the portal of entry and heals spontaneously); (b) secondary syphilis (a generalised skin eruption appears after about 6 weeks later; and (c) early latent syphilis. The *late non-infectious stage* comprises: (a) late latent syphilis; (b) benign late syphilis; (c) cardiovascular syphilis; and (d) neurosyphilis. This stage occurs about a decade after the occurrence of the primary lesion. Degenerative and irreversible necrotic lesions that do not contain treponemes are characteristic of late non-infectious stage.

Direct demonstration of *T. pallidum* by dark-field microscope is the simplest and the most rapid method of diagnosing syphilis. Smears obtained from chancre or mucous patches are stained by Fontana's stain or by silver impregnation technique. *T. pallidum* can also be demonstrated in biopsies of tissues by silver impregnation or by immunofluorescence. However, the organisms can not be detected after the primary stage of syphilis, when the lesions have healed. Though PCR is available, it has not yet been standardised. Therefore, the main method for diagnosing syphilis is by testing a patient's serum for antibodies to *T. pallidum*.

10.7.1 – Collection of Blood Sample

Materials Required: sterile gloves, tourniquet, disposable syringe and needle no. 21 or 22 gauge, sterile cotton wool, discarding jar containing 1 per cent sodium hypochlorite solution, spirit, centrifuge, sterile dry test tubes, and Pasteur pipettes.

10.7.1.1 – Steps for collecting blood sample

1. Apply tourniquet above the cubital fossa.
2. Ask the patient to clench his or her fist, with the thumb facing inside, so that the veins become prominently visible.
3. Clean the region around the cubital fossa with spirit swab.
4. Wear gloves and feel the antecubital veins.
5. With the help of sterile disposable needle and syringe, collect 3–5 mL of venous blood.
6. Open the tourniquet, remove the needle from the vein and apply pressure on the site with spirit swab.
7. Transfer the collected blood in a properly labelled sterile test tube and allow it to clot for 1 hour.
8. Discard the syringe and needle in the discarding jar containing 1 per cent sodium hypochlorite solution.
9. Remove the supernatant serum into another sterile test tube with the help of Pasteur pipette.
10. Centrifuge the test tube at 1500 revolutions per minuted (RPM) for 15 minutes. Any red cells transferred to this test tube will settle down after centrifugation.
11. Remove the serum with a Pasteur pipette in to a sterile screw cap bottle.
12. Label the vial and store in the refrigerator at 4°C, till further use.
13. Store the vial in the freezer of the refrigerator, if needed in future.

10.7.1.2 – Precautions during blood collection

1. During blood collection, avoid needle-stick injuries.
2. Any breach in the integrity of the skin such as abrasion, wound, or injury should be covered with waterproof adhesive tape and sterile gloves should be worn.
3. Hot (autoclaved) sterile syringes may haemolyse the blood; and should not be used.
4. The test tube should be carefully labelled with the patient's name, age, sex, date of collection, and registration number.

10.7.2 – Serological Tests for Syphilis

Serological tests can be used for the diagnosis of late primary, secondary and late syphilis. Two major groups of antibodies are produced by the immune system of the host, who is infected by *T. pallidum*.

10.7.2.1 – Non-specific antibody

About 1–3 weeks after the appearance of the primary lesion, a substance called *reagin* (an antibody complex) appears in the serum. The presence of reagin is detected by serological tests that use non-treponemal antigen i.e. antigens from beef heart. Two types of non-treponemal tests are used to detect and measure *reagin* in the serum.

(a) Flocculation Tests: VDRL test and rapid plasma reagin (RPR) card test.

(b) Complement Fixation Test: Wassermann reaction.

False positive results may be obtained with sera from healthy individuals or patients suffering from other diseases because these tests detect non-specific antigens that are shared by treponemes and mammalian tissues. "Acute" biological false positive results turn "negative" within 6 months in viral infections such as measles, infectious mononucleosis, mycoplasma pneumonia, and malaria. "Chronic" biological false positive results persist for 6 months or longer and may occur in leprosy and autoimmune diseases.

10.7.2.2 – Specific treponemal antibody

This reacts with antigens prepared from live or dead treponemes, or those prepared from extracts of virulent Nichol's strains of *T. pallidum* maintained in rabbit testicle. The treponemal tests include:

1. Fluorescent treponemal antibody absorption test (FTA-ABS).
2. Micro-haemagglutination assay for *T. pallidum* (MHA-TP).
3. *T. pallidum* immobilization (TPI) test.
4. Enzyme immunoassay (EIA).
5. *T. pallidum* haemagglutination (TPHA) test.
6. *T. pallidum* particle agglutination (TP-PA) test.

Venereal Disease Research Laboratory (VDRL) and TPHA are used as *screening tests* in pregnant women, blood donors, and "at risk" patients. VDRL is more sensitive than TPHA in early syphilis, while the converse is true in latent and late stages of the disease. FTA-ABS and TPPA are *confirmatory tests*, to be used when one screening test is positive.

10.7.2.3 – VDRL test

The acronym VDRL stands for "Venereal Disease Research Laboratory". In this test, *reagin* antibodies in patient's serum are detected by "cardiolipin antigen" (an alcoholic extract of bovine heart muscle, to which lecithin and cholesterol are added).

Materials and Equipment Required: round bottle (30 mL capacity), water bath, micropipettes to deliver 20–60 µL, VDRL slide with depressions (14 mm in diameter), antigen (0.5 mL vial), buffered saline, and mechanical rotator.

Preparation of Fresh Antigen: Transfer 0.4 mL of buffered saline to a 30 mL round bottle by micropipette. Add 0.5 mL of the antigen (available commercially) directly to the saline while gently rotating the bottle on a flat surface. Add antigen over a period of 6 seconds and continue rotating for 10 seconds. Add 4.1 mL of buffered saline and mix well by repeatedly inverting the bottle about 30 times in 10 seconds.

Procedure for Qualitative VDRL Test: Inactivate the patient's serum by heating in a water bath at 56°C for 30 minutes. Add 60 µL of inactivated serum in VDRL slide. Add 20 µL of fresh antigen to the inactivated serum in VDRL slide. Mix and rotate the slide for 4 minutes in a mechanical rotator (to be set at 180 RPM). Read the test microscopically. Report the results as: (a) "reactive"; (b) "weakly reactive"; or (c) "non-reactive".

Procedure for Quantitative VDRL Test: Dilute the serum in geometric progression and titrate these dilutions with freshly prepared antigen. The quantitative result is reported as the highest dilution in which the test is fully reactive. VDRL titre is between 1:8 and 1:16 in primary syphilis, and between 1:16 and 1:128 in secondary stage, cardiovascular and neurosyphilis.

10.7.2.4 – RPR card test

The acronym RPR stands for *rapid plasma reagin*. The RPR card test is commercially available as a ready-to-use kit. The test is performed according to the manufacturer's instructions supplied with the kit. In this test, finely divided carbon (charcoal) particles and choline chloride are added to cardiolipin (i.e. VDRL antigen).

Advantages over VDRL Test
1. Simpler than the VDRL test, it can also be performed at the peripheral health centres (Cheesbrough, 2000).
2. Addition of finely divided carbon particles enables visual reading of results and the reactivity of the antigen is enhanced.
3. There is no need for heat inactivation of the sample.
4. Plasma as well as serum can be used in the RPR test and blood from finger prick is sufficient.
5. Can be tested on plastic or paper cards.

Its disadvantage is that it can not be used with CSF. RPR card test kit contains the following materials: (a) plastic-coated card with circles on them; (b) antigen suspension in an unbreakable container; (c) 20 gauge needle without bevel; and (d) disposable plastic stirrer.

Other materials needed, but not provided with the kit are: (a) normal (physiological) saline; (b) pipettes (automatic or glass) – 1 mL, 2 mL, and 0.5 mL (with 0.01 mL calibrations); and (c) mechanical rotator adjusted to rotate forming a circle, 2 cm in diameter. Speed should be adjusted to the number of RPM, as per kit specifications.

Steps: Take one test-card and with the help of a pipette, place 0.05 mL of the *unheated* serum on one of the circles. Spread the serum sample on the circle, using the disposable plastic stirrer. Gently shake the RPR card test antigen, and put one drop (1/60 mL) of this antigen in the serum sample of the circle. Rotate the card on the mechanical rotator, after adjusting the speed. The duration and the number of RPM should be as per the instructions supplied with the kit. Remove the card from the rotator, and see the results *immediately* with naked eye in bright light.

Results and Interpretation: Formation of small to large clumps indicates that the test is "reactive". Lack of clumps or presence of slight roughness indicates a "non-reactive" test.

Caution: False positive results may be obtained in conditions like viral pneumonia, malaria, leptospirosis, tuberculosis, and connective tissue disorders such as disseminated lupus erythematosus (Gupta & Mahajan, 2003).

10.7.2.5 – Fluorescent treponemal antibody absorption test

In the FTA-ABS test, the patient's serum is absorbed with an autoclaved supernate from cultures of treponemes in order to remove group-specific antibody. Binding of *T. pallidum*-specific antibody is demonstrated by indirect immunofluorescence method. This test can detect IgG and IgM antibodies. It is the earliest serological test that becomes positive and remains positive for many years, even after treatment.

10.7.2.6 – Micro haemagglutination assay

In the MHA-TP, tanned sheep erythrocytes are sensitised with extract of Nichol's strain of *T. pallidum* and then mixed with patient's serum. If the serum contains treponemal antibodies, the erythrocytes clump together.

10.7.2.7 – Treponema pallidum IMMOBILIZATION (TPI)

When serum of a syphilitic patient and complement are added to actively motile Nichol's strain of *T. pallidum*, these spirochetes are immobilized. Although this is the most specific test for diagnosing syphilis, it is difficult, time consuming, and expensive.

10.7.2.8 – Enzyme immuno assay (EIA)

This assay detects the antigen-antibody complex by using a tracer complex (that contains horse radish peroxidase conjugated monoclonal antibody to *T. pallidum*).

10.7.2.9 – Treponema pallidum Haemagglutination (TPHA)

Antibodies in patient's serum agglutinate sheep erythrocytes (coated with extract of *T. pallidum*). TPHA is often negative in early syphilis but may become positive at low titres (1:80 to 1:320) towards the end of the primary stage. The titre rises sharply during the secondary stage.

Table 1. Interpretation of serological tests for syphilis

VDRL	TPHA	FTA-ABS	Interpretation
Positive	Negative	Negative	False positive reaction – repeat to exclude primary infection
Positive	–/+	Positive	Primary infection
Positive	Positive	Positive	Untreated or recently treated
Negative	Positive	Positive	Fully or partially treated
Negative	Positive	Negative	Past history of treated syphilis

10.7.2.10 – Treponema pallidum Particle Agglutination (TP-PA)

TP-PA is a specific serological test for the detection of antibodies in various species and subspecies of pathogenic treponemes, which cause syphilis, yaws, pinta, bejel, and endemic syphilis. This test is used to confirm the reactive results of a non-treponemal screening test for syphilis such as VDRL test. TP-PA is based on passive agglutination wherein serum samples containing antibodies to pathogenic treponemes react with gel particles (sensitised with sonicated antigens of Nichol's strain of *T. pallidum*) to form a smooth mat of agglutinated gel in the microfilter tray well. If antibodies are not present, the gel particles settle to the bottom of the tray well, forming a characteristic compact button of unagglutinated particles. The control well containing unsensitized gel should also show this compact button or absence of agglutination (Pope & Fears, 2000). SERUM SAMPLE: Collect *fresh* serum sample (about 0.5 to 1 mL, *never* less than 0.4 mL) using regular red top or serum separator vaccutainers and allow the specimen to clot at room temperature and centrifuge. Transfer to polypropylene screw-capped vial with capacity of 2 mL. Serum samples are stable up to 72 hours at 4°–8°C. For longer periods of storage, the recommended temperature is −20°C or lower. Repeated freezing and thawing may compromise the integrity of the specimen. Excessively haemolysed, contaminated or lipaemic sera may give atypical results and should not be used. A serum sample is considered too haemolysed to be tested if printed material can not be read through it. Heat-inactivated (56°C for 30 minutes) serum may be used. Excessive inactivation time or temperature may cause equivocal results.

10.7.3 – Interpretation of Serological Tests

Non-treponemal (non-specific) tests become negative or show decline in titres with effective therapy, while treponemal tests usually remain positive even after successful therapy.

REFERENCES

Aggarwal A.K. and Kumar R., 1999, Community based study of reproductive tract infections among ever-married women of reproductive age in a rural area of Haryana. J Commun Dis 31: 223–228.

Amsel R., Totten P.A., Spiegel C.A., *et al.* 1983, Non-specific vaginitis: Diagnostic criteria and microbial and epidemiological associations. Am J Med 74(1): 14–22.

Bang R.A., Bang A.T., Baitule M., *et al.* 1989, High prevalence of gyaenocological diseases in rural Indian women. Lancet I: 85–88.

Beigi R.H., Weisendfeld H.C., Hillier S.L., *et al.* 2005, Factors associated with absence of H_2O_2-producing lactobacillus among women with bacterial vaginosis. J Infect Dis 191(6): 924–929.

Burtin P., Taddio A., Ariburnu O., *et al.* 1995, Safety of metronidazole in pregnancy: a meta-analysis. Am J Obstet Gynecol 1995 172: 525–529.

Cheesbrough M, 2000, District laboratory practice in tropical countries. Part 2. Cambridge: Cambridge University Press.

Cohn J.A., Hashemi F.B., Camarca M., *et al.* 2005, HIV-inducing factor in cervicovaginal secretions is associated with bacterial vaginosis in HIV-1 infected women. AIDS 39(3): 340–346.

Coli E., Bertulessi C., Landoni M., Parazzini F., 1996, Bacterial vaginosis in pregnancy and preterm birth: evidence from the literature. J Int Med Res 24(4): 317–324.

Crowley T., Low N., Turner A., *et al.* 2001, Antibiotic prophylaxis to prevent post-abortal upper genital tract infection in women with bacterial vaginosis: Randomised control trial. Br J Obstet Gynaecol 108(4): 396–402.

Das A., Jana S., Chakraborty A.K., *et al.* 1994, Community based survey of STD/HIV infection among commercial sex workers in Calcutta (India). Part III: Clinical findings of sexually transmitted diseases (STD). J Commun Dis 26: 192–196.

Eschenbach D.A., Hillier S., Critchlow C., *et al.* 1988, Diagnosis and clinical manifestation of bacterial vaginosis. Am J Obstet Gynecol 158(4): 819–828.

Goyal R., Sharma P., Kaur I., *et al.* 2004, Bacterial vaginosis and vaginal anerobes in preterm labour. JIMA 102 (10): 548–550; 553.

Gupta M.C. and Mahajan B.K., 2003, Textbook of preventive and social medicine. 3rd edn. New Delhi: Jaypee Brothers Medical Publishers.

Hillier S.L., Krohn M.A., Robe L.K., *et al.* 1993, The normal vaginal flora, H_2O_2-producing lactobacilli, and bacterial vaginosis in pregnant women. Clin Infect Dis 16 Suppl 4: S273–S281.

Joesosef M.R., Korundeng A., Runtupalit C., *et al.* 2001, High risk of bacterial vaginosis among women with intrauterine devices in Manado, Indonesia. Contraceptive 64(3): 169–172.

Kambo I.P., Dhillon B.S., Singh P., *et al.* 2003, Self reported gynecological problems from twenty three districts of India: an ICMR Task Force Study. Indian J Community Med 28: 67–73.

Larsson P.G. and Carlsson B., 2002, Does pre- and post-operative metronidazole treatment lower the vaginal cuff infection rate after abdominal hysterectomy among women with bacterial vaginosis? Infect Dis Obstet Gynecol 10(3): 133–144.

Mardh P.A. and Westrom L., 1970a, Antibodies to *Mycoplasma hominis* in patients with genital infections and healthy contacts. Br J Vener Dis 46(5): 390–397.

Mardh P.A. and Westrom L., 1970b, Tubal and cervical cultures in acute salpingitis with special reference to *Mycoplasma hominis* and T-strain *mycoplasma*. Br J Vener Dis 46(3): 179–186.

Mardh P.A., Lind J., Svensson L., *et al.* 1981, Westrom L, Moller BR. Antibodies to *Chlamydia trachomatis, Mycoplasma hominis,* and *Neisseria gonorrhoeae* in serum from patients with acute salpingitis. Br J Vener Dis 57(2): 125–129.

Marrazzo J.M., Koutsky L.A., Eschenbach D.A., *et al.* 2002, Characterization of vaginal flora and bacterial vaginosis in women who have sex with women. J Infect Dis 185(9): 1307–1313.

McDoland H., Brocklehurst P., and Parson J., 2005, Antibiotic for treating bacterial vaginosis in pregnancy. Cochrane Database Systematic Review 25(1): CD 000262.

Misra T.N., Chawla S.C., and Bajaj P., 1997, Gynecological disease in women of reproductive age group – Unmet needs in MCH care. Indian J Community Med 22: 104–109.

Nandan D., Gupta Y.P., Krishnan V., *et al.*, 2001, Reproductive tract infections in women of reproductive age group in Sitapur/Shahjehanpur district of Uttar Pradesh. Indian J Public Health 45: 8–13.

Palai P., Singh A., and Pallai V., 1994, Treating vaginal discharge in slum women. Bull PGI Chandigarh 28: 107–110.

Piper J.M., Mitchel E.F., and Ray W.A., 1993, Prenatal use of metronidazole and birth defects: no association. Obstet Gynecol 82(3): 348–352.

Platx-Chistensen J.J., Sundstrom E., and Larsson P.G., 1994, Bacterial vaginosis and cervical intraepithelial neoplasia. Acta Obstet Gynecol Scand 73(7): 586–588.

Pope V. and Fears M.B., 2000, Serodia *Treponema pallidum* passive particle agglutination (TP-PA) test. In: A manual of tests for syphilis (S.A. Larsen, V.Pope, R.E. Johnson, E.J. Kennedy Jr., eds.). Suppl. Washington DC: American Public Health Association, pp 363–378.

Ram R., Bhattacharyya K., Goswami D.N., *et al.* 2006, Syndromic approach for determination of reproductive tract infections among adolescent girls. JIMA 104(4): 178–181.

Ranjan R., Sharma A.K., and Mehta G., 2003, Evaluation of WHO diagnostic algorithm for reproductive tract infections among married women. Indian J Comm Med 28: 81–84.

Silver H.M., Speeling R.S., St. Clair P.J., and Gibbs R.S., 1989, Evidence relating bacterial vaginosis to intraamniotic infection. Am J Obstet Gynecol 161: 808–812.

Thakur J.S., Swami H.M., and Bhatia S.P.S., 2002, Efficacy of syndromic approach in management of reproductive tract infections and associated difficulties in a rural area of Chandigarh. Indian J Community Med 27: 77–79.

SECTION THREE
CLINICAL ASPECTS

CHAPTER 11

ACCIDENTAL OCCUPATIONAL EXPOSURE

Abstract

Virtually all health care providers and hospital workers who are likely to come in contact with infected blood or body fluids are at high risk of accidental occupational exposure to HBV or HIV. Preventive measures include hand washing, using personal protective barriers, safe handling of sharps and their decontamination and disposal, and immunisation against hepatitis B. The guiding principle is to presume that all specimens, all patients or clients are potentially infected, unless proved otherwise. In case of accidental occupational exposure, the site of exposure is treated, spills are decontaminated, the incident is reported to the concerned authorities and the exposed person is tested for hepatitis B/C and HIV infection, after counselling and informed consent. Post-exposure prophylaxis (PEP) is available against HBV and HIV, while there is no specific PEP for exposure to hepatitis C virus (HCV). During the follow-up period, persons exposed to HBV or HIV must not donate blood, semen, or body organ for transplant, should avoid pregnancy and abstain from sexual intercourse or use latex condom every time during sexual intercourse. Persons receiving PEP ought to be monitored for drug toxicity. Specialist opinion on PEP is available on the Internet.

Key Words

Exposure code, HIV status code, Post-exposure prophylaxis, Risk of HIV infection, Universal biosafety precautions, Window period

11.1 – INTRODUCTION

As compared to the general public, health care providers and hospital workers are occupationally exposed to an over-abundance of pathogens. Diverse type of pathogens can be potentially transmitted through infected blood and body fluids. These include hepatitis viruses B, C, D, and G, HIV-1 and HIV-2, cytomegalovirus (CMV), Epstein–Barr virus (EBV), *Plasmodium*, *Treponema pallidum*, viruses causing haemorrhagic fevers such as dengue, *Brucella*, *Yersinia pestis*, *Mycobacteria*, and slow viruses (NACO, Training Manual for Doctors). Currently, the main concern is the prevention of accidental occupational exposure to HIV because of fear, stigma, social and economic cost associated with HIV infection, and non-availability of an effective preventive vaccine or cure creates apprehension in the mind of an exposed health care provider. Even if a patient's ELISA report is "negative", he or she may still harbour the virus during the "window period". Normally, antibodies that are produced by the human body in response to other infections try to neutralise the pathogen. But, in the case of HIV infection, anti-HIV antibodies do not neutralise HIV. Hence, HIV-positive individuals will remain infectious to others for the rest of their lives (Mehta & Rodrigues, 1996; WHO, 1996).

The sources of exposure are blood and body fluids, infected human organs, tissues, specimens, *percutaneous* accidental injuries with contaminated sharp instruments or needle stick injuries leading to inoculation of blood or body fluids, and contaminated equipment. *Mucocutaneous* accidental occupational transmission may occur by contamination of mouth, conjunctiva, wound (breached skin), abrasion, scratches, and dermatitis (NACO, 2002).

Virtually all health care providers and hospital workers who are likely to come in contact with infected blood or body fluids are at high risk of accidental occupational exposure. This high-risk group includes:

(a) All health care providers such as physicians, surgeons, nurses, and midwives
(b) Persons handling infected, or potentially infected blood, body fluids, excretions, and secretions (microbiologists, pathologists, laboratory workers, and sweepers)
(c) Those who handle or embalm dead bodies, perform autopsies (autopsy workers, forensic experts, pathologists, anatomists, and embalmers)

11.2 – RISK OF HIV INFECTION

In case of exposure through wounds, skin, mouth, or conjunctiva, called *mucocutaneous exposure*, the risk of HIV infection is 0.3 per cent. Exposure through needle stick, sharps, called *percutaneous exposure* carries a risk between 0.25 and 0.3 per cent. Among HIV-infected individuals, *typical progressors* are relatively more infectious in seroconversion stage and terminal stage of the disease. In comparison, the risk of acquiring hepatitis B infection varies from 9 to 30 per cent, since carriers of HBV may have 10 million to 1 billion infectious virons per mL of blood. The risk of hepatitis C virus (HCV) infection is 3–10 per cent for accidental occupational exposures. This lower risk of HIV infection is because HIV-infected persons have only about 100–10,000 infectious virons per mL of peripheral blood (NACO, 2002; Mehta & Rodrigues, 1996). Longitudinal studies on 1,074 accidental needle stick exposures revealed three seroconversions. Contamination of mucous membrane or skin did not result in seroconversion in 104 cases. In a progressive study following 870 needle stick injuries or cutting injuries, four persons (0.46 per cent) tested HIV seropositive (Mandlebrot *et al.*, 1990).

11.2.1 – Model for Calculating Risk

The probability of acquiring HIV infection during a surgical operation can be calculated by a hypothetical model:

P_1 = probability of surgeon acquiring HIV infection = 0.46 per cent
P_2 = probability of surgeon getting an injury = 10 per cent
P_3 = probability of patient having HIV infection = 1 per cent
P = probability of acquiring HIV in one surgical operation = $P_1 \times P_2 \times P_3$ = 0.0000046 per cent
If the surgeon performs 200 surgeries in 1 year, the probability = $1(1 - P) \times 200$ = 0.09 per cent
The cumulative probability in a career of 30 years would be 2.72 per cent (Raahave & Bremmelgaard, 1991).
In a study conducted in Amsterdam, the 30-year cumulative risk was estimated at 0.0012 for a surgeon who performs about 500 surgeries (Leentvaar *et al.*, 1990).

11.2.2 – Reliability of Risk Estimates

Risk estimates are based on studies conducted in industrialised countries and therefore, extrapolation of risks must be treated with extreme caution. In the *developing countries*, accidental occupational exposure may be relatively more common due to lack of training of health care providers in "universal biosafety precautions", carelessness, even among those who have been trained, presence of many undiagnosed HIV-infected persons who are not routinely screened for HIV particularly before emergency surgeries, and likelihood of false negative HIV tests during the "window period" (NACO, 2002; Mehta & Rodrigues, 1996). The prevalence of HIV infection varies geographically. Injuries occur in up to 10 per cent of surgical procedures. During emergency surgeries, there is greater likelihood of accidental injury occurring to any member of the surgical team (Mehta & Rodrigues, 1996).

11.3 – PREVENTION

Practising "universal biosafety precautions" (also called "universal work precautions") can prevent accidental occupational exposure. The guiding principle is to *presume that all specimens, all patients or clients are infected (or potentially infected), unless proved otherwise.* Prevention holds the key to avoiding occupational HIV transmission (NACO, 2002; WHO, 1996; CDC, 1997, 1998; UK Health Department, 1990). The methods for prevention are discussed in Chapter 7.

11.4 – MANAGEMENT OF ACCIDENTAL EXPOSURE

Accidental occupational exposure should be treated as an *emergency*. This is because some persons who have had such an exposure may require post-exposure prophylaxis (PEP) with antiviral drugs. Since PEP should be started within 2 hours of exposure and *not* later than 72 hours of exposure, sufficient stocks of "starter packs" of PEP drugs should be available for all 24 hours.

11.4.1 – Immediate First Aid: Dos and Don'ts

Needle Stick or Percutaneous Exposure: (a) Do not put the pricked finger into your mouth. This may be done as a reflex and can be dangerous; (b) Encourage bleeding from the wound by squeezing; (c) Wash with soap and plenty of water; and (d) Apply any antiseptic. It is not necessary to use antiviral agents for wound care. Caustic agents such as household bleach should *never* be used for wound care (www.uchsc.edu/sm/aids; NACO, 1999).

Splashes to Nose, Mouth, or Skin: Wash area around the splash with plenty of water.

Splashes to Eyes: Irrigate eyes with clean water, saline, or sterile irrigating fluids.

11.4.2 – Immediate Decontamination of Spills

11.4.2.1 – Procedure

(a) Wear latex, vinyl, or India rubber gloves before decontamination.
(b) Cover the spill with absorbent material (cotton, gauze, absorbent tissue paper).
(c) Pour disinfectant *over* absorbent material and *around* the spill, leave for 30–60 min.
(d) Clean surface with fresh absorbent material, and dispose it off in a special container for contaminated waste.
(e) Sweep broken glass, etc. with a brush into special container for contaminated waste.
(f) Wipe the surface (spillage area) once again with disinfectant (NACO, 1999; CDC, 1997).

11.4.3 – Immediate Reporting

The health care providers should report all the spills of blood or body fluids and accidental occupational exposures immediately to supervisory staff, who will inform the Hospital Infection Control Officer (HICO) and HICC. All cases of accidental occupational exposure should be reported by hospital authorities to designated state health authorities. Designated hospital officials should maintain written record of all accidental exposures in the format given below:

(a) *Details for identification* – name, age, gender, designation, and employee number.
(b) *Details of accidental occupational exposure* – date, time, place of exposure, exposure code (EC), and HIV status code (HIV SC).
(c) *History of the incident* – date of injury, date of reporting, site and depth of injury, nature of injury (needle stick, laceration, sharp cut, splash of fluids, splattered glass, etc.).

(d) *Action taken report* – action taken in emergency or casualty, dates of immunisation with hepatitis B vaccine or immunoglobulin, dates of estimation of anti-hepatitis B antibody titre with titre levels, dates of estimation of hepatitis B antigen titre with titre levels, and dates of HIV tests by ELISA technique with reports.

11.4.4 – Evaluating the Exposure

Evaluation of exposure is based on EC and HIV SC.

11.4.4.1 – Exposure Code (EC)

NO EXPOSURE: Intact skin only.
EC-1: Breached skin or mucous membrane, *low-volume* exposure (few drops of body fluid, short duration of exposure).
EC-2: Breached skin or mucous membrane, *high-volume* exposure (many drops of body fluid, major splash, long duration of exposure).
EC-2: Percutaneous, *less severe* exposure (solid needle, superficial scratch).
EC-3: Percutaneous, *more severe* exposure (hollow bore needles, deep puncture, visible blood on sharp instrument, needle used on patient's artery or vein).

11.4.4.2 – HIV Status Code (HIV SC)

HIV-Negative: HIV-negative patient
HIV SC-1: HIV-positive patient, *low-titre* exposure (asymptomatic, high CD4 count).
HIV SC-2: HIV-positive patient, *high-titre* exposure (advanced AIDS, high viral load, low CD4 count).
HIV SC UNKNOWN: HIV status of patient or source is unknown.

11.4.4.3 – Base-line tests

Susceptibility of the exposed person to blood-borne pathogens is determined by baseline tests for hepatitis B surface antibody, anti-HCV, and HIV antibody, preferably within 72 hours (www.uchsc.edu/sm/aids).

11.4.4.4 – Factors determining risk of infection

Factors Related to Exposure: Type of exposure (mucocutaneous or percutaneous), depth of injury, quantity of blood or body fluids involved, duration of exposure, viral load in patient's blood at the time of exposure, and timeliness and dosage of PEP (if determined to be required).

High-risk Departments: haemodialysis unit, pathology or microbiology laboratory, surgical or trauma or emergency or intensive care units, blood bank, oral surgery or dentistry, obstetrics and gynaecology, and skin and STIs department.

Factors Related to Procedures: per vaginal (PV) and per rectal (PR) examinations; invasive diagnostic and therapeutic procedures; wound dressing; operation theatre procedures; handling blood, body fluids, and tissues; waste disposal, cleaning and house keeping; faulty procedures in Central Sterile Supplies Department (CSSD), post-mortem examination, and embalming.

Unless blood is *visible*, exposure to nasal discharge, saliva, sputum, stool, sweat, tears, urine, and vomitus does *not* pose a risk of transmission of blood-borne pathogens. Exposure to the following body fluids may pose a *significant* risk of transmission of blood-borne pathogens – blood, CSF, amniotic fluid, semen, cervicovaginal secretions, synovial fluid, peritoneal fluid, pleural fluid, and pericardial fluid.

11.4.4.5 – Step 5 Evaluate the exposure source

1. *If the source patient is known and can be tested* – After obtaining the source patient's informed consent and pre-test counselling, test for markers of HBV, HCV, and HIV infection.
   (a) If HBsAg positive, consider test for presence of HbeAg.
   (b) If HCV antibody positive, consider testing for HCV viral load.
   (c) And if positive for HIV antibody, HIV viral load and clinical status should be considered (www.uchsc.edu/sm/aids).
2. *If the source patient is known, but cannot be tested* – consider medical diagnosis, clinical symptoms, and history of high-risk behaviour in patients receiving haemodialysis, or blood transfusion, IDU, MSM, prison inmates, refugees, immigrants from highly endemic areas and other vulnerable groups such as mentally handicapped persons.
3. *If the source patient is not known* (e.g. exposure from needle in sharps container) – evaluate the likelihood of high-risk exposure based on the prevalence of blood-borne pathogens in the community or in the health facility. Do not test used needles and other sharp instruments for blood-borne pathogens since the reliability of these findings is not known (www.uchsc.edu/sm/aids).

11.4.4.6 – Step 6 Counselling and testing

Pre-test counselling should be provided *before* collecting the first sample of blood for laboratory tests and/or initiating PEP (NACO, 2002). The possible risks and benefits of PEP and details about follow-up are to be explained to the exposed person. Possible risks include side effects of ARV, and likelihood of seroconversion despite early initiation of PEP (Mitchell *et al.*, 1997; Wig, 2002).

11.4.4.7 – Step 7 Disease-specific Post-Exposure Prophylaxis (PEP)

PEP AGAINST HEPATITIS B INFECTION: Exposed persons *not* previously vaccinated against hepatitis B, should receive hepatitis B immunoglobulin 0.06 mL per kilogram body weight, administered intramuscularly along with a primary course of hepatitis B vaccine. Exposed persons who have been previously vaccinated against hepatitis B should receive a booster dose of hepatitis B vaccine (NACO, Training Manual for Doctors).

PEP against HCV: At present, there are no recommendations for PEP against HCV. Exposed persons to receive counselling, testing, and follow-up as for HIV exposures (www.uchsc.edu/sm/aids).

PEP against HIV: Most occupational exposures do not lead to HIV infection. The physician should carefully consider risks of acquiring HIV infection and possible side effects (toxicity) due to antiviral drugs used in PEP. For exposures with lower risk of infection, it is not advisable to start PEP (NACO, 2002). Delay in obtaining information on the source patient should not delay initiation of PEP since modifications can be made later, if necessary (www.uchsc.edu/sm/aids). Ideally, PEP should be started within 2 hours of exposure and not later than 72 hours of exposure. Hence, adequate stocks of PEP drugs (called "starter packs") should be available all 24 hours. If PEP is recommended, the following baseline investigations must be carried out *within* 72 hours and should be repeated 2 weeks later – complete blood count, urine analysis (for those receiving indinavir), renal function tests, and liver function tests (www.uchsc.edu/sm/aids).

The *basic* or *expanded* regimen may be prescribed, based on EC and HIV SC to select cases. The optimal duration of treatment is *unknown*. If the exposed person tolerates the antiviral drugs, the PEP is given for 4 weeks (NACO, 2002; www.uchsc.edu/sm/aids). The exposed person is informed about signs and symptoms of acute retroviral syndrome (flu-like syndrome), the need to report for additional tests at the onset of symptoms (www.uchsc.edu/sm/aids), and possible side effects of ARV drugs used in PEP. Only a physician or specialist in HIV medicine should determine the need for PEP (Mitchell *et al.*, 1997).

The Centers for Disease Control and Prevention, Atlanta, USA, has recommended schedules for PEP for health care providers accidentally exposed to HIV. The basic two-drug regimen comprises ZDV 200 mg thrice daily for 4 weeks with lamivudine (3TC) 150 mg twice daily for 4 weeks. The expanded three-drug regimen (advised if the source patient has advanced HIV disease) consists of ZDV 200 mg thrice daily for 4 weeks, 3TC 150 mg twice daily for 4 weeks, and nelfinavir 750 mg thrice daily for 4 weeks. In India, indinavir is used in the expanded regimen in a dose of 800 mg thrice daily for 4 weeks in place of nelfinavir.

11.4.4.8 – Follow-up

After Exposure to HBV: Test for anti-hepatitis B surface antigen 1–2 months after the last dose of vaccine. Anti-hepatitis B surface antigen cannot be determined up to 6–8 weeks after administration of hepatitis B immunoglobulin (www.uchsc.edu/sm/aids). During follow-up, the exposed person is to receive psychological counselling if needed. He or she must observe the following precautions strictly: (a) not to donate blood, semen, or body organ for transplant; (b) to avoid pregnancy; and (c) to abstain from sexual intercourse, or use latex condom every time during sexual intercourse (www.uchsc.edu/sm/aids; NACO, 2002; Mitchell *et al.*, 1997).

After Exposure to HCV: (a) Tests for anti-HCV antibody and liver enzymes are to be repeated for at least 4–6 months after exposure. Anti-HCV enzyme immunosorbent assays are to be confirmed by supplemental tests. (b) HCV RNA is tested for at least 4–6 weeks post-exposure. Caution should be exercised due to occurrence of false positive results. (c) During the 4–6 month follow-up period, the exposed person must refrain from donating blood, semen, or body organ for transplant. Changes in sexual activity, pregnancy, breastfeeding, or professional activities are *not* recommended. Mental health counselling is to be offered, if necessary (Mitchell *et al.*, 1997).

After Exposure to HIV: Blood test for HIV antibody should be done *immediately* after exposure, 6 weeks later, followed by 12 weeks, 6 months, and 12 months after exposure (NACO, 2002). The exposed health care provider ought to be followed up for the next 6 months for fever, pharyngitis, malaise, skin rash, lymph node enlargement, myalgia, and arthralgia (NACO, 2002; Burke *et al.*, 1993). The occurrence of illness being suggestive of acute retroviral syndrome, test for HIV viral load (www.uchsc.edu/sm/aids). Extended follow-up for 12 months is recommended if the source patient is co-infected with HIV and HCV (www.uchsc.edu/sm/aids). During the entire follow-up period, the exposed person is to be counselled to observe the following precautions strictly: (a) not to donate blood, semen, or body organ for transplant; (b) to avoid pregnancy; and (c) to abstain from sexual intercourse, or use latex condom every time during sexual intercourse (www.uchsc.edu/sm/aids; NACO, 2002; Mitchell *et al.*, 1997). If the person is HIV negative, 1 year after the accidental exposure, it means that he or she is *not* infected. PCR can give results even at the end of second or fourth week of exposure (NACO, 2002).

11.5 – SEEKING SPECIALIST OPINION

In addition to seeking opinion of specialists in HIV medicine, doctors can avail of information from the Internet (www.uchsc.edu/sm/aids). Some of these sources are
• National Clinicians' PEP Hotline (PEPline) – www.ucsf.edu/hivcntr
• Mountain Plains E-mail Clinical Consultation Service for HIV Infection – hiv-consultation@uchsc.edu
• HIV/AIDS Treatment Information Service – www.hivatis.org
• Needlestick! – www.needlestick.mednet.ucla.edu
• Hepatitis Hotline – www.cdc.gov/hepatitis

Indications for seeking specialist opinion
1. *Delayed reporting after exposure* – later than 24–36 hours (www.uchsc.edu/sm/aids).
2. *Unknown source of infection* – for example, needle in sharps container. Use of PEP is determined on case-by-case basis, considering severity of exposure and epidemiological likelihood of HIV exposure. Needles or other sharp instruments should not be tested for HIV (www.uchsc.edu/sm/aids).

3. *Known or suspected pregnancy in exposed person* – PEP must not be denied solely on the basis of pregnancy. The ARV agents that are currently approved for use in pregnancy may be prohibited if new information emerges (www.uchsc.edu/sm/aids). ZDV should be cautiously used in first trimester of pregnancy (Lewin *et al.*, 1997). The physician should carefully consider the risks before recommending PEP to a pregnant health care provider (NACO, 2002). It is better to consult a standard protocol.

4. *Known or suspected resistance of source patient's virus to certain ARV* – Select alternate drugs. The influence of drug resistance on the risk of transmission is not known. At the time of exposure, drug resistance testing of source patient's virus is not recommended (www.uchsc.edu/sm/aids).

5. *Toxicity of PEP regimen* – Nausea, diarrhoea, and headaches are common symptoms, which can be managed without changing the PEP regimen. Seek specialist opinion if the adverse effects are difficult to manage. Usually, dosage intervals may be modified (e.g. giving a lower dose of the drug more frequently) to relieve the symptoms (www.uchsc.edu/sm/aids).

6. *Expanded regimens* – Use of nevirapine, a protease inhibitor, has been associated with severe toxicity in exposed health care providers. It is advisable to seek expert opinion when using this drug or when considering dual protease inhibitor therapy (www.uchsc.edu/sm/aids).

REFERENCES

Burke R.A., Garvin G.M., and Sulis C.A., 1993, Infection control and risk reduction for health care workers in HIV infection – a clinical manual, 2nd edn. Boston: Little Brown.

Centers for Disease Control and Prevention (CDC), 1997, Perspectives in disease prevention and health promotion. update – universal precautions for prevention of transmission of HIV, hepatitis B and other blood borne pathogens in health care settings. Morb Mort Wkly Rep Suppl 37(24): 377–388.

Centers for Disease Control and Prevention (CDC), 1998, Recommendations for prevention of HIV transmission in health care settings. Morb Mort Wkly Rep Suppl (2S).

Leentvaar K.A., *et al.*, 1990, Needle stick injuries, surgeons, and HIV risks. Lancet 335: 546–547.

Lewin S.R., Crowe S., Chambers D.E., and Cooper D.A., 1997, Antiretroviral therapies for HIV. In: Managing HIV (G.J. Stewart ed.). North Sydney: Australasian Medical Publishing Company Limited.

Mandlebrot D.A., *et al.*, 1990, A survey of exposures, practices, and recommendations of surgeons in the care of patients with human immuno-deficiency virus. Surg Gynae Obs 171(2): 99–106.

Mehta A. and Rodrigues C., 1996, HIV and the surgeon. New Mediwave 5: 38–44.

Mitchell D.H., Sorrell T.C., and McDonald P.J., 1997, HIV control in medical practice. In: Managing HIV (G.J. Stewart, ed.). North Sydney: Australasian Medical Publishing.

Mountain Plains AIDS Education and Training Center. A quick guide to post-exposure prophylaxis in the health care setting. University of Colorado, Denver, USA. www.uchsc.edu/sm/aids

National AIDS Control Organisation (NACO), 1999, Manual for control of hospital associated infections – standard operative procedures. New Delhi: Government of India.

National AIDS Control Organisation (NACO), 2002, Specialists' training and reference module. New Delhi: Government of India.

National AIDS Control Organisation (NACO), Training manual for doctors. New Delhi: Government of India.

Raahave D. and Bremmelgaard A., 1991, New operative technique to reduce surgeons' risk of HIV infection. J Hosp Infect 18 Suppl A: 177–183.

UK Health Department, 1990, Guidance for clinical health care workers: protection against infection with blood-borne viruses. Recommendations of the Expert Advisory Group. London: Her Majesty's Stationery Office. www.uchsc.edu/sm/aids

WHO, 1996, Guidelines for preventing HIV, hepatitis B virus and other infections in the health care setting. New Delhi: SEARO.

Wig N., 2002, Anti-retroviral therapy – are we aware of adverse effects? JAPI 50: 1163–1171.

CHAPTER 12

HIV-RELATED NEUROLOGICAL DISORDERS

Abstract

Headache, fever, and neck stiffness occurring during the stage of seroconversion are due to aseptic meningitis. In advanced disease, patients may have generalised headache with photophobia for many weeks or months. Seizures and transient neurological deficits may occur. Underlying illnesses may precipitate AIDS-related dementia, where markers of immune activation may be present in CSF. The clinical severity exceeds the neuropathological abnormalities and viral load.

Key Words

AIDS dementia, HIV-related headache, Mononeuritis multiplex, Myopathy, Peripheral neuropathy, Seizures, Sensory neuropathy, Vacuolar myelopathy, ZDV-associated myopathy

12.1 – AIDS DEMENTIA COMPLEX

Before the advent of strong and effective ARV drugs, about 15–20 per cent of HIV-infected persons developed dementia and another 20–25 per cent had cognitive or motor dysfunction, when the CD4 cell count dropped below 100 cells per μL. The risk factors for development of dementia are anaemia and elevated levels of beta-2-microglobulin in CSF (McArthur *et al.*, 1993). Underlying illnesses that may precipitate this condition include cerebral lymphoma or toxoplasmosis, cryptococcal meningitis, depression, metabolic disorders, progressive multifocal leucoencephalopathy (PML), and systemic opportunistic illnesses. ARV drugs such as ZDV, stavudine, abacavir, and nevirapine easily cross the blood-brain barrier.

Clinical Manifestations: It is initially a subcortical dementia. Stage 0 (normal) to Stage 4 (severely impaired) have been described (Sidtis & Price, 1990). In the initial stages, the reflexes are brisk and symmetrical. Primitive reflexes may be seen in later stages. The clinical manifestations include poor concentration, disturbed short-term memory, slowing of thought processes and psychomotor slowing, impairment of rapid alternating and repetitive movements, motor incoordination, tandem (heel-to-toe) gait, bradykinesia, social apathy and withdrawal (rare), and abnormal ocular saccades (Currie *et al.*, 1988; Brew & Currie, 1993). The clinical severity exceeds the neuropathological abnormalities and

171

viral load (Glass *et al.*, 1993), and is associated with the presence of markers of immune activation in the CSF (Brew, 1992). Neuropsychological assessment reveals impairment of cognitive domains of memory, executive function, psychomotor speed, reaction time, and complex attention (Maruff *et al.*, 1994). Patients may need extra help in remembering to take their medications, making simple arithmetic calculations in daily life, and remembering commonly used telephone numbers. They may get lost in familiar places.

Investigations: CSF is to be examined for protein, glucose, culture, syphilis, and cryptococcal antigen. Levels of beta-2-microglobulin and neopterin are to be estimated. Serological tests are essential to rule out syphilis and cryptococcal infection. Computerised axial tomography (CAT) of the cranium shows cortical atrophy, enlargement of ventricles, widened sulci, and attenuation of white matter. Magnetic resonance imaging (MRI) shows similar changes as the cranial CAT scan, but is more sensitive in detecting illnesses such as PML.

Chemoprophylaxis: Early initiation of ZDV prophylaxis (at least 600 mg per day) has been recommended, but its efficacy is unclear.

Management: In the early stages of the disease, the patient and family should be counselled and medical power of attorney recommended (Wright *et al.*, 1997). The response to ARV drugs takes 6–8 weeks to be clinically apparent. Improvement is seen in 50 per cent of cases. Total blood counts are to be done every 2 weeks when patients are on high-dose ZDV therapy (Wright *et al.*, 1997). The clinical assessment is repeated every 4–6 weeks to determine response to treatment. Of the new drugs being studied, selegiline appears promising. Some studies have found ongoing brain damage even in patients taking ARV drugs (Fact Sheet 505, 2006).

12.1.1 – Progressive Multifocal Leucoencephalopathy (PML)

Leucoencephalopathy is a disease of the white matter of the brain. As its name denotes, the disease gets worse in a short time and occurs in multiple sites at the same time. The disease is difficult to diagnose and before strong ARV drugs became available, most patients died within 2 years. There is no approved treatment for PML, though several treatments may be helpful. Cytosine arabinoside, a toxic drug that can damage the bone marrow, seemed to be effective against PML in one study, but not in others. Acyclovir, dexamethasone, cidofovir, beta interferon, heparin, peptide-T, n-acetyl cysteine, and topotecan are among the drugs that have been studied, with varying degrees of success (Fact Sheet 516, 2006).

12.2 – VACUOLAR MYELOPATHY

This condition is rare in HIV-infected children (Brew & Currie, 1993; Glass *et al.*, 1993). It affects about 10 per cent of AIDS patients (Dal Pan *et al.*, 1994). It is due to vacuolar degeneration of the posterior and lateral columns of the spinal cord (Wright *et al.*, 1997). Toxins, metabolic disorders, or excessive production of cytokines may cause vacuolar myelopathy. It manifests as a progressive spastic paraparesis, with impaired perception of vibration and proprioception in feet and legs. Sensory loss of posterior-column type may be present, which lacks a definite sensory level. There may be loss of sphincter control in late stages of the disease. This condition responds poorly to ARV drugs. Addition of another ARV drug may be helpful. Management includes occupational therapy, physiotherapy, using walking aids, and tackling incontinence (Wright *et al.*, 1997).

12.3 – MYOPATHIES

ZDV may cause myositis. The inflammatory type of myopathy is similar to polymyositis, while non-inflammatory type is not yet understood (Simpson & Wolfe, 1991). Both HIV-related polymyositis and ZDV-associated myopathy manifest as progressive proximal muscle weakness, myalgia, and loss of more than 10 per cent of body weight. ZDV-associated myopathy mainly affects the lower limbs and causes wasting of buttocks. Serum creatine kinase (CK) levels are elevated. Electromyography (EMG) and muscle biopsies are abnormal. A trial of corticosteroids or intravenous Ig is indicated for HIV-related polymyositis. Patients with ZDV-associated myopathy usually recover in 6–12 weeks after withdrawal of the drug. Once the myopathy has resolved, ZDV may be reintroduced at a lower dose. If relapse occurs, another ARV drug should be used.

12.4 – PERIPHERAL NEUROPATHY

HIV-1 Sensory Neuropathy: This type of neuropathy is rare in asymptomatic HIV-infected persons but occurs in up to 35 per cent of AIDS patients (So *et al.*, 1988); disease is due to axonal degeneration with macrophage infiltration that affects all types of nerve fibres. This condition has been associated with infection by MAC. An identical and indistinguishable type of neuropathy is caused by ARV drugs (didanosine, zalcitabine, and stavudine). The neuropathy is distal, symmetrical, and predominantly sensory. Its manifestations include symmetrical numbness, hyperaesthesia, impaired perception of pain, temperature, light touch, vibration, and proprioception. Ankle jerks may be decreased or absent. Burning sensation in soles is a rare manifestation. There is no known specific treatment (Simpson & Wolfe, 1991). Management of HIV-1 sensory neuropathy involves withdrawal of all neurotoxic drugs (ARV drugs, isoniazid, alcohol, diabetes, high-dose metronidazole, thalidomide, and dapsone) and replacement with another non-neurotoxic drug. Mild to moderate neuropathic pain is treated with a *combination* of paracetamol and codeine. Tricyclic anti-depressants (amitriptyline 10 mg at night or sodium valproate 200 mg thrice daily) are to be administered and the dosage may be increased if pain is not controlled. For severe neuropathic pain, narcotic analgesics such as morphine are indicated. Clinical improvement may take up to 12 weeks (Wright *et al.*, 1997).

Inflammatory Demyelinating Polyneuropathy: This condition occurs during sero-conversion or asymptomatic phase. The acute manifestations of this condition are similar to that of Guillian–Barré syndrome. It is probably an autoimmune disorder (Simpson & Wolfe, 1991). It manifests as progressive weakness, loss of reflexes and rarely, mild sensory disturbances. EMG and nerve conduction studies are essential for confirming the diagnosis. Corticosteroids are to be used cautiously. Plasmaphresis and intravenous administration of Igs have been successfully tried. Physiotherapy and long-term rehabilitation are essential components of management (Wright *et al.*, 1997).

Mononeuritis Multiplex: During the asymptomatic or early symptomatic phases of HIV infection, acute nerve palsies may occur. One or more nerves may be involved. This mild form of mononeuritis is probably an autoimmune disorder (Simpson & Wolfe, 1991). This condition may resolve without specific treatment (So & Olney, 1991). In advanced disease, widespread neuropathy may occur, which requires hospitalisation. Nerve conduction studies are indicated. Since cytomegalovirus has been linked to severe form of mononeuritis multiplex, a trial of ganciclovir may be necessary (Wright *et al.*, 1997).

12.5 – HIV-RELATED HEADACHES

During the stage of seroconversion, patients may present with headache, fever, and neck stiffness due to aseptic meningitis. Examination of CSF may or may not show pleocytosis. Headaches are commonly seen in advanced disease. Patients may have generalised headache with photophobia for many weeks or months. The patient should be explained that most headaches resolve within several weeks. HIV-related headaches need to be differentiated from that due to sinusitis, migraine, depression, cryptococcal meningitis, cerebral toxoplasmosis, and cerebral lymphoma. Investigations include CAT or MRI of the cranium with sinus views, CSF examination (cytology, culture, cryptococcal antigen) and serology for cryptococcal antigen. Headache is treated with a combination of paracetamol and codeine. Tricyclic antidepressant (amitriptyline 50–75 mg at night) may also be used.

12.6 – SEIZURES

Seizures are seen in a variety of conditions that include cerebral space-occupying lesion, subclinical HIV-associated CNS involvement or dementia, metabolic disturbances, and cryptococcal meningitis (Holtzman *et al.*, 1989). The cause cannot be determined in up to 46 per cent of patients (Wong *et al.*, 1990). A complete history of the seizure should be recorded and withdrawal effects from alcohol or drugs are to be ruled out. CAT or MRI of the cranium may be undertaken to exclude cerebral lymphoma, toxoplasmosis, and PML. If CAT or MRI is normal, CSF examination (protein, glucose, cytology, culture, cryptococcal antigen) is essential. Haematological and serological investigations include total blood count, blood sugar, liver function tests, serum electrolytes, serum calcium and magnesium, and cryptococcal antigen. Electroencephalogram (EEG) helps to exclude focal seizures. If left untreated, seizures may recur. Clonazepam or sodium valproate is the recommended front-line drug. Use of phenytoin and carbamazepine is associated with adverse drug reactions such as skin rash, leukopenia, and abnormal liver function tests (Holtzman *et al.*, 1989).

12.7 – CEREBROVASCULAR DISEASE

Transient Neurological Deficits (TNDs) may occur in advanced HIV disease. These are similar to transient ischaemic attacks. This condition is associated with thrombocytopenia, deficiency of protein S, and presence of anticardiolipin antibodies. Some patients have responded to ARV agents. Cardiac disease, cerebral lesions, cryptococcal meningitis, and neurosyphilis are among the causes of cerebral infarction or haemorrhage. The patient must be hospitalised (Wright *et al.*, 1997). Investigations include CAT or MRI of the cranium, carotid doppler studies (for recurrent TNDs), EEG to exclude focal seizures, echocardiogram, chest radiography, total blood count, clotting profile (bleeding time, clotting time, prothrombin time, estimation of activated C, protein C, and protein S), and serological tests – cardiolipin antibody, syphilis serology, and cryptococcal antigen titre. Underlying cardiac or cerebral disease must be treated. If TNDs are recurrent, without underlying pathology, switch to a new ARV drug. Some patients may require anticoagulation therapy with low dose aspirin or warfarin (Wright *et al.*, 1997).

REFERENCES

Brew B.J., 1992, Medical management of AIDS patients: Central and peripheral nervous system abnormalities in HIV-1 infection. Med Clin North Am 76: 63–81.

Brew B.J. and Currie J., 1993, HIV-related neurological disease. Med J Aust 53: 104–108.

Currie J., Benson E., Ramaden B., *et al.*, 1988, Eye movement abnormalities as a predictor of the acquired immunodeficiency syndrome dementia complex. Arch Neurol 45: 949–953.

Dal Pan G.J., Glass J.D., and McArthur J.C., 1994, Clinico-pathologic correlations of HIV-1 associated vacuolar myelopathy: an autopsy-based case-controlled study. Neurology 44: 2159–2164.

Glass J.D., Wesselingh S.L., Selnes D.A., and McArthur J.C., 1993, Clinical-neuropathologic correlation in HIV-associated dementia. Neurology 43: 2230–2237.

Holtzman D.M., Kaku D.A., and So Y.T., 1989, New-onset seizures associated with human immunodeficiency virus infection: causation and clinical features in 100 cases. Am J Med 87: 173–177.

McArthur J.C., Hoover D.R., Bacellar M.A., *et al.*, 1993, Dementia in AIDS patients: incidence and risk factors. Neurology 43: 2245–2252.

Maruff P., Currie J., Malorie V., *et al.*, 1994, Neuropsychological characterization of the AIDS dementia complex and rationalization of a test battery. Arch Neurol 51: 689–695.

New Mexico AIDS Education and Training Center, 2006, Fact Sheet 505. Dementia and nervous system problems. University of New Mexico Health Sciences Center. www.aidsinfonet.org. Revised May 15.

New Mexico AIDS Education and Training Center, 2006, Fact Sheet 516. Progressive multifocal leucoencephalopathy. University of New Mexico Health Sciences Center. www.aidsinfonet.org. Revised February 21.

Sidtis J.J. and Price R.W., 1990, Early HIV-1 infection and the AIDS dementia complex. Neurology 40: 323–326.

Simpson D.M. and Wolfe D.E., 1991, Neuromuscular complications of HIV infection and its treatment. AIDS 5: 917–926.

So Y.T. and Olney R.K., 1991, The natural history of mononeuropathy multiplex and simplex in patients with HIV infection. Neurology 41(Suppl 1): 375.

So Y.T., Holtzman D.M., Abrams D.I., and Olney R.K., 1988, Peripheral neuropathy associated with acquired immunodeficiency syndrome: prevalence and clinical features from a population-based survey. Arch Neurol 45: 945–948.

Wong M.C., Suite M.D., and Labar D.R., 1990, Seizures in human immunodeficiency virus infection. Arch Neurol 47: 640–642.

Wright E.J., Brew B.J., Currie J.N., and McArthur J.C., 1997, HIV-induced neurological disease. In: Managing HIV (G.J. Stewart, ed.). North Sydney: Australasian Medical Publishing, pp 76–79.

CHAPTER 13

HIV-RELATED PSYCHOLOGICAL DISORDERS

Abstract

The psychological reactions experienced by an individual to the news of HIV positive test result depends on factors such as available social support systems and pattern of coping with major stresses in the past. In order to avoid anticipated social stigma, some individuals deny reality or bear the additional burden of secrecy from family, friends, and colleagues. With the onset of symptoms and visible signs of illness, denial, and secrecy cannot be sustained as coping mechanisms. Some individuals strive to achieve a purposeful life, and associate themselves with self-help groups of HIV-infected persons. There is a need for ongoing guidance after pre- and post-test counselling sessions. Counselling sessions are also necessary when patients develop new symptoms. Anxiety and stress should be treated. The individual may be referred to organisations that provide counselling and support services. Persons involved in care of HIV patients may also face psychological stress due to the emotionally demanding task of giving bad news, fear of accidental occupational exposure to HIV, distressing manifestations of the disease, and a feeling of powerlessness at the limitations of some treatments. Hence, HIV care providers should take regular breaks, develop a range of professional and personal interests and relationships outside the workplace, recognise early signs of excessive stress and seek support, when necessary.

Key Words

Burnout, Closet homosexuals, Denial, Depression, Dichotomy, Frustration, Gays, Guilt, Impact on providers, Isolation, Secrecy, Social stigma, Withdrawal

13.1 – PSYCHOLOGICAL REACTIONS TO DIAGNOSIS

The reaction of an individual to the news of HIV positive test result involves a sequence of psychological reactions that may require specialist intervention. Diagnosis of HIV positive status precipitates a great amount of changes in life, which has been rated as equivalent to the death of a spouse or a prison sentence. Thus there is a need for pre- and post-test counselling sessions. Counselling sessions are also necessary when patients develop new symptoms. The individual's response may depend on factors such as social support system available to the individual and pattern of coping with major stresses in the past. These factors are also predictors of both physical and psychological illnesses (Cohen, 1988). The stages of psychological reactions have been studied in asymptomatic homosexual men, but these are similar in heterosexuals and persons infected by non-sexual routes. All individuals do not go through these stages (Ross *et al.*, 1989).

On receiving the news of HIV positive test result, the individual experiences emotional shock, leading to feeling of "guilt" (self-blame) or "denial" (refusal to accept the diagnosis). Due to denial, the individual may get the HIV test repeated at different laboratories hoping for a negative test result. Guilt about their high-risk behaviour (promiscuity or use of injecting drugs) is another source of psychological stress. Once the individual realises the truth, *frustration* sets in. If the individual knows others who indulge in the same type of high-risk behaviour but are HIV negative, the reactions vary from anger, distress, feeling of powerlessness, and blaming one's fate or destiny. "Dichotomy" is an issue to be tackled. People are seen as infected ("us") and non-infected ("them") and responded accordingly. For such persons, HIV status becomes the central defining issue in life. On realising the gravity of the situation and the ultimate outcome of the condition, the person may go into *depression*. Worry about social stigma leads to *withdrawal* and *isolation*. The individual may start worrying about his or her spouse and family members and may have fear of infecting others.

13.2 – DENIAL

Some individuals cope with early HIV infection by denying its reality. The onset of symptoms and visible signs of illness creates a situation of *disclosure* and denial is difficult to sustain as a coping mechanism. Dealing with loss of denial is equivalent to dealing with emotional shock at the time of diagnosis. If the diagnosis of HIV infection has not been revealed to family members and close contacts, it can create psychosocial problems, both at home as well as at the workplace, and aggravate the stress experienced by the patient.

13.3 – EFFECT OF SOCIAL STIGMA

Family members and close friends may not be aware of sexual habits of some male homosexuals (also called "gays"). These persons have been called "closet homosexuals". Such persons may not be able to share the news of their HIV positive test result with their family members and close friends. This leads to lack of social support that aggravates their stress. The social reality is that people generally react to HIV-infected persons with fear and prejudice. Homosexuals and drug users are socially marginalised and stigmatised. Irrespective of mode of infection, HIV positive individuals suffer from the prejudiced assumption that they may have been infected through "deviant" or illegal activities. This assumption challenges the social identity of persons who may have been infected by heterosexual contact or through medical interventions such as blood transfusion. Anticipation of social discrimination complicates the psychological response to the news of HIV positive test and affects the person's ability to cope.

It may damage self-esteem of HIV-infected persons and may cause or worsen feeling of depression. In order to avoid anticipated social stigma, some individuals bear the additional burden of *secrecy* from family, friends, and colleagues (Miller, 1987).

13.4 – LIVING WITH HIV

A great amount of courage is required to face a disease that seems to have no cure so far. Many HIV-infected people experience fear, anxiety, hopelessness, loneliness, and depression. Some patients accept their diagnosis and come to terms with their condition. During the time period between diagnosis and the onset of serious illness, some strive to achieve a purposeful life. They may test the reactions of others; look for other HIV positive persons for sharing and positive reinforcement (Miller, 1987). Group cohesiveness leads to altruistic behaviour and a feeling of belonging to the group. These groups usually have a positive attitude to being HIV positive and also have role models that are open about their HIV positive status. Some individuals select their sexual and other partners only from within the community of infected persons. This solves problems like fear of rejection by a prospective partner and of transmitting HIV to an uninfected partner (Miller, 1987). Though yoga cannot replace professional counselling, yoga techniques are known to help in reducing excessive fear and anxiety, and learning stress-coping skills. Meditation helps in self-awareness and in building inner strength through relaxation.

13.5 – ROLE OF THE DOCTOR

The infected persons experience considerable anxiety about becoming dependent on others for basic physical care and stress related to actual or anticipated discrimination or abandonment. Psychological reactions such as denial, anger, and depression that occur in HIV-infected individuals may result in lack of trusting relationship with health care providers and missed opportunities for prevention and treatment of opportunistic infections. Psychological reactions such as shock, denial, and depression also occur in family members and may affect their ability to care for the HIV-infected person. Hence, family members also need to be considered while managing the HIV-infected individual. It is the responsibility of the doctor to provide accurate HIV-related information, address and treat anxiety and stress, and refer the patient to organisations that provide counselling and support services (Miller, 1987).

It is essential that the patient be fully informed so that both the doctor and the patient make their decisions with full knowledge of the circumstances. Lack of rapport between the doctor and patients results in poor compliance with prescribed treatment, visits to multiple doctors who may use multiple drug regimens, and self-medication with traditional or alternative medicines that lead to interactions.

13.6 – SYMPTOMATIC HIV INFECTION

The emotional and psychological conflicts are related to changes in life circumstances caused by symptomatic illness and can be as varied as the symptoms themselves. The social support systems that worked well before the onset of symptoms may become irrelevant. The responses may include fear, anxiety, and uncertainty about dependence on others, depression, and suicidal ideas. The psychological state can affect the quality of life beyond the impact of physical illness. The patients and their family members usually experience these psychological problems during the phase of recovery, when investigations and medical interventions do not distract them. These problems may also manifest when the patient is discharged from the hospital. Hence, the family physician may have to provide patients and their family members with support and interventions that include: (a) information on likely opportunistic infections, (b) information on legal issues (Will, pensions, and power of attorney), and occupational issues (sick leave, loss of capacity to work), (c) psychiatric interventions including specific counselling, and (d) spiritual support. If the patient and family members are mentally primed to deal with symptomatic illness, the onset of symptoms may have a positive impact because they can prepare themselves for further significant changes in life. When HIV seropositivity is discovered late in the course of an illness, the psychological stress may be aggravated and coping becomes more difficult (Kelly *et al.*, 1997).

13.7 – PSYCHOLOGICAL IMPACT ON PROVIDERS

13.7.1 – Psychological Stress

Persons caring for HIV patients may face psychological stress due to the emotionally demanding task of giving bad news such as HIV positive status and likelihood of death, fear of contracting HIV infection through accidental occupational exposure, the distressing nature of manifestations of the disease, increased scrutiny by consumer rights organisations and AIDS activists, and a feeling of powerlessness at the limitations of some treatments (Kelly *et al.*, 1997).

13.7.2 – Burnout

"Burnout" refers to physical and emotional symptoms caused by task-related stress (Ross & Seeger, 1988). This condition is caused by working harder for long periods of time, without being able to set reasonable time limits. This leads to development of indifference and negative attitudes. Burnout may lead to feeling of lack of personal accomplishment; emotional exhaustion or emotional numbness, depression, and anxiety; antagonised relationship with patients; and estranged family relationships. Health care personnel should be aware of these problems and be able to take preventive measures and seek assistance if required.

13.7.3 – Coping Strategies: Dos and Don'ts

- Set limits to occupational commitment that one can take.
- Take regular breaks (clearly defined work-free time, holidays).
- Maintain personal interests such as hobbies, physical exercise, and relationships outside the workplace.
- Recognise early signs of excessive stress, and seek advice and support from friends and colleagues.
- Avoid using self-medication such as sedatives or alcohol to cope with symptoms of stress.
- Develop a range of professional interests to balance the intensity of occupational commitment and avoid professional isolation.
- Participate in institutional activities that promote healthy coping – continuing medical education (CME), in-service training, and team meetings (Kelly *et al.*, 1997).

REFERENCES

Cohen L.H., 1988, Life events and psychological functioning. Newburg Park, CA: Sage. Cited in: Kelly *et al.*, 1997.

Kelly B.J., Todhunter L., and Raphael B., 1997, HIV care: impact on the doctor. In: Managing HIV (G.J. Stewart, ed.). North Sydney: Australasian Medical Publishing.

Miller D., 1987, Living with AIDS and HIV. London: Macmillan.

Ross M.W. and Seeger V., 1988, Determinants of reported burnout in health professionals associated with the care of patients with AIDS. AIDS 2: 395–397.

Ross M.W., Tebbie W.E.M., Vilunas D., *et al.*, 1989, Staging of psychological reactions to HIV infections in asymptomatic homosexual men. J Psychol Human Sexol 2: 93–104.

CHAPTER 14

CO-INFECTION WITH HIV AND TUBERCULOSIS

Abstract

As compared to their HIV negative counterparts, HIV positive individuals infected by *Tubercle bacillus*, have about 10 times higher risk of progression from latent to clinically active tuberculosis. In India, the strategy of HIV testing of all newly diagnosed tuberculosis patients is neither feasible nor cost-effective, due to the large number of new cases of tuberculosis that are detected each year. The standard treatment regimens used in the Revised National Tuberculosis Control Programme (RNTCP), are equally effective in HIV positive patients. Directly Observed Treatment, Short Course (DOTS) strategy has been shown to improve survival of patients with co-infection. Concurrent treatment with protease inhibitors can either reduce activity or prolong half-life of rifampicin. Chemoprophylaxis against tuberculosis is currently not recommended in India.

Key Words

Adult tuberculosis, ART in tuberculosis, BCG vaccine, Childhood tuberculosis, DOTS, Ghon focus, Granuloma, Immune reconstitution syndrome, Langhan's giant cells, Mantoux test, Multi-drug resistance, Post-primary infection, Primary infection, Primary complex, RNTCP, Tubercle

14.1 – MAGNITUDE

14.1.1 – Tuberculosis

Currently, tuberculosis is the single biggest infectious disease that kills about 2–3 million persons worldwide each year. Annually, about 8 million new cases are diagnosed. The global incidence of tuberculosis is increasing at the rate of 0.4 per cent per year. Globally, the economic cost of tuberculosis is estimated at US$12 billion. Despite the availability of effective drugs, tuberculosis remains a global health challenge. In 1993, the WHO declared tuberculosis a global emergency (Health Action Information Network – HAIN, 2003). The re-emergence of tuberculosis in the developed countries in the 1990s is attributed to the HIV epidemic and environmental and social changes. The advent of HIV epidemic and multi-drug resistant (MDR) strains of tuberculosis has made the re-emergence of tuberculosis even more dangerous (HAIN, 2003). In 2001, the WHO drafted a "Global Plan to Stop TB". The plan seeks to expand the DOTS approach and to improve the existing tools for diagnosis, treatment, and prevention. In its first phase, the plan envisages a detection rate of 70 per cent and a cure rate of 85 per cent by the year 2005 (HAIN, 2003).

Of the world's tuberculosis patients, 40 per cent live in South and South-East Asia. Of the world's annual 700,000 tuberculosis-related deaths, 95 per cent occur in Bangladesh, India, Indonesia, Myanmar, and Thailand. Though India leads in the total number of world's cases, Philippines has a higher rate of cases per 100,000 population (HAIN, 2003). Tuberculosis remains the leading cause of infectious death in India, killing close to 500,000 persons each year. There is an additional burden of two million new cases every year. Since most victims are aged between 18 and 45 years (the most economically productive age group), the disease causes losses in family income and national productivity. Studies show that on an average, a patient loses 3–4 months of work with lost earning amounting to 20–30 per cent of a family's annual income (HAIN, 2003). The economic burden of tuberculosis in India has been estimated to be Rs. 148.5 billion (about US$3 billion) per year (Pathni & Chauhan, 2003). The overall prevalence of HIV infection among tuberculosis patients was 9.0 per cent in 2005 (NACO, 2006).

14.1.2 – HIV-Tuberculosis Co-infection

Sub-Saharan Africa: Tuberculosis is the most common opportunistic infection in Sub-Saharan Africa. The HIV seroprevalence in tuberculosis patients is up to 75 per cent. It is the most frequent cause of death among HIV-infected persons in sub-Saharan Africa (Harries *et al.*, 2004). Tuberculosis-related deaths are expected to double by the year 2010 as the immune system (of currently HIV-infected persons) becomes more vulnerable to active disease (HAIN, 2003). In much of Africa, the spread of HIV is primarily responsible for driving the parallel epidemic of tuberculosis, often at the rate of 6 per cent per year (Corbett *et al.*, 2003).

Thailand: A case control study in Northern Thailand between 1990 and 1998 has found that 72 per cent of males and 66 per cent of females had tuberculosis that was directly attributable to HIV infection (Tansuphasawadikul *et al.*, 1999). Both cohort and case control studies have shown that the relative risk of acquiring active tuberculosis among HIV-infected persons varied from 5 to 20 per cent (Glynn, 1998). Seropositivity rate of 40 per cent has been reported among tuberculosis patients in Northern Thailand (Yanai *et al.*, 1996).

India: Since India has a high prevalence of tuberculosis, the problem of HIV-tuberculosis co-infection is overwhelming. An estimated 40 per cent of adults in India are already infected with *Mycobacterium tuberculosis* (Pathni & Chauhan, 2003). According to HIV sentinel surveillance report for 2005, released by NACO in April 2006, the overall prevalence of HIV infection among tuberculosis patients was 9.0 per cent. The HIV prevalence among tuberculosis patients in four sentinel sites was – Davangere district, Karnataka: 9.5 per cent; Guntur district, Andhra Pradesh: 16 per cent; Nashik district, Maharashtra: 4.3 per cent; and Tiruvannamalai district, Tamil Nadu: 6.3 per cent (NACO, 2006). A study conducted in New Delhi hospital during 1994–1999 reported that out of 555 patients with tuberculosis, 9.4 per cent were HIV-positive while the overall seropositivity rate at the same hospital was 0.4 per cent (Sharma *et al.*, 2003). Seropositivity rate of 30 per cent has been reported among tuberculosis patients in Mumbai (Mohanty & Basheer, 1995).

14.2 – SOURCES OF TUBERCULOUS INFECTION

The predominant source of infection is an individual with pulmonary tuberculosis whose sputum smear is positive. Coughing, talking, sneezing, spitting, and singing by such an individual produce droplet nuclei that contain *Tubercle bacilli*. Droplet nuclei are infectious particles of respiratory secretions usually less than 5 µ in diameter. Owing to their small size, they can directly lodge in the terminal alveoli of the lungs by avoiding the mucociliary defences of the bronchi. A single cough can produce up to 3,000 droplet nuclei that remain suspended in air for prolonged periods. Transmission by droplet nuclei generally occurs indoors and in the dark because direct sunlight can kill *Tubercle bacilli* within 5 minutes (Harries *et al.*, 2004). Milk-borne tuberculosis is spread by consumption of milk from cattle infected by *M. bovis*. Infection of the tonsils presents as cervical lymphadenitis ("scrofula"). The intestinal tract may also be infected (Harries *et al.*, 2004). Since milk is boiled before consumption in India, milk-borne tuberculosis is not a public health problem.

14.3 – HOST FACTORS

Some host factors associated with increased risk of HIV infection also predispose to tuberculosis. These include poverty, migration, and gender and they share a symbiotic relationship (HAIN, 2003).

Poverty: Poverty forces people to live in overcrowded conditions that increase the risk of transmission of the disease. Poverty is also accompanied by under-nutrition, lack of access to health care and poor living conditions such as poor ventilation, lack of safe drinking water, and inadequate sanitation. Workers exposed to silica-containing dust are also vulnerable. These factors compromise the body's ability to fight infections. Therefore, poor people are more vulnerable to tuberculosis, as compared to their relatively affluent counterparts. Thus, tuberculosis control programmes can succeed only if the socio-economic condition of the target population is improved. Due to losses in earnings, many families are forced to sell their land or livestock, or take their children out of school. These school dropouts help out with domestic chores or work outside their homes as child labourers. In India, 300,000 children from tuberculosis-afflicted households are forced to leave school every year (HAIN, 2003).

Migration: Migrants are pushed to poverty if they are unable to find work. Actual or perceived discrimination and social maladjustment prevents these migrants from seeking health care. Illegal international migrants do not seek health care services for fear of detection and deportation. Single migrants also constitute a high-risk group for HIV infection.

Gender: Female tuberculosis patients face more social stigma and discrimination as compared to their male counterparts. For fear of social ostracism, many families in male dominated societies do not seek tuberculosis treatment for their womenfolk. In India alone, more than 100,000 tuberculosis-afflicted wives are abandoned by their husbands, each year (HAIN, 2003). HIV-infected women also face a similar situation.

14.4 – NATURAL HISTORY OF TUBERCULOSIS

14.4.1 – Risk of Infection

The risk of infection of a new host is determined by: (a) concentration of infected droplet nuclei in the inhaled air, (b) duration of exposure to inhaled droplet nuclei, and (c) susceptibility of the new host to infection (HAIN, 2003). The risk of infection is high with prolonged, close, indoor exposure to a person with sputum positive pulmonary tuberculosis. The risk of transmission of infection from a person with sputum negative pulmonary tuberculosis is low. The risk is even lower from a person with extrapulmonary tuberculosis (Harries *et al.*, 2004).

14.4.2 – Risk of Progression of Infection to Disease

Infection with *M. tuberculosis* can occur at any age. Once infected, the person may stay infected probably for life. In India, an estimated 40 per cent of the adult population is already infected with *M. tuberculosis*. (Pathni & Chauhan, 2003). About 90 per cent of persons who are infected with *M. tuberculosis* (but without HIV infection) do not develop tuberculosis disease. In such asymptomatic but infected individuals, the only evidence of infection may be a positive tuberculin skin test. Infected persons can develop tuberculosis disease at any time. Infants, children, the elderly, and persons with immune suppression (due to malignancy, malnutrition, measles or pertussis infection, corticosteroid therapy, HIV infection) are more vulnerable to develop tuberculosis. Infants and children have an immature immune system and usually develop the disease within 2 years of exposure and infection. The disease is more likely to spread from the lungs to the other parts of the body in this age group. Those who do not develop the disease in this age group may do so later in life. Physical or emotional stresses may also trigger progression of infection to disease (Harries *et al.*, 2004).

14.4.3 – Untreated Tuberculosis

HIV infection, *per se*, is not fatal. HIV destroys the immune system and makes the infected person vulnerable to multiple infections. If left untreated, by the end of 5 years, 50 per cent of patients with pulmonary tuberculosis will be dead; 25 per cent will be self-cured by strong immune defence; and the remaining 25 per cent will remain ill with chronic infectious tuberculosis (Harries *et al.*, 2004).

14.5 – PATHOGENESIS AND IMMUNOPATHOLOGY

14.5.1 – Primary Infection

Primary infection occurs in persons who have not had any previous exposure to *Tubercle bacilli*. Droplet nuclei, which are usually less than 5 µm in diameter, can directly lodge in the terminal alveoli of the lungs by avoiding the mucociliary defences of the bronchi. Infection begins with multiplication of *Tubercle bacilli* in the lungs and the resulting lesion is called *Ghon focus*. The lymphatics drain the bacilli to hilar lymph nodes. The Ghon focus, together with the enlarged hilar lymphadenopathy, forms the *primary complex* (Harries *et al.*, 2004).

From the primary complex, bacilli may spread via the blood stream, throughout the body. Rapid progression to intra-thoracic disease is more common in children under 5 years of age. The immune response (delayed-type hypersensitivity (DTH) and cellular immunity) develops about 4–6 weeks after primary infection. The ensuing events are determined by the quantum of infecting dose and the strength of the immune response.

14.5.2 – Immunopathology of Primary Infection

M. tuberculosis multiplies and within about 3 weeks, the population of the bacilli reaches 10^3–10^4 (the number required to trigger an immune response). Once this number is reached, the mycobacterial multiplication suddenly stops.

If the tubercular antigens are represented by *class-I MHC molecules*, cell mediated immunity (CMI) is developed by activating the CD4 cells. However, if class-II MHC molecules represent the tubercular antigens, CD8 cells – cytotoxic cells concerned with DTH – are activated. CD8 cells also have the ability to recognise infected macrophages and destroy them by direct cytotoxic action. Various immunological chemicals activate the resting macrophages to engulf, ingest, and digest the mycobacteria. This process is further enhanced by vitamin D3, which is tuberculostatic.

The activated macrophages release chemicals from the cell wall of the digested mycobacteria, which results in converting monocytes into epithelioid cells and Langhan's giant cells, forming a granuloma (called the *Tubercle*). The centre of the granuloma has low pO_2 and low pH, which is unfavourable for the growth of *M. tuberculosis*. Thus, during this stage, immunologic control can be achieved, by "walling off" the infection. The granuloma may get *fibrosed* (in adult tuberculosis), or *calcified* (in childhood tuberculosis). As long as the infection is "walled off", the person remains infected, but clinically asymptomatic. About 90 per cent otherwise healthy persons, infected with the *Tubercle bacillus* do not develop clinical tuberculosis.

14.5.3 – Outcome of Primary Infection

In about 90 per cent of cases, the immune response stops the multiplication of bacilli, but a few dormant bacilli may persist. A positive tuberculin (Mantoux) test would be the only evidence of infection. In some individuals, the immune response is not vigorous enough to prevent the multiplication of bacilli and various manifestations of the disease may occur after a latent period of months or years. Some may develop hypersensitivity reactions such as phylctenular conjunctivitis, erythema nodosum, and dactylitis. Intra-thoracic disease (lung infiltrates, pneumonia, consolidation, collapse, or pleural effusion) may occur. Lymphadenopathy (particularly in cervical region), meningitis, or miliary tuberculosis constitute disseminated type of tuberculosis (Harries *et al.*, 2004).

14.5.4 – Post-Primary Tuberculosis

Post-primary tuberculosis occurs either by: (a) *reactivation* – dormant bacilli acquired from a primary infection begin to multiply due to trigger factors like weakening of immune system by HIV infection, or (b) *reinfection* – occurrence of repeat infection in an individual who has previously had a primary infection (Harries *et al.*, 2004).

14.5.5 – Immunopathology of Post-Primary Tuberculosis

In the later stages of primary infection, gamma and delta lymphocytes become responsive to antigens of *M. tuberculosis*. There seems to be a balance between CMI and DTH. If the immunologic control by CMI and DTH is not balanced, then the tissue-damaging action of DTH may predominate. This leads to liquefaction necrosis of the tubercular granuloma and subsequent activation of the disease. In pulmonary tuberculosis, this leads to development of cavities. Thus, DTH is more harmful, and not helpful, to the host. The immune response results in a pathological lesion that is localised, often with extensive tissue destruction and cavitation.

14.5.6 – Outcome of Post-Primary Tuberculosis

Though post-primary tuberculosis usually affects the lung, it can affect any part of the body. *Pulmonary* tuberculosis may manifest as cavities, upper lobes infiltrates, progressive pneumonitis, endobronchial tuberculosis, fibrosis, and pleural effusion. The common clinical features of *extrapulmonary* tuberculosis are lymphadenopathy (usually cervical), meningitis, cerebral tuberculoma, pericardial effusion, constrictive pericarditis, ileo-caecal and peritoneal tuberculosis, and involvement of the skeletal system (spine, bone, and joints). Involvement of skin (lupus vulgaris, tuberculids), miliary tuberculosis, involvement of kidney and adrenals, tuberculous epididymitis and orchitis, tubo-ovarian or endometrial tuberculosis and empyema are rare manifestations of extrapulmonary disease. The hallmarks of post-primary tuberculosis are – extensive destruction of lung tissue with cavitation, involvement of upper lobe, positive sputum smear, usually with absence of intra-thoracic lymphadenopathy. Patients with these lesions can spread infection in the community (Harries *et al.*, 2004).

14.6 – DIAGNOSTIC TECHNIQUES

Clinical manifestations such as cough, fever, and chest pain are also seen in other diseases and are thus not typical for diagnosis of tuberculosis. Sputum microscopy is not very accurate and is also subject to observer errors. Interpretations of chest radiographs are also subjective. Though *M. tuberculosis* was identified in 1882, and its entire genome was completed only in 2002, there is no modern diagnostic kit for tuberculosis till date. On the other hand, within a few months of outbreak of Severe Acute Respiratory Syndrome (SARS), the entire genome of the pathogen was identified and a diagnostic kit was developed. PCR that amplifies specific DNA sequences of the *Tubercle bacillus* is available, but very expensive (Bezbaruah, 2004).

In the developed countries, all newly diagnosed tuberculosis patients are tested for HIV serostatus. However, in India, this strategy is neither feasible nor cost-effective, due to the large number of new cases of tuberculosis that are detected each year. Pulmonary tuberculosis can be diagnosed in HIV positive individuals by sputum examination. Diagnosis of extrapulmonary and disseminated forms of tuberculosis is possible by histopathology and various sophisticated imaging techniques. Based on the genome sequence of *M. tuberculosis* (which was completed in 2002), sensitive DNA-based diagnostic tests are being developed. The Foundation for Innovative New Diagnostics (FIND), a WHO-funded organisation, is working on cheap innovative diagnostic tests. A colour-based assay (wherein cultures of *M. tuberculosis* will light up bright green) is being developed. The All India Institute of Medical Sciences (AIIMS), New Delhi, has developed a specific PCR technique and a solution for spotting mycobacteria more clearly (Bezbaruah, 2004).

14.7 – DIRECTLY OBSERVED TREATMENT, SHORT COURSE

The WHO and the International Union Against Tuberculosis and Lung Diseases (IUATLD) recommend "DOTS" as the most effective and affordable strategy to control tuberculosis. The DOTS strategy involves: (a) diagnosis of cases by sputum microscopy from among patients with symptoms suggestive of tuberculosis, (b) free-of-cost intermittent therapy with a standardised drug regimen, (c) direct observation of drug consumption by a trained health worker to ensure adherence, (d) reliable and regular drug supply, (e) adequate health infrastructure and trained health personnel, (f) political commitment, and (g) monitoring and evaluation of the programme (HAIN, 2003).

Treatment supporter is a health worker or a community volunteer who provides encouragement and support for the person taking antitubercular treatment. DOTS strategy was first introduced in 1991. Since then, incidence rates have decreased in high-burden countries. High cure rates have been reported in selected areas, but these areas may be isolated islands of excellence. An adequate public health infrastructure is a pre-requisite for starting DOTS. Critics point out that the DOTS programme will only be as efficient as the public health services of the country where it is being implemented. In countries where the basic health services are inadequate, the long-term sustainability of DOTS is endangered. The current treatment regimens involve the use of four or five antitubercular drugs, to be taken for at least 6 months. Studies reveal that most patients stop treatment after the initial 2–3 months since the symptoms are relieved. Poor compliance to treatment increases the risk of multi-drug resistance. If the treatment is stopped after 2 months, the estimated risk of relapse is 70 per cent, while after 4 months it is 40 per cent (HAIN, 2003).

14.7.1 – Fixed-Dose Combinations

These contain two or more drugs within the same tablet. Currently, fixed-dose combinations are more expensive than the total cost of single drugs. The situation is likely to change with increase in production of fixed-dose combinations. The WHO and IUATLD recommend the use of fixed-dose combinations for the following reasons:
1. Increase in patient adherence: the probability of patients forgetting to take a particular medication is reduced since they have fewer pills to swallow.
2. Improves adherence of health care personnel to standardised regimens.
3. Drug management becomes easier (fewer items with a single expiry date).
4. Managing drug supplies becomes easier since fewer drugs and lower volumes need to be procured and delivered to rural and remote areas.
5. Reduces possibility of drug resistance by reducing the likelihood of prescription errors or use of wrong drug combinations (HAIN, 2003).

14.7.2 – DOTS in India

The government of India introduced DOTS strategy under the RNTCP in 1997. By the end of 2001, the population covered by DOTS increased to 45 per cent and the number of DOTS-notified smear positive cases nearly doubled. But at the national level, the total numbers of smear positive cases (both DOTS and non-DOTS taken together) changed little. In order to reach the targeted case detection rate of 70 per cent, DOTS must be extended geographically and at the same time, the proportion of cases detected under DOTS programme must be increased. The cure rate for patients registered in the year 2000 was 84 per cent (HAIN, 2003). The World Bank has provided a loan of US$142.5 million to the government of India. Danish International Development Agency (DANIDA), Global Fund for AIDS, Tuberculosis and Malaria (GFATM), and Global Development Fund financially support the DOTS expansion programme (HAIN, 2003).

Despite sound financing, the key problem in the DOTS programme continues to be access to drugs, information, and treatment. Many patients and even medical practitioners are still not aware of the DOTS programme (HAIN, 2003). Given the low Indian health budget, the long-term sustainability of the programme is uncertain (Bezbaruah, 2004).

14.8 – HIV-TUBERCULOSIS CO-INFECTION

14.8.1 – Effect of HIV Epidemic on Tuberculosis

HIV infection expedites the spread of tuberculosis by re-activating latent tubercular infection, accelerating progression of recently acquired tubercular infection, and by predisposing to exogenous re-infection by *M. tuberculosis* (Pathni & Chauhan, 2003). Among those infected by *M. tuberculosis*, it is estimated that the *lifetime risk* of progression from latent to clinically active tuberculosis is 50 per cent (i.e. about 10 times higher) in HIV positive individuals, as compared to the *lifetime risk* of 5–10 per cent faced by their HIV negative counterparts (Harries *et al.*, 2004). HIV infection along with active tuberculosis leads to depletion of CD4 lymphocytes, increased multiplication of HIV, and elevated plasma levels of HIV-RNA. In HIV positive individuals, progressive depletion of CD4 cells results in predominance of CD8 cells and DTH, which helps the re-activation of primary tubercular infection and dissemination of the disease. In pulmonary tuberculosis, HIV also impairs the innate resistance of alveolar macrophages. Thus, HIV and *M. tuberculosis* form a *deadly alliance* (WHO, 2002). Besides complicating the diagnosis of tuberculosis, HIV contributes to increase in incidence of tuberculosis (Narain *et al.*, 1992; Raviglione *et al.*, 1992). The deadly alliance between HIV and tuberculosis, each potentiating the impact of the other, has been documented and is currently obvious in Africa (Narain *et al.*, 1992; Raviglione *et al.*, 1992; WHO, 2002; Dye *et al.*, 1999).

Since the rate of progression from infection with *M. tuberculosis* to clinical tuberculosis is accelerated in persons who are HIV-infected, an increase in incidence of tuberculosis can be expected in areas with high incidence of HIV seropositivity. In Africa, countries with high HIV prevalence rates also have a high prevalence of tuberculosis (Godfrey-Faussett & Ayles, 2003). The linear relationship between prevalence of HIV seropositivity and tuberculosis indicate that rapid spread of HIV infection would increase the case load of tuberculosis (Godfrey-Faussett & Ayles, 2003; WHO, 2001; Tripathy & Narain, 2001).

Pathogenesis: Pathogenesis of both HIV and tuberculosis relates directly to CMI. HIV infection, which depletes CD4 lymphocytes, also causes defective immunological response to *M. tuberculosis*. HIV infection can alter the pathogenesis of tuberculosis by reactivation of latent tuberculous infection to active disease, which is more common, or by causing rapid progression from recent infection to clinical tuberculosis.

Diagnostic Challenges: As compared to HIV-negative patients, a lesser proportion of HIV-positive patients with pulmonary tuberculosis will have sputum positive smears. This will *reduce* the *sensitivity* of sputum smear examination (mainstay for diagnosis of tuberculosis). Chest radiographic findings that are not specific for tuberculosis in HIV-negative patients are even more non-specific in the HIV-infected. Patients with HIV-tuberculosis co-infection have frequent illnesses with pulmonary involvement caused by organisms other than *M. tuberculosis* (Narain & Lo, 2004).

Chest Radiographic Findings: No chest radiographic finding is *typical* of pulmonary tuberculosis, especially with concomitant HIV infection. The chest radiographic abnormalities in patients with concomitant HIV infection are indicative of the degree of immune suppression. If the immune suppression is *mild*, the appearance is often "classical" (i.e. presence of cavitations and upper lobe infiltrations). Thus in patients with early HIV infection, the chest radiographic findings are indistinguishable from that in seronegative patients. In *severe* immune suppression, the appearance is often "atypical": (a) less often cavitating and smear positive, (b) lesions are bilateral, diffuse, and reticular or reticulonodular, (c) miliary pattern is more common, (d) lower lobe infiltration, and (e) mediastinal lymphadenopathy. If mediastinal lymphadenopathy is seen, it is necessary to rule out bronchial carcinoma, lymphoma, and sarcoidosis (very rare in India).

Response to Antitubercular Treatment: The response to antitubercular treatment is similar among HIV-positive and HIV-negative patients. HIV-induced immune suppression does not seem to interfere with the effectiveness of antitubercular treatment (Alwood et al., 1994). Since the treatment and response to treatment are similar for HIV-positive and HIV-negative patients, there is no justification for carrying out HIV tests in clinical settings (Narain & Lo, 2004).

Adverse Reactions: Adverse drug reactions are more common in HIV-positive patients, as compared to their HIV-negative counterparts. Most reactions occur in the first 2 months of treatment. Skin rash and hepatitis is attributed to rifampicin. Rifampicin can reduce the activity of several drugs used in HIV-infected patients. Thiacetazone is usually associated with exfoliative dermatitis, Stevens–Johnson syndrome, and toxic epidermal necrolysis. Therefore, thiacetazone should *never* be given to HIV-positive tuberculosis patients and should not be prescribed in areas where HIV prevalence is high (Narain & Lo, 2004).

14.8.2 – Effect of Tuberculosis on the HIV/AIDS epidemic

Up to 40 per cent of deaths in HIV-infected patients are due to tuberculosis (Corbett *et al.*, 2003). Tuberculosis accounts for one-third of deaths among HIV-infected individuals worldwide (HAIN, 2003; WHO, 2004). The degree of immune suppression is the most important predictor of survival of HIV-infected patients with tuberculosis and low CD4 counts are associated with high mortality (Narain & Lo, 2004). Tuberculosis occurs earlier in the course of progression of HIV infection, when CD4 counts are around 400 cells per µL. Extrapulmonary (especially lymphadenititis) and disseminated forms of tuberculosis are more common (HAIN, 2003). More than 2–3 million HIV-infected persons in India are also afflicted with tuberculosis (HAIN, 2003).

Accelerated Progression of HIV: Tuberculosis intensifies the HIV/AIDS epidemic by accelerating progression of HIV infection and by shortening the survival of HIV positive patients. There is a six- to sevenfold increase in viral load in patients with co-infection, as compared to HIV-positive patients without tuberculosis.

Accelerated HIV-Induced Immune Suppression: Active tuberculosis is associated with transient depletion of CD4 lymphocytes. Tuberculosis increases production of cytokines like tumour necrosis factor (TNF) which increases replication of HIV in vitro. HIV-infected persons with tuberculosis appear to have higher risk of opportunistic infections and death, as compared to their counterparts with similar CD4 counts, but without tuberculosis.

14.8.3 – Clinical Manifestations of Co-infection

Though disseminated and extrapulmonary tuberculosis is relatively more common in patients with co-infection, pulmonary tuberculosis is still the most common manifestation. The manifestations depend on the degree of immune suppression (Harries *et al.*, 2004). The most frequently seen forms of extrapulmonary tuberculosis in HIV-infected persons are pleural effusion, widespread tuberculous lymphadenopathy, miliary tuberculosis, pericardial involvement, meningitis, and disseminated tuberculosis with mycobacteriemia (Harries *et al.*, 2004).

The clinical manifestations in HIV-positive children are similar to that in their HIV-negative counterparts in early stages of HIV infection. In late stages, disseminated forms of tuberculosis may occur. In the *early stage*s of HIV infection in adults, the clinical manifestations of pulmonary tuberculosis often resembles that of *post-primary* pulmonary tuberculosis, the sputum smears are frequently positive, and chest radiographs often show cavitary lesions or localised parenchymal involvement in the upper lobes (HAIN, 2003). In the *late stage*s of HIV infection in adults, disseminated forms of tuberculosis are seen, the sputum smears are frequently negative for acid-fast bacilli, and chest radiographs are not "typical" and often show diffuse infiltrates, with no cavities.

14.8.4 – DOTS for Co-Infection

The principles of control of tuberculosis are the same even for patients with co-infection. The standard treatment regimens used in the RNTCP, are equally effective in HIV positive patients (Espinal *et al.*, 2000). The DOTS strategy has improved survival of patients with co-infection (HAIN, 2003; NACO). The DOTS strategy depends primarily on *passive case finding*. In view of the symbiotic association between HIV and tuberculosis, active case finding for tuberculosis may be a helpful approach in areas with high HIV prevalence (Narain & Lo, 2004).

In areas with high prevalence of co-infection, the rise in number of tuberculosis patients may increase the workload of public health facilities, with the following possible outcomes: (a) excess laboratory workload leading to over- or under-diagnosis of pulmonary tuberculosis, (b) inadequate supervision of antitubercular chemotherapy, (c) low cure rates and high rate of recurrence, (d) high morbidity and mortality during treatment, (e) poor adherence of patients due to adverse drug reactions, and (f) increased transmission of multidrug-resistant tuberculosis among HIV-infected patients (Harries *et al.*, 2004).

14.9 – MULTIDRUG-RESISTANT TUBERCULOSIS (MDR-TB)

Multidrug-resistant tuberculosis (MDR-TB) is defined as "disease caused by *M. tuberculosis*, which is resistant to at least two first-line antitubercular drugs – isoniazid (INH) and rifampicin" (NACO). Some bacilli are inherently resistant to some drugs. If a single drug is used to treat a patient, only those bacilli that are sensitive to that drug are killed, allowing resistant bacilli to multiply. Multiple drugs are used in the intensive phase of tuberculosis treatment so that the number of viable *M. tuberculosis* is greatly reduced (HAIN, 2003). There are two specific types of drug resistance.

Primary Resistance: occurs when someone who harbours drug-resistant forms of *M. tuberculosis* infects another individual. Sometimes, a patient may withhold information on previous treatment with antitubercular drugs. Such cases may be wrongly labelled as that of primary drug resistance (HAIN, 2003).

Secondary (or Acquired) Resistance: is due to the emergence of drug resistant strains as the dominant population. Secondary drug resistance is attributable to –
1. Use of correct combinations for inadequate duration – This can occur due to interruptions in drug supply. Another reason is poor adherence of patients to the prescribed drug regimen due to ignorance, poverty, or relief of symptoms after partial treatment. DOTS programme requires biweekly visits to a designated health facility. During each visit, the patient loses the day's wages and has to bear the additional cost of to and fro travel. Ignorance of patients is because some health personnel do not care to provide information about the disease and its treatment.

2. Use of wrong combinations – Non-adherence of doctors to current recommendations on treatment of tuberculosis is either due to ignorance or recalcitrance (HAIN, 2003).

The best way to prevent MDR-TB is to ensure that patients with drug-sensitive disease are given the correct drug regimens and that they continue treatment for the prescribed duration, till they are declared cured. When drug resistance occurs, the treatment has to be individualised. A combination of reserve second-line drugs (amikacin, capreomycin, ciprofloxacin, cycloserine, ethionamide, kanamycin, levafloxacin, ofloxacin, and prothionamide) is prescribed. Second-line drugs are expensive (therefore, not available in many developing countries) and cause serious side effects. They need to be taken for up to 2 years to prevent relapse. DOTS-PLUS is a pilot project for treating MDR-TB with second-line drugs (HAIN, 2003).

14.10 – ANTIRETROVIRAL THERAPY IN CO-INFECTION

14.10.1 – Drug Interactions

Drug interactions can result in ineffectiveness of ARV therapy or antitubercular drugs or increase risk of drug toxicity. Protease inhibitors (PIs) and non-nucleoside reverse transcriptase inhibitors (NNRTIs), used in ARV therapy, should be administered cautiously to patients with co-infection because rifampicin stimulates the activity of the hepatic microsomal enzyme cytochrome P450. This leads to decreased blood levels of PIs and NNRTIs. These drugs can also stimulate or inhibit the same enzyme system and cause altered blood levels of rifampicin. Isoniazid can interact with abacavir and cause peripheral neuropathy. NsRTIs (didanosine, zalcitabine, and stavudine) may also cause peripheral neuropathy (Harries *et al.*, 2004).

In patients with co-infection (particularly those with sputum positive pulmonary tuberculosis), the priority is to treat tuberculosis in order to stop transmission of the disease. With careful evaluation and management, patients with co-infection can be administered ARV therapy and antitubercular drugs at the same time. In a patient with risk of dying (low CD4 count and disseminated tuberculosis) it may be necessary to start ARV therapy and antitubercular drugs simultaneously. In HIV-infected patients with smear-positive tuberculosis, who are not at risk of dying, ARV therapy may be deferred until the completion of intensive phase of antitubercular treatment. This decreases the risk of immune reconstitution syndrome and prevents risk of interaction between antitubercular and ARV drugs (Harries *et al.*, 2004).

14.10.2 – Immune Reconstitution Syndrome

Patients with HIV-tuberculosis co-infection may develop temporary worsening of clinical or radiographic manifestations of tuberculosis, after initiation of antitubercular treatment. This paradoxical reaction is believed to be due to immune reconstitution that occurs as a result of simultaneous administration of antitubercular and ARV drugs. Features include high fever, lymphadenopathy, expanding lesions of central nervous system, and worsening of findings on chest radiography. Such patients should be thoroughly investigated and tuberculosis treatment failure should be ruled out. Patients with severe paradoxical reactions may be given prednisolone in the dose of 1–2 mg per kg body weight for 1–2 weeks followed by gradually tapering doses. However, the efficacy of prednisolone is not clinically proven (Harries *et al.*, 2004).

14.11 – PREVENTION

Chemoprophylaxis: In various studies abroad, different drug regimens of varying duration have been tried for reducing the risk of developing active tuberculosis. The possible target groups include HIV seropositive individuals, household contacts of tuberculosis patients, and health care providers exposed to tuberculosis patients (WHO, 2004). It is difficult to recommend this strategy in the Indian situation. It is necessary to study the drug regimens, dosage, and duration of chemoprophylaxis, and their efficacy in reducing the risk of developing active tuberculosis. The *NACO Technical Resource Group on Chemoprophylaxis* has deferred antitubercular chemoprophylaxis till the availability of more scientific data in the Indian setting (NACO).

Immunoprophylaxis: BCG (*Bacille Calmette-Guérin*) vaccine was developed over a 13-year period from 1908 to 1921. The mother (original) vaccine was released in 1921. Currently, this is the only vaccine available for preventing tuberculosis. BCG is still used for routine immunisation of infants because it is thought to protect against life-threatening forms of childhood tuberculosis. It is not effective against adult forms of tuberculosis. More than 5 billion doses have been administered but tuberculosis continues to be rampant in most regions of the world. It is not effective against adult forms of tuberculosis. The disease cannot be controlled by BCG vaccine and a new vaccine needs to be developed (Bezbaruah, 2004). Prolonged use of BCG vaccine is one of the selective forces implicated in the spread of Beijing serotype of *M. tuberculosis* in East Asia (Van Soolingen *et al.*, 1995). If this hypothesis is proved, the use of live vaccines against tuberculosis may need reconsideration (Deivanayagam, 2003).

14.12 – RESEARCH

14.12.1 – New Antitubercular Drugs

A total of eight new drugs are undergoing trials worldwide. In India, the Tuberculosis Research Centre, Chennai is conducting clinical trials on ofloxacin, which can shorten the duration of treatment to 4 months. A combination of gatifloxacin and moxifloxacin is also under trial. The GFATM has developed a new drug called PA-824, which appears to be effective against multidrug-resistant tuberculosis (MDR-TB). This drug acts like a "Trojan Horse" and "fools" the mycobacterium to come out of its dormant state. The drug then destroys the mycobacterial cell wall (Bezbaruah, 2004).

The Council of Scientific and Industrial Research (CSIR) coordinates New Millennium Indian Technology Leadership Initiative (NMITL). NMITL is a public-private sector partnership, which includes Lupin Laboratories, Central Drug Research Institute (Lucknow), Indian Institute of Chemical Technology (Hyderabad), National Chemical Laboratory (Pune), and University of Hyderabad. Currently, 33 research projects are underway, for developing drug delivery systems and bio-enhancers for tuberculosis treatment. In September 2004, NMITL announced the discovery of a new antitubercular drug named *sudoterb*. Pre-clinical studies show that the drug is relatively less toxic and is compatible with the currently used antitubercular drugs. A multi-drug regimen that includes sudoterb may bring down the treatment period from 6–12 months to just 2 months. Patent protection has been secured in India and the United States. The team has applied to the Drug Controller of India for permission to start clinical tests. Three phases of clinical trials may be conducted over a 4-year period after which, the drug can be marketed (Kashyap, 2004).

14.12.2 – New Vaccines

The WHO-funded GFATM has allocated US$2 billion, while Bill and Melinda Gates Foundation has pledged US$89 million for research on tuberculosis vaccine. In 2003, the pharmaceutical company AstraZeneca established a Tuberculosis Research Centre in Bangalore (WHO, 2004). In 2003, Central Drug Research Institute (Lucknow) reported the development of a new vaccine against tuberculosis, based on a related non-pathogenic mycobacterium. This vaccine is undergoing clinical trials (Bezbaruah, 2004).

A genetically modified BCG vaccine (containing new tuberculosis-specific genes) has been developed at Delhi University. Pre-clinical studies of this vaccine at Tuberculosis Research Centre (Chennai) have shown promising results. DNA vaccines (based on two genes found in *M. tuberculosis*) are under development at the Indian Institute of Science, Bangalore. The US Biotechnology Company Aeras has also developed two vaccines (Bezbaruah, 2004).

Invitrogen, also an American Biotechnology Company, has used DNA that codes for a *M. leprae* antigen to prepare a vaccine against tuberculosis. This new DNA vaccine produces an immunogenic peptide, which stimulates the T-cells of the host's immune system to produce gamma-interferon. This vaccine was found to be effective in killing the *Tubercle bacilli* in mice that were heavily infected with *M. tuberculosis*. Scientists believe that if this vaccine is used along with antitubercular drugs, it could produce faster cure in tuberculosis-afflicted patients (Nature, 1999).

REFERENCES

Alwood K., Keruly J., Moore-Rice K., *et al.*, 1994, Effectiveness of supervised, intermittent therapy for tuberculosis in HIV-infected patients. AIDS 8: 1103–1108.

Bezbaruah S., 2004, The new weapons. India Today. 31 May, pp 60–62.

Corbett E.L., Watt C.J., Walker N., *et al.*, 2003, The growing burden of tuberculosis: global trends and interactions with the HIV epidemic. Arch Int Med 163: 1009–1021.

Deivanayagam C.N., 2003, Tuberculosis in the twenty-first century – the Indian response. JIMA 11(3): 139.

Dye C., Scheele S., Dolin P., *et al.*, 1999, Global burden of tuberculosis – estimated incidence, prevalence and mortality by country. WHO Global Surveillance and Monitoring Project. JAMA 282: 677–696.

Espinal M.A., Gupta R., and Raviglione M.C., 2000, Working to decrease the costs of anti-tubercular drugs. Essent Drugs Monit 28, 29: 12–13.

Glynn J.R., 1998, Resurgence of tuberculosis and the impact of HIV infection. Bull 54: 579–593.

Godfrey-Faussett P. and Ayles H., 2003, Can we control tuberculosis in high prevalence settings? Tuberculosis (Edinb) 83: 68–76.

Harries A., Maher D., and Graham S., 2004, TB/HIV – a clinical manual, 2nd ed. Geneva: World Health Organization, pp 210.

Health Action Information Network (HAIN), 2003, Racing against tuberculosis. Health Alert Asia-Pacific Edition 2: 2–11.

Kashyap B.C., 2004, Fighting TB (Editorial). Science Reporter, October, p 5.

Mohanty K.C. and Basheer P.M.M., 1995, Changing trend of HIV infection and tuberculosis in a Bombay area since 1988. Indian J Tuberc 42: 117–120.

Narain J.P. and Lo Y-R., 2004, Epidemiology of HIV-TB in Asia. Indian J Med Res 120(10): 277–289.

Narain J.P., Raviglione M.C., and Kochi A., 1992, HIV associated tuberculosis in developing countries: epidemiology and strategies for prevention. Tuberc Lung Dis 73: 311–321.

National AIDS Control Organisation (NACO), 2006, HIV/AIDS epidemiological surveillance and estimation report for the year 2005. New Delhi: Government of India; April, p 4.

National AIDS Control Organisation (NACO), Training manual for doctors. New Delhi: Government of India.

Nature 1999; 400: 269–272. Cited in: Essential Drugs Monitor 2000; 28 & 29.

Pathni A.K. and Chauhan L.S., 2003, HIV/TB in India – a public health challenge. JIMA 101(3): 148–149.

Raviglione M.C., Narain J.P., and Kochi A., 1992, HIV associated tuberculosis in developing countries: clinical diagnosis and treatment. Bull WHO 70: 515–526.

Sharma S.K., Aggarwal G., Seth P., and Saha P.K., 2003, Increasing HIV sero-positivity among adult tuberculosis patients in Delhi. Indian J Med Res 117: 239–242.

Tansuphasawadikul S., Amornkul P.N., Tanchanpong C., *et al.*, 1999, Clinical presentation of hospitalized adult patients with HIV infection and AIDS in Bangkok, Thailand. AIDS 21: 326–332.

Tripathy S. and Narain J.P., 2001, Tuberculosis and human immunodeficiency virus infection. In: Tuberculosis (S.K.Sharma and A. Mohan, eds.). New Delhi: Jaypee Brothers Medical Publishers, pp 404–412.

Van Soolingen D., Qian L., de Haas P.E., *et al.*, 1995, Predominance of a single genotype of mycobacterium tuberculosis in countries of East Asia. J Clin Microbiol 33: 3234–3238.

WHO, 2001, Global tuberculosis control. WHO Report 2001. www.who.int/gtb/publications/globrep01/contents.htm/

WHO, 2002, A deadly partnership: tuberculosis in the era of HIV. Consensus Statement. Geneva: WHO/TB/96.204.

WHO, 2004, Report on the global tuberculosis epidemic since 1996. www.who.int/gtb/publications/tbrep_96

Yanai H., Uthaivarovit W., Panich V., *et al.*, 1996, Rapid increase in HIV-related tuberculosis, Chiang Rai, Thailand, 1990–1994. AIDS 10: 527–531.

CHAPTER 15

SEXUALLY TRANSMITTED INFECTIONS

Abstract

Five diseases (syphilis, chancroid, gonorrhoea, Lymphogranuloma venereum, and granuloma inguinale) are known as "classical STIs". The "second generation STIs" include infections where sexual transmission is epidemiologically important and where sexual transmission is possible, and HIV infection. In the syndromic approach, the main etiological agents are classified into a group of symptoms and signs. Six major syndromes have been identified: genital ulcer (in both sexes), urethral discharge (in males), scrotal swelling (in males), vaginal and cervical discharge (in females), pelvic inflammatory disease (in females), and inguinal swelling (in both sexes). Since these six syndromes are easy to identify, flow charts have been devised for each syndrome. Each flow chart is user-friendly and depicts the decisions and actions one has to take, in a step-by-step manner. Therefore, even non-specialists at any rural or urban health care facility can initiate treatment promptly. This approach includes only those syndromes that are treatable and would lead to severe complications, if left untreated. Other STI syndromes, such as genital warts and dysuria in women are not included in this approach.

Key Words

Acyclovir, Aetiological approach, Chancre, Chancroid, Clinical approach, Genital warts, Gonorrhoea, Granuloma inguinale, Herpes genitalis, Lymphogranuloma venereum, Reproductive tract infections, Sexually transmitted infections, Syndromic approach, Syphilis

15.1 – INTRODUCTION

STIs are a group of communicable diseases that are predominantly transmitted by sexual contact. More than 20 diseases are listed in this group. These diseases were formerly known as "venereal diseases" (after Venus, the Roman goddess of love) and as "STDs". Change in nomenclature has helped in including other diseases, which are seldom transmitted by the sexual route. However, not every disease that is transmitted sexually can be considered STI. One partner may acquire an infection non-sexually (e.g. vaginal candidiasis following antimicrobial therapy) and the infection may be sexually transmitted to the other partner. Five diseases (syphilis, chancroid, gonorrhoea, lymphogranuloma venereum, and granuloma inguinale) are known as *classical STIs*, while those recently

199

recognised are known as *second generation STIs*. The second generation STIs comprise: (a) infections such as non-gonococcal urethritis, herpes progenitalis, genital warts, trichomoniasis, and moniliasis, where sexual transmission is epidemiologically important, (b) diseases where sexual transmission is possible (genital scabies, pediculosis, molluscum contagiosum, and hepatitis B), and (c) HIV infection (Ahuja, 1981).

The term RTI includes STIs, infections due to overgrowth of commensals in the reproductive tract of women (also called "super-infection"), and iatrogenic infections associated with insertion of intrauterine devices (IUDs) and medical procedures. The true incidence of STIs or RTIs is not known since patients tend to conceal these conditions. Moreover, many of these diseases are *not* notifiable and reporting to the health care system is inadequate. The available data suggest a rising incidence of gonorrhoea and syphilis. Many agents causing STIs are developing antimicrobial resistance. The incidence of *second generation* STIs is increasing, as compared to that of *classical* STIs.

15.2 – CO-INFECTION OF HIV WITH STIs

HIV epidemic has refocussed attention on STIs because patterns of high-risk behaviour, social contacts, and social network are similar for HIV and other STIs.

15.2.1 – Effect of STIs on HIV Infection

The risk of transmission of HIV is enhanced by the presence of other ulcerative and non-ulcerative STIs (Kreiss *et al.*, 1989; Quinn *et al.*, 1987; Greenblatt *et al.*, 1988; Piot *et al.*, 1988; Laga *et al.*, 1993; Wasserheit, 1992). STIs act as a cofactor for HIV infection and there is a definite linkage between the presence of STIs and the risk of developing HIV infection. HIV has been isolated from genital secretions, tissue and mononuclear cells in patients with STIs. These cells are present in increased number in inflammatory conditions. Ulcerative STIs disrupt the integrity of the genital mucosa. HIV has been isolated from genital ulcers (Kreiss *et al.*, 1989). *Epidemiological synergy* between viral STIs and HIV can be explained by STI-related molecular events (Quinn *et al.*, 1987; Kreiss *et al.*, 1989).

After entering the genome of the host cell, the viral nucleic acid (called "provirus") remains dormant for a long time. The dormant provirus may be activated to start rapid replication by certain infections like syphilis, gonorrhoea and diseases caused by CMV and herpes simplex virus (HSV). Once a large number of viral particles are produced, they lyse the host cell and produce immunological damage (Malaviya, 1990). Many of the AIDS-related malignancies, such as carcinoma of the uterine cervix, are the end result of STIs.

15.2.2 – Effect of HIV Infection on STIs

The usual clinical presentation and the natural history of some STIs may be altered, leading to problems in clinical diagnosis (Wasserheit, 1992). The serological tests may not be reliable in an individual with immune suppression. Vaccination against hepatitis B may not be effective if the individual has HIV-induced immune suppression. Moreover, HIV-positive individuals with STIs may not respond to the standard treatment prescribed for their HIV negative counterparts (Wald *et al.*, 1993).

15.3 – SYNDROMIC MANAGEMENT OF STIs

The traditional method of diagnosing STI is through the *etiological approach*, wherein laboratory diagnosis is established by identification of the causative organisms in smears and/or cultures. Its advantage is that when the etiological agent is identified by laboratory tests, the disease or condition can be treated cost-effectively. However, this approach is expensive (needs skilled personnel and a network of advanced laboratories) and time consuming (since one has to wait for the results). Moreover, the requisite laboratory facilities are not available at primary health care level, in rural as well as urban areas, in most developing countries.

The *clinical approach* relies upon clinical diagnosis alone and does *not* involve laboratory tests. The vast majority of patients with STIs seek treatment in private or public sector clinics, which lack the required facilities and skilled personnel. It is difficult to clinically differentiate between various types of STIs, especially in the presence of mixed infections. For example, it is not possible to clinically differentiate between gonococcal and chlamydial urethritis. Studies have shown that clinical diagnosis is reliable only in 5 per cent of cases. Since the above-mentioned approaches have their own limitations, a third approach, called the *syndromic approach* has been recommended so that patients with STI can be treated quickly and cost-effectively, by non-specialists at primary health care level (NACO, 1998).

15.3.1 – Rationale

HIV epidemic has brought STIs into focus because they increase the risk of HIV transmission. Several pathogens that cause STIs have developed resistance to antimicrobials and consequently, some low-cost treatment regimens have become ineffective. The syndromic approach is based on the presumption that laboratory facilities are not available or affordable (WHO, 1997). Use of *standardised protocols* for diagnosis, treatment, and follow up ensures adequate treatment at all levels of the health care system, facilitates training and supervision of health care providers, and delays development of antimicrobial resistance.

15.3.2 – Features of Syndromic Approach

A *syndrome* is a group of symptoms (complained of by patients) and signs (found on examination). In this approach, the main etiological agents are classified into a group of symptoms and signs. Though STIs are caused by a variety of organisms, they give rise to a limited number of clinical syndromes (AIDS Prevention and Control Project, 1998). Six major syndromes have been identified

1. Genital Ulcer (in both sexes) – causative organisms are *Treponema pallidum, Chlamydia trachomatis, Calymmatobacterium granulomatis*, and Herpes simplex.
2. Urethral Discharge (in males) – due to *C. trachomatis* and *Neisseria gonorrheae*
3. Scrotal Swelling (in males) – caused by *C. trachomatis, N. gonorrheae* and viruses. Surgical conditions may also cause scrotal swellings.
4. Vaginal and Cervical Discharge (in females) – causative organisms are *C. trachomatis, N. gonorrheae, Candida albicans, Gardnerella vaginalis*, and *Trichomonas vaginalis*.
5. Pelvic Inflammatory Disease (in females) – due to *C. trachomatis, N. gonorrheae*, and anaerobic organisms.
6. Inguinal Swelling (in both sexes) – Causative organisms are *Hemophilus ducreyi* and *C. trachomatis*.

Since these six syndromes are easy to identify, flow charts have been devised for each syndrome. Each flow chart is user-friendly and depicts the decisions and actions one has to take, in a step-by-step manner. Therefore, non-specialists at any rural or urban health care facility can initiate treatment promptly (NACO, 1998; AIDS Prevention and Control Project, 1998).

15.3.3 – Components of Syndromic Approach

Diagnosis and Treatment of Specific Syndromes: Flow charts are used for diagnosis. If the condition is highly refractory to treatment (e.g. genital herpes, vulvo-vaginal candidiasis), the patient is to be informed likewise. However, in case of patients with low abdominal pain and scrotal swelling it is important to ensure that there is no surgical emergency (WHO, 1997).

Patient Education on Risk-reduction: During every visit to the health care facility, the patient is advised about regular treatment; follow-up, safer sex practices, partner notification, and genital hygiene. Education for prevention is an essential part of management of STIs (WHO, 1997).

Promotion of Condoms: Clients need adequate knowledge about correct use and disposal of condoms. Condoms are provided free of cost, or at an affordable price, at the health care facility.

Counselling: Counselling, if done in confidence, facilitates modification of high-risk behaviour, helps in assessing chances of acquiring HIV infection due to high-risk behaviour, and permits discussion of possibility of problems like *incurability* (herpes genitalis), *infertility* (gonorrhoea), and *infection of progeny* (congenital syphilis) (WHO, 1997).

Partner Management: The patient is encouraged to voluntarily inform his or her partners of the infection and the need for clinical evaluation and treatment. Telling the spouse or sexual partner about the diagnosis of STI may be emotionally painful issue for many patients and the counsellor is required to tackle this issue (WHO, 1997). Partner notification and management has an important role in STI risk-reduction.

15.3.4 – Advantages

The syndromic approach is a scientific and simple method of managing STIs with good compliance (NACO, 1998; WHO, 1997; AIDS Prevention and Control Project, 1998). The cure rate can be as high as 95 per cent (NACO, 1998). It is free from errors in clinical judgement even in mixed infections and HIV-positive individuals in whom the usual clinical presentation of STIs may be altered. Increase in cost-effectiveness through money saved on laboratory tests (NACO, 1998; WHO, 1997; AIDS Prevention and Control Project, 1998). Compliance is increased since it obviates the need to wait for results of laboratory investigations. Offering prompt treatment at the first visit renders the patient non-infectious quickly and can control the spread of STI. This approach is feasible for both urban and rural areas at primary health care level.

15.3.5 – Disadvantages

This approach includes only those syndromes that are treatable and would lead to severe complications, if left untreated. The other STI syndromes, such as genital warts and dysuria in women, are not included in this approach. There is a potential risk of overtreatment.

15.4 – MANAGING STIs IN HIV-INFECTED INDIVIDUALS

STIs may be asymptomatic. Therefore, a detailed sexual and life history ought to be taken and specific questions suggestive of STI-related signs and symptoms should be asked. Privacy and confidentiality be respected while eliciting sexual history. Patients with problems pertaining to genitalia tend to be cautious and evasive in giving a history (WHO, 1997). Serological tests (for syphilis and hepatitis B/C) are recommended for each case of STI. Pre- and post-test counselling is essential. A person may have more than one STI. Conversely, STI always involves more than one person. A STI may involve multiple anatomical sites. Women are to be examined gynaecologically for *Neisseria* and *Chlamydia* infection and recommend bacteriological examination of cervical or vaginal secretions and Pap smear.

15.5 – CLINICAL ASPECTS

15.5.1 – Syphilis

The causative organism is a spirochaete *T. pallidum* (subspecies *pallidum*). Though usually spread sexually, the disease can also be transmitted by blood transfusion and transplacentally to offspring. Syphilis can affect any organ in the body and can mimic any disease. The incubation period varies from 9 to 90 days, with an average of 3–4 weeks.

The disease has an *early stage*, which occurs within the first 2 years of infection, and includes – primary syphilis, secondary syphilis, and early latent syphilis. The *late stage* comprises late latent syphilis, benign late syphilis, cardiovascular syphilis, and neurosyphilis.

The *primary stage* represents the local tissue reaction at the site of entry of the organism and lasts for 3–8 weeks. The *secondary stage* represents the generalised tissue reaction, which occurs after the organism has spread to all the tissues in the body. This stage lasts for 3–9 months. In the *latent stage*, serological tests are positive, but there are no clinical manifestations. The CSF is non-reactive.

Primary Syphilis: A small, red, painless, non-tender papule is seen on the genitalia or extragenital sites such as lips, tongue, and fingers about 3–4 weeks after exposure (incubation period varies from 9 to 90 days). It develops into a well-defined hard sore (or *chancre*) in 2–3 weeks. The ulcer, which is usually single, painless, and non-tender, has an indurated base. The untreated sore heals in 6–8 weeks with a thin atrophic scar and is followed by a painless bubo in the groin. The lymph nodes in the neck, axilla, and epitrochlear region become bilaterally enlarged with rubbery consistency and are painless and non-tender. The primary sore, blood, and body fluids are *infective* in this stage (Benenson, 1990).

Secondary Syphilis: This stage is characterised by sore throat, anaemia, skin rashes, lymphadenitis, and swelling in bones and joints. Arthralgia may be worse at night. Systemic symptoms like headache, fever, and malaise may be present. *Cutaneous lesions* are symmetrically distributed, polymorphic, non-pruritic and are abundant on the central part of the body, as compared to the extremities (called "centripetal distribution"). These skin lesions may manifest as rose-coloured macules, coppery papules, pustules, or ulcers. *Condyloma lata* (Latin: *lata* = broad; plural: condylomata) are flat, moist, raised, warty papule found on moist surfaces like groin, anus, vulva, and infra-mammary region. All the lymph nodes are enlarged, but painless and non-tender. *Snail-track ulcers* (white mucous patches or plaques which erode and form ulcers) may develop in the throat, mouth, prepuce and vulva. In this stage of syphilis, blood, body fluids, condylomata, and "snail-track" ulcers are *infective* (Benenson, 1990).

Early Latent Syphilis: The visible manifestations disappear with or without treatment. If the patient has had the disease for less than 2 years, it is called "early latent syphilis". In the latent stage, blood and body fluids are *not* infective (Benenson, 1990).

Tertiary Syphilis: This stage is seen in inadequately treated or untreated patients. *Benign tertiary syphilis* occurs about 8–10 years after the primary stage. *Gumma* (plural: gummata) is the characteristic lesion. Gummata are painless nodules, which may progress slowly, suppurate and form ulcers on the skin. Cutaneous gummata are usually asymmetrical. Gummatous ulcers have punched-out edges and are covered with "wash leather" (or "chamois leather") slough. They heal with a thin non-contractile scar. Gummata may be seen on tongue, soft palate, pharynx, lips, nose, bones (sternum, clavicle, vertebrae, long bones), and viscera (stomach, liver, spleen, intestines, testes). In the tertiary stage of syphilis, blood and body fluids are *not* infective. A broken and ulcerating gumma in the throat or skin may be infective ((Benenson, 1990).

15.5.1.1 – Diagnosis

Dark-field microscopy is the simplest and most rapid method of diagnosing syphilis. However, the organisms cannot be detected after the primary stage, when the skin lesions have healed. Hence, the main method for diagnosing syphilis is by testing a patient's serum for antibodies to *T. pallidum*. Serological tests (RPR and Treponema pallidum particle agglutination (TPPA)) are used for the diagnosis of late primary, secondary, and late syphilis. The *RPR test* is advised for any patient with STI to identify asymptomatic cases. If positive, the test should be repeated after 3–6 months of treatment.

15.5.1.2 – Syphilis and HIV infection

HIV infection (the most recent STI) and syphilis (the oldest known STI) share common features: (a) multiple modes of transmission, including the tragedy of mother-to-child (transplacental) transmission, (b) uncertain and prolonged asymptomatic phase, and (c) unpredictable clinical manifestations of varying severity affecting every organ system (Ramachandran, 1990). The primary lesion is an ulcer, which increases the risk of transmission of HIV. In HIV-infected individuals, syphilitic ulcers may be multiple, larger and may persist for longer periods. The symptoms of tertiary syphilis (e.g. meningitis, ocular complications, and gummata) may occur during the primary, secondary, and latent phases (Wald *et al.*, 1993). Serological tests for syphilis may not be reliable. Seroconversion for syphilis occurs more rapidly in HIV-infected individuals (Wald *et al.*, 1993). The effectiveness of past treatment determines the occurrence of reactivation or relapse. Secondary and tertiary stages of syphilis may occur earlier in HIV-positive patients (Wald *et al.*, 1993). Neurological syphilis is also more frequently seen (Musher *et al.*, 1990). For treatment of syphilis in HIV-positive individuals, injection procaine penicillin should be preferred over benzathine penicillin.

15.5.2 – Chancroid

(Synonym: SOFT SORE). A slender rod-shaped organism, *Haemophilus ducreyi* (or Ducrey's bacillus), causes this disease. The average incubation period is 2–5 days, up to 14 days. The incubation period is only about 24 hours if there are abrasions on the genitalia. Though the sexual route primarily transmits the disease, it can also spread by auto-inoculation in persons with poor personal hygiene. The disease is infective till the lesions heal. The lesion appears on the genitalia as a papule or vesicle, which ulcerates rapidly. The ulcers are painful, tender, multiple, shallow, and ragged with undermined edges (unlike the hard sore of syphilis). The floor of the ulcer is covered with necrotic slough. Extragenital lesions may be seen on lips, tongue, chin, breast, and umbilicus. There is unilateral enlargement of regional lymph nodes, which are tender, firm, and matted (unlike that of syphilis). The lymph nodes soften, suppurate, forming "buboes" and get fixed to the skin. The bubo ruptures with a single opening on the skin (unlike the bubo of LGV). Phimosis is a common complication in males. Extensive ulceration may lead to destruction of prepuce and shaft of penis. Swabs taken from the base of the ulcer are used for microscopic examination and culture.

Chancroid and HIV Infection: Being an ulcerative STI, it increases the risk of transmission of HIV. In HIV-positive patients, the ulcers tend to be larger, more prominent and do not respond well to the standard single dose regimen (Wald et al., 1993).

15.5.3 – Gonorrhoea

This disease is caused by *N. gonorrhoeae*. The incubation period is 2–5 days. Non-sexual routes, such as sharing infected towels, can also transmit this disease. The acute stage is characterised by inflammation of the urethra (in males) and of urethra, cervix, and vagina (in females). Usual manifestations in *males* are – thick, purulent, greenish yellow discharge from external urethral meatus, severe pain during micturition, and frequency and urgency of micturition. Symptoms are more marked if the posterior urethra is inflamed. In *females*, the symptoms are dysuria, frequency, and urgency. Epidemiologically, females are more liable to transmit the disease since they have few symptoms or obvious signs. In *female children*, the lesion is essentially *vulvo-vaginitis* and the infection may be acquired from infected towels, infected parents, or sexual assault. Babies born to infected mothers may develop gonococcal eye infection, which is acquired during passage through the birth canal. The condition is known as *ophthalmia neonatorum*. The eyelids are oedematous and stuck together with a purulent discharge. The conjunctiva is red. In severe cases, involvement of the cornea leads to keratitis and corneal ulceration.

Complications: In males, the infection may extend to prostate, seminal vesicles, bladder, renal pelvis, or rectum by contiguity or by lymphatics. Stricture may occur at bulbous urethra due to fibrosis of the urethral mucous membrane. In females, the infection may spread to uterus, fallopian tubes, peritoneum, and Bartholin glands. About 30 per cent of women with salpingitis may develop secondary sterility. In both sexes, blood-borne spread to joints, muscles, tendons, and eyes may occur. Meningitis and endocarditis are rare.

Laboratory Diagnosis: Smears taken from urethral discharge (in males) or cervix uteri (in females) are used for microscopic examination under Gram's stain to demonstrate Gram-negative kidney-shaped intracellular organisms. Enriched media such as chocolate agar are used for culturing *N. gonorrhoeae*.

Gonorrhoea and HIV Infection: Inflammation and formation of micro-ulcers in the genital mucosa may increase the risk of HIV transmission (Wald *et al.*, 1993). The disease may be co-transmitted with HIV in case of unprotected sexual intercourse. In HIV-infected women, gonorrhoea may progress to painless pelvic inflammatory disease or tubo-ovarian masses. These women are less likely to have leukocytosis or abdominal pain (Wald *et al.*, 1993; Hoegsberg *et al.*, 1990).

15.5.4 – Granuloma Inguinale

C. granulomatis (or *Donovania granulomatis*) causes this disease. The disease is also known as "donovanosis" after Major Donovan, who discovered *Donovan bodies* in India in 1905. The incubation period is unknown, and is probably between 8 and 80 days. The lesion starts as a hard papule or nodule, which soon ulcerates. The floor of the ulcer has "beefy red" or "velvety red" granulations and bleeds on touch. The edge is rolled out. Satellite ulcers develop in the periphery of the initial lesion due to auto-inoculation and gradually the edges overlap. The regional lymph nodes are *not* enlarged. However, the lymphatics may get blocked subsequently due to fibrosis resulting in pseudo-elephantiasis of the genitals. The lesion is usually seen on the genitalia. Extragenital lesions may occur on moist and warm areas like folds between the scrotum and thighs (in males) or labia and vagina (in females). Blood-borne spread may occur to liver, spleen, bones, and lymph nodes. The disease is infectious till wet lesions are present on the skin or mucosa. Smears are taken by scraping the edge of the ulcer. Giemsa stain reveals *Donovan bodies* (Gram-negative, large, oval, intracytoplasmic bodies inside large mononuclear cells).

15.5.5 – Lymphogranuloma Venereum (LGV)

This disease, caused by *C. trachomatis* (types L-1, 2, and 3), affects the lymphatics and lymph nodes. The incubation period varies from 7 to 21 days. The lesion begins as a vesicle on the external genitalia, which ulcerates rapidly and heals without a scar. The primary lesion may not be noticed. After 1–3 weeks, the disease recurs as "climatic bubo" (so called because it is found mainly in tropics and subtropics). The inguinal lymph nodes, which are unilaterally enlarged, firm, and tender, get fixed due to periadenitis. When both the femoral and inguinal lymph nodes are enlarged, the inguinal ligament that separates the two masses appears as a groove. This is called the "grooving sign". This mass suppurates and breaks down, with seropurulent discharge from multiple sinuses (unlike the bubo of chancroid). During recurrence, systemic manifestations (fever, bodyache, malaise, joint pains, and splenomegaly) may be present. The disease is infective till the lesions are clinically active.

Complications: In both sexes, multiple abscesses may form along the lymphatics resulting in their destruction. The lymphatics are destroyed over a period of few weeks to 20 years or longer and their destruction results in elephantiasis of genitalia. Anorectal stricture may occur due to polypoid growth in rectum. This is more common in females. Urethral fistulae may occur.

Laboratory Diagnosis: Demonstration of intracellular "Halberstaedler- Prowazek inclusion bodies" with Giemsa stain.

15.5.6 – Herpes Genitalis

HSV type II (HSV-2) causes recurrent episodes of sores or ulcers in the genital area. Prevalence of HSV-2 infection is generally higher in women than in men and is associated with increasing age and sexual activity in both the developing and developed countries (Smith & Robinson, 2002). In the developed world, HSV-2 has emerged as the leading cause of genital ulcers (Maynaud & McCormick, 2001). The usual pattern of manifestation consists of a first episode, followed by recurrences. The incubation period varies from 7 days to months. The disease is *infective* even in the asymptomatic stage. The disease starts with appearance of very small vesicles, which break down forming superficial ulcers. These ulcers heal by themselves and recur again. There is intense burning sensation over the vesicles and ulcers.

Laboratory diagnosis involves the demonstration of IgG and IgM antibodies to HSV-2. The *first episode* of herpes genitalis infection is treated by oral administration of acyclovir 200–400 mg, six times a day, for 7 days or until remission occurs. Topical application of 5 per cent acyclovir cream is also advised. *Frequent* recurrent episodes are treated with orally administered acyclovir 400 mg twice a day (or 200 mg orally thrice a day). The WHO recommends acyclovir as a first-line treatment in countries where the prevalence of HSV-2 infection is greater than 30 per cent. Daily suppressive treatment with acyclovir is advised for patients who experience more than six episodes of genital herpes per year (WHO, 2003). Severe disease requires intravenous administration of the drug. Acyclovir-resistant infection usually responds to intravenous administration of foscarnet.

Herpes Genitalis and HIV Infection: Herpes genitalis is an independent risk factor for transmission of HIV. Activated lymphocytes including CD4 cells, are often recruited to sites of inflammation and are primed to receive or present HIV at the site of ulceration. Thus, genital ulcers can be a portal of entry or exit for HIV. HIV infection increases the frequency of reactivation of HSV-2 and the episodes become more extensive and persistent with the decline in CD4 cell counts. HSV-2 increases HIV replication by directly affecting HIV transcription (Margolis *et al.*, 1992) or by indirect mechanisms involving cytokines (Clouse *et al.*, 1989). In HIV-positive individuals, ulcers may persist indefinitely and the recurrences may be more frequent. Atypical sites, such as oesophagus may be involved. Chronic mucocutaneous infection may manifest as painful ulcerating lesions in genital, perianal, or perioral regions. Disseminated infections, myelitis and encephalitis are rare (Wald *et al.*, 1993). Persistence of herpes for more than 1 month is considered as AIDS-defining illness. HIV-infected patients may not respond to treatment with acyclovir.

15.5.7 – Genital Warts

(Synonym: Condyloma accuminata). Caused by HPV, the infection may be subclinical or may present with a broad spectrum of clinical manifestations from warts to neoplasms. This disease is not included in syndromic management.

Genital Warts and HIV Infection: The treatment is more difficult and the lesions may recur frequently (Wasserheit, 1992; Hoegsberg *et al.*, 1990). Neoplastic changes may occur in the lesions in cervix uteri or anus. Hence, Pap smear should be done annually in all women suffering from genital warts. If the CD4 count drops below 200 cells per µL, the Pap smears should be taken six monthly.

REFERENCES

Ahuja M.M.S., 1981, Progress in clinical medicine, 4th series. New Delhi: Arnold Heinemann.

AIDS Prevention and Control Project, 1998, Quality STD care – training module for private medical practitioners. Chennai: Voluntary Health Services.

Benenson A.S., 1990, Control of communicable diseases in man, 15th edn. Washington, DC: American Public Health Association.

Clouse K.A., *et al.*, 1989, Monokine regulation of human immunodeficiency virus-1 expression in a chronically infected human T-cell clone. J Immunol 142: 431–438.

Feingold A.R., Pehrson P.O., Peclersen C., *et al.*, 1990, Cervical cytological abnormalities and papilloma virus infection in women infected with human immunodeficiency virus. J AIDS 3: 986–993.

Greenblatt R.M., Lukfehart S.A., Plummer F.A., *et al.*, 1988, Genital ulceration as a risk factor for human immunodeficiency virus infection. AIDS 2: 47–50.

Hoegsberg B., Abulatia O., Sedis A., *et al.*, 1990, STDs and HIV among women with PID. Am J Obs Gynecol 163: 1135–1139.

Kreiss J.K., Coombs R., Plummer F.A., *et al.*, 1989, Isolation of HIV from genital ulcers in Nairobi prostitutes. J Inf Dis 160: 380–384.

Kreiss J.K., Koech D., Plummer F.A., *et al.*, 1986, AIDS virus infection in Nairobi prostitutes. N Engl J Med 314: 414–418.

Laga M., Manoka A., Kivuvu M., *et al.*, 1993, Non-ulcerative sexually transmitted diseases as risk factors for HIV-1 transmission in women: results of a cohort study. AIDS 7: 95–102.

Malaviya A.N., 1990, AIIMS manual of AIDS and HIV infection. New Delhi: Department of Medicine, All India Institute of Medical Sciences.

Margolis D.M., *et al.*, 1992, Transactivation of HIV-1 LTR by HSV-1 immediate-early genes. Virology 186: 788–791.

Maynaud P. and McCormick D., 2001, Interventions against sexually transmitted infections to prevent HIV infection. Br Med Bull 58: 129–153.

Musher D.M., Harnill R.J., Baughin R.E., *et al.*, 1990, The effect of HIV infection on the course of syphilis and on the response to treatment. Ann Int Med 113: 872–881.

National AIDS Control Organisation (NACO), 1998, Simplified STI and RTI treatment guidelines. New Delhi: Government of India.

Piot P., Plummer F.A., Mhalu F.S., *et al.*, 1988, AIDS: an international perspective. Science 239: 573–579.

Quinn T.C., Piot P., McCormick J.B., *et al.*, 1987, Serologic and immunologic studies in patients with AIDS in North America and Africa: the potential role of infectious agents as co-factors in human immunodeficiency virus infection. JAMA 257: 2617–2621.

Ramachandran P., 1990, HIV infection in women. ICMR Bull 20(11/12): 111–119.

Smith J.S. and Robinson N.J., 2002, Age-specific prevalence of infection with herpes simplex virus types 2 and 1: a global review. J Infect Dis 186(Suppl 1): S3–S28.

Wald A., Corey L., Hunter H.H., and Holmes K.K., 1993, Influence of HIV infection on manifestations and natural history of other sexually transmitted diseases. Ann Rev Pub Health 14: 19–42.

Wasserheit J., 1992, Epidemiological synergy: interrelationships between HIV and other STDs. Sex Transm Dis 19: 61–77.

World Health Organization (WHO), 1997, Management of sexually transmitted diseases at district and PHC levels. Regional publication no. 25. New Delhi: SEARO; July.

World Health Organization (WHO), 2003, Guidelines for the management of sexually transmitted infections. www.who.int. Revised 2003.

CHAPTER 16

ANTIRETROVIRAL THERAPY

Abstract

ARV is indicated in specified categories of HIV-infected individuals, prevention of mother-to-child transmission of HIV, and for PEP. ARV therapy has reversed the progress of illness in many patients with advanced disease, and prevented the progression of disease in those who are asymptomatic or relatively healthy. Keys to successful adherence include pretreatment patient education, minimum number of pills, packaging of pills, use of fixed-drug combinations, avoidance of food precautions, adherence-friendly frequency of dosing (not more than twice daily), fitting ARV therapy into the patient's lifestyle, and involvement of friends, relatives, and community members to support adherence. Pretreatment and periodic clinical and laboratory monitoring is mandatory for ARV therapy. In case of accidental occupational exposure to HIV, a specialist should explain the possible risks such as side effects of ARV drugs and benefits of PEP, and determine the likelihood of the exposed person's adherence to the prescribed regimen. Emergence of drug-resistant strains of HIV is a widespread and growing problem.

Key Words

Abacavir, Adherence, Access to treatment, Antiretroviral treatment, Darunavir, Efavirenz, Enfuvirtide, Fusion inhibitor, Indinavir, Lamivudine, Nelfinavir, Nevirapine, NNRTI, Non-nukes, Nukes, Post-exposure prophylaxis, Protease Inhibitors, Reverse transcriptase inhibitors, Ritonavir, Saquinavir, Tenofovir, Zalcitabine, Zidovudine

16.1 – INTRODUCTION

When HIV enters the blood stream through any one of the routes of transmission, it is attracted by lymphocytes that have matured in thymus and bear CD4 receptors on their surface. HIV binds to the CD4 cell receptor via its outer glycoprotein (gp 120) cover and enters the cytoplasm of the lymphocyte, where it sheds its outer coat, envelope, viral RNA, and unique enzyme *reverse transcriptase*. This enzyme gets activated and facilitates conversion of RNA into provirus DNA. This provirus DNA then creates its own mirror image and with the help of another enzyme *integrase*, integrates with the host cell genome and becomes an integral part of the host cell. This DNA copy enters the nucleus of the infected cells and multiplies and produces messenger RNA along with the multiplication of the nucleus of the host cell and is always immunologically active.

Messenger RNA directs the production of new viral particles which form into small virions with the help of another enzyme, *protease*. These small virions then bud out of the host cell and affect other cells that bear the CD4 receptor. Thus, each infected host cell becomes a "factory" of HIV that produces billions of viruses (NACO, National Guidelines).

Reverse transcriptase enzyme dominates genomic RNA. If a provirus DNA copy is not made promptly, viral replication stops. Since reverse transcriptase enzyme exists only in HIV, drugs that inhibit this enzyme will affect the virus without affecting the normal host cells. For these reasons, reverse transcriptase continues to be the *prime target* of drugs used in ARV therapy (WHO, 2003a).

16.1.1 – Public Health Aspects

The introduction of ARV treatment in 1996 in the developed countries dramatically improved morbidity and mortality rates, improved quality of life and transformed the perception of HIV/AIDS from a plague to a manageable chronic illness (Palella *et al.*, 2003). ARV therapy neither destroys the virus nor cures HIV infection. The objective of ARV treatment is to retard the progress of illness in many patients with advanced disease and prevent onset of symptomatic HIV disease in asymptomatic or relatively healthy HIV-positive individuals (Harries *et al.*, 2004; Wig, 2002).

ARV treatment improves quality of life and offers hope to HIV-affected individuals and encourages voluntary disclosure of HIV infection in order to receive ARV treatment. Appropriate ARV treatment is a *life-saving tool* at the individual level and also an important *public health intervention* for retarding the progress of the HIV epidemic (John, 2001; Clinton, 2003).

16.1.2 – Art: Principles and Perspectives

The cost of treatment of tuberculosis, leprosy, and malaria is borne by the Government. Treatment for these diseases is a *public health intervention* to reduce the incidence of disease. The same principle should also apply to ARV treatment for HIV/AIDS. It is necessary to decide what clinical conditions signal the need for starting ARV treatment. As long as an HIV-infected individual has good quality of life, ARV treatment can be withheld. This makes sound medical and economic sense. When opportunistic infections, such as tuberculosis are diagnosed, the immediate step is to initiate specific treatment against these diseases and not to start ARV treatment. Currently the decision-making process for starting ARV treatment is left to an individual physician's own criteria. A mechanism, such as a decision by a panel of three physicians, is necessary to ensure that strict adherence to criteria for starting ARV treatment are developed and validated through research. Such a provision already exists in India: for MTP between 12 and 20 weeks gestation, the opinion of two specialists is necessary (John, 2004).

Since CD4 cell counts may not be available everywhere, it is necessary to decide on the use of simple tests, such as haemoglobin, hematocrit, total lymphocyte count, for monitoring ARV treatment. Currently it is possible to computerise clinical and laboratory data of each patient who is started on ARV treatment, with a programmed alert for timely follow-up. ARV treatment protocols and guidelines for monitoring should be made available to all physicians in private and public sector to prevent the "therapeutic anarchy" that currently prevails in case of antitubercular treatment. Likewise, every medical college and postgraduate training centre in the country ought to receive ARV treatment protocols and guidelines for monitoring to ensure nationwide uniformity of teaching curriculum for tomorrow's doctors (John, 2004).

16.1.3 – Prerequisites

Successful implementation of ARV treatment requires political commitment, diagnosis and registration of patients, standardised treatment regimens, regular and reliable supply of ARV drugs, and provision for monitoring and evaluation of the programme. ARV drugs must be used in standardised multidrug combinations to prevent development of drug resistance (Harries *et al.*, 2004).

16.1.4 – Classification of Antiretroviral Drugs

None of the currently available ARV drugs can kill HIV. ARV drugs belong to the following main classes; they act at different stages of the life cycle of HIV
• Reverse transcriptase inhibitors (RTIs or "nukes") – Appendix 2
• Non-nucleoside RTIs (NNRTIs or "non-nukes") – Appendix 6
• Protease Inhibitors (PIs) – Appendix 4
• Attachment and Fusion Inhibitors – Appendix 5
• Immune stimulators – Appendix 6
• Integrase inhibitors – Appendix 6
• Antisense drugs – Appendix 6
• Maturation inhibitors – Appendix 6
• Zinc finger inhibitors or zinc ejectors – Appendix 6
RTIs or "nukes" are further subdivided into two groups: (a) Nucleoside analogue RTIs (NsRTIs), and (b) Nucleotide analogue RTIs (NtRTIs).

16.2 – CLINICAL AND LABORATORY MONITORING

16.2.1 – Baseline Pretreatment Clinical Assessment

For all the four recommended first-line regimens (the interested reader may read details from reference: WHO, 2003a), the baseline clinical assessment should include:
• Documentation of past medical history.
• Identification of coexisting medical conditions such as tuberculosis, pregnancy, and major psychiatric illnesses, and documentation of concomitant

medications including traditional therapies. Active tuberculosis is to be managed in accordance with national tuberculosis control programmes.

• Recording the patient's weight and assessing his or her readiness for ARV treatment.
• Identifying current symptoms and physical signs.
• Staging of HIV disease (WHO, 2003a).

16.2.2 – Periodic Clinical Assessment During Treatment

This includes assessment for signs and symptoms of potential drug toxicities, patient's adherence to therapy, response to therapy, recording body weight, and basic laboratory monitoring (see below).

16.2.3 – Laboratory Monitoring

The WHO has recommended a three-tiered system of laboratory monitoring in resource-limited settings (WHO, 2003a).

16.2.3.1 – Primary health care centres (Level 1)

Rapid HIV antibody testing, assessment of haemoglobin levels (if ZDV is being considered for use), pregnancy testing, and sputum smear microscopy for tuberculosis.

16.2.3.2 – District hospitals (Level 2)

Rapid HIV antibody testing and capability to resolve indeterminate rapid HIV antibody test report by second serological method, complete and differential blood count, CD4 cell count, serum alanine amino transferase (ALT), pregnancy testing, and sputum smear microscopy for tuberculosis.

16.2.3.3 – Regional referral centres (Level 3)

Rapid HIV antibody testing, complete and differential blood count, CD4 cell count, serum ALT, pregnancy testing, sputum smear microscopy for tuberculosis, assessment of viral load, and full serum chemistry including electrolytes, renal function, liver enzymes, and lipids (WHO, 2003a).

16.3 – ARV THERAPY FOR ADULTS AND ADOLESCENTS

As per recommendations of the WHO, in resource-limited settings, adults and adolescents should be started on ARV treatment when they have confirmed HIV infection and one of the following conditions.

16.3.1 – In Settings where CD4 Testing is Available

• WHO Stage IV HIV disease, irrespective of CD4 cell count.
• WHO Stage III HIV disease, with a CD4 cell count of less than 350 cells per μL to assist decision-making. This includes HIV wasting, chronic diarrhoea of

unknown aetiology, prolonged fever of unknown aetiology, pulmonary tuber-culosis, recurrent invasive bacterial infections or recurrent/persistent mucosal candidiasis.

• WHO Stage I, or Stage II HIV disease, with a CD4 cell count of less than 200 cells per μL.

The precise CD4 level above 200 cells per mm^3, at which ARV treatment should be initiated, has not yet been established (WHO, 2003a). However some recom-mendations suggest that the treatment should not be started until the CD4 T-lymphocyte count falls below 350 cells per mm^3 or viral load exceeds 50,000 copies per mL. These recommendations are based on the risk of developing AIDS within 6 years without treatment (Kelly, 2002).

16.3.2 – In Settings where CD4 Testing Facilities are not Available

The presence of a *total lymphocyte count* of 1,200 cells per mm^3 or below can be used as a *substitute indication* for treatment in the presence of symptomatic HIV disease. The total lymphocyte count is not useful indicator in an asymptomatic patient. Thus, in the absence of CD4 testing facilities, asymptomatic HIV patients (WHO Stage I) should not be started on ARV treatment because currently no other reliable marker is available in resource-limited settings.

• WHO Stage IV HIV disease, irrespective of total lymphocyte count.
• WHO Stage III HIV disease, irrespective of total lymphocyte count. This includes HIV wasting, chronic diarrhoea of unknown aetiology, prolonged fever of unknown aetiology, pulmonary tuberculosis, recurrent invasive bacte-rial infections or recurrent/persistent mucosal candidiasis.
• WHO Stage II HIV disease, with a total lymphocyte count of less than 1,200 cells per mm^3.

Assessment of viral load (plasma HIV-1 RNA levels) is not necessary to start therapy and is not recommended as a routine test by the WHO (2003a).

16.3.3 – First-Line Regimens (Adults and Adolescents)

In order to develop accessible ARV treatment programmes, countries with resource limitations need to standardise treatment regimens, select a first-line regimen, and select a limited number of second-line regimens. In such settings, patients who cannot tolerate or fail the first-line and second-line regimens will be referred (for individualised care) to specialists in HIV medicine (WHO, 2003a).

16.3.4 – Criteria for Selecting First-Line Regimens

These include availability, storage requirements, and cost of ARV drugs, espe-cially availability in fixed-dose combinations (FDCs) or as co-blister packs, availability of potency and profile of adverse reactions, laboratory monitoring requirements, potential for maintaining future treatment options, availability of data on expected patient adherence, coexisting conditions such as metabolic diseases and infections, pregnancy or risk of pregnancy, concomitant use of medications and potential drug interactions, and potential for infection with a strain of HIV with reduced susceptibility to one or more ARV drugs (WHO, 2003a). "Holding regimens" contain drugs such as lamivudine (3TC) help in keeping the virus "handicapped" with multiple mutations, so that it multiples slowly (Fact Sheet 408, 2006).

16.3.5 – Changes in First-Line Regimen

ARV drugs may need to be changed due to treatment failure or drug toxicity. This has been termed "salvage therapy" (Fact Sheet 408, 2006). When toxicity is related to an identifiable drug in the regimen, the offending drug can be replaced with another drug that does not have the same adverse effects. Protease inhibitor-based regimens are primarily reserved for second-line therapy. Though these regimens have proven clinical efficacy, they have significant interactions with other drugs such as rifampicin. A functioning cold chain is essential for ritonavir (RTV)-boosted regimens. Co-formulations of PIs with NNRTIs are not available. Therefore, protease inhibitor-based regimens may be considered for first-line regimens where prevalence of NNRTI resistance in the community is more than 5–10 per cent (WHO, 2003a).

Viral Load: Within 2–4 weeks after starting ARV therapy, the HIV-RNA should be preferably less than 10,000 copies per mL, i.e. one log reduction in viral load or more. If HIV-RNA is more than 100,000 copies per mL or the reduction in viral load is less than 0.5 logs, ARV therapy should be adjusted by adding or switching drugs. Viral loads should be repeated every 4–6 months during periods of clinical stability. If viral load returns to 0.3–0.5 logs of pretreatment levels, then ARV treatment is no longer working and should be changed.

CD4 Cell Count: Within 2–4 weeks of starting ARV treatment, CD4 cell count should increase by at least 30 cells per mm^3. If this is not achieved, ARV treatment should be changed. CD4 cell counts should be obtained every 3–6 months during periods of clinical stability and more frequently in case of symptomatic HIV disease. If CD4 cell count drops to base line (or below 50 per cent of increase from pretreatment levels, then ARV treatment should be changed.

16.3.6 – Treatment Failure

This can be assessed *clinically* by disease progression, *immunologically* by CD4 cell counts, and *virologically* by measuring viral loads. The WHO recommends the use of clinical criteria, and, where possible, CD4 count criteria to define treatment failure. Testing for drug resistance will not be routinely available in resource-limited settings in the foreseeable future. In developing countries, recognition of treatment failure will be delayed when based on clinical criteria and/or CD4 criteria alone. Such a delay will provide greater opportunity for evolution of drug resistant mutations before the ARV drug regimen is changed (WHO, 2003a).

16.3.7 – Clinical Signs of Treatment Failure

Occurrence of new opportunistic infection or malignancy denotes clinical progression of disease. This should be differentiated from *immune reconstitution syndrome* (IRS). This condition is characterised by the appearance of signs and symptoms of an opportunistic disease as an inflammatory response to a previously subclinical opportunistic infection. IRS is seen a few weeks after initiation of ARV treatment in patients with advanced immune deficiency. Its occurrence may lead to development of atypical presentations of some opportunistic infections. *Recurrence* of a previous opportunistic infection is also a sign of treatment failure. However, recurrence of tuberculosis may not represent progression of HIV disease since reinfection may occur. Hence clinical evaluation is essential. Onset or recurrence of WHO *Stage III conditions* may also indicate treatment failure. If the patient is asymptomatic and treatment failure is defined using CD4 cell count alone, a confirmatory CD4 cell count should be considered (WHO, 2003a).

16.4 – ARV THERAPY FOR WOMEN

While choosing a regimen for women with childbearing potential or who are pregnant, there is likelihood that the ARV drugs may be administered in the first trimester of pregnancy, before pregnancy is diagnosed. Efavirnez (EFV), a NNRTI, should be avoided in such women because of its potential for teratogenicity. However, EFV remains a viable option in those women for whom effective and reliable method of contraception can be assured (WHO, 2003a).

The recommended first-line regimen for women with childbearing potential or pregnant women is *Stavudine (d4T) or Zidovudine (ZDV) + Lamivudine (3TC) + Nevirapine (NVP)*. For pregnant women, it is desirable to start ARV treatment after the first trimester. However, for severely ill women, the benefit of early ARV treatment outweighs any potential foetal risks and ARV treatment should be started in such cases. Women receiving efavirenz (EFV)-containing regimens, who become pregnant, should continue their treatment; with the exception that EFV is replaced by NVP (WHO, 2003a).

The dual NRTI combination of d4T/didanosine or di deoxy inosine (ddI) should be avoided in pregnancy due to its potential for lactic acidosis in pregnant women. Symptomatic NVP-associated hepatotoxicity or serious rash is more frequent in women (as compared to men) and is more likely in women with high CD4 cell counts (more than 250 cells per mm^3). PIs can lower the blood levels of oral contraceptives. Therefore, women receiving PIs should use additional or alternative contraceptive methods, such as consistent use of condoms to avoid pregnancy (WHO, 2003a).

16.4.1 – Prevention of Mother-to-Child Transmission

ARV regimen for HIV positive pregnant women should be started before and during delivery, and also within 48 hours of the delivery. ARV drugs act mainly by reducing viral load in the mother, so less quantity of virus is transferred to the infant, and preventing fixation of HIV in the child's tissues. An issue of concern is the potential impact of NVP prophylaxis used for PMTCT on the subsequent treatment of the mother and her infant. Until more information is available, women who have received single dose prophylaxis with NVP or 3TC for PMTCT should be considered eligible for NNTRI-based regimens (WHO, 2003a).

16.5 – ARV THERAPY FOR INFANTS AND CHILDREN

It is difficult to make a laboratory diagnosis of HIV infection in infants and children aged *below 18 months* due to persistence of maternal antibodies. The WHO recommends initiation of ARV treatment if the infant/child has virologically proven infection (using either HIV-DNA PCR, HIV-RNA assay, or immune-complex dissociated p24 antigen) and has:

- WHO Paediatric Stage III Disease: AIDS-defining opportunistic conditions, severe failure to thrive, progressive encephalopathy, malignancy, meningitis, irrespective of CD4 cell count.
- WHO Paediatric Stage II Disease: chronic diarrhoea, severe persistent or recurrent candidiasis, weight loss, persistent fever, severe recurrent bacterial infections, generalised lymphadenopathy, with consideration of using CD4 cell count less than 20 per cent to assist in decision-making.
- WHO Paediatric Stage I: (i.e. asymptomatic) and CD4 cell count less than 20 per cent. WHO Stage I should be treated *only if* facilities for CD4 cell count are available (WHO, 2003a).

The current WHO staging system for paediatric HIV infection was developed many years ago. Many of the clinical symptoms in Paediatric Stage II and Stage III are not specific for HIV infection. These symptoms may overlap with those seen in HIV-negative children in resource-limited settings. Till the revision of the classification system, the existing classification system is to be used in resource-limited settings for defining parameters for initiating ARV treatment (WHO, 2003a).

Breastfed infants are at risk of HIV infection during the entire period of breastfeeding. A negative virologic or antibody test at one age does not exclude the child becoming infected at a later time if breastfeeding is continued. The penetration of ARV drugs into human breast milk has not been quantified for most of these drugs. Some ARV drugs like NVP are present in breast milk, but the quantity of the drug that would be ingested by the infant through breast milk would be less that the required therapeutic levels. Ingestion of *subtherapeutic levels* of ARV drugs through breast milk may lead to development of resistance. Therefore, if a breastfed infant is sufficiently ill to require ARV treatment, standard paediatric doses of ARV drugs should be started, irrespective of whether the mother is receiving ARV treatment (WHO, 2003a).

16.5.1 – Recommended First-Line ARV Regimens in Infants and Children

Dosing in children is based on *body surface area* or *body weight* in order to avoid the risk of underdosing and the development of resistance. The formula (Harries *et al.*, 2004) for calculating BODY SURFACE AREA (in m²) is – *BODY SURFACE AREA (in m²)* = $\sqrt{\{(HEIGHT \ in \ centimetres \times WEIGHT \ in \ kg)/3600\}}$.

The doses must be adjusted as the child grows. Non-expert personnel need to be provided with a table of drug doses to ensure administration of correct doses. Regimens chosen for children should be similar to those used by the parents in order to avoid different timings and thus improve adherence (WHO, 2003a).

Some ARV drugs are available in specially designed formulations for paediatric use. But, these formulations may not be widely available in resource-limited settings. Use of tablets that require cutting up (especially unscored tablets) can result in under- or overdosing, leading to either drug resistance or toxicity (WHO, 2003a).

16.5.2 – Clinical and Laboratory Monitoring in Infants and Children

The parameters for laboratory monitoring for infants and children on ARV treatment are the same as for adults and adolescents. In addition to the clinical parameters recommended for adults, the *clinical monitoring* in infants and children should include growth and nutritional status, developmental milestones, and neurological symptoms (WHO, 2003a). Salient *clinical signs of response* to ARV treatment in children include improvement in growth and achievement of developmental in children, decreased frequency of bacterial infections, oral thrush, other opportunistic infections, and/or improvement in neurological symptoms.

16.5.3 – Changing ARV Therapy Regimens in Infants and Children

The indications and principles for changing ARV treatment in infants and children is the same as for adults and adolescents. Management of drug toxicity is also the same. If toxicity is related to an identifiable drug, the offending drug is replaced with one that does not have the same side effects (WHO, 2003a).

16.5.4 – Clinical Signs of Treatment Failure

Clinically, treatment failure is determined by lack of growth (or decline in growth) in children, who show initial response to treatment; loss of neurodevelopmental milestones or development of encephalopathy; occurrence of new opportunistic infection or malignancy indicating disease progression; and recurrence of previous opportunistic infections, such as oral candidiasis, that is refractory to treatment (WHO, 2003a).

16.5.5 – CD4 Cell Criteria for Treatment Failure

• Return of CD4 cell percentage (for children older than 6 years of age, of absolute CD4 cell count) to pretreatment baseline levels, or below, in absence of other concomitant infection to explain transient decrease in CD4 counts.

• More than 50 per cent decrease of CD4 cell percentage (for children older than 6 years of age, of absolute CD4 cell count) from peak treatment levels, in absence of other concomitant infection to explain transient decrease in CD4 cell counts (WHO, 2003a).

16.6 – ARV THERAPY FOR TUBERCULOSIS PATIENTS

Treatment of tuberculosis is a *central priority* and should not be compromised by ARV treatment. The optimal time for initiating ARV treatment in tuberculosis patients is *not known*. ARV treatment may be *life saving* in patients with advanced HIV disease, during the first 2 months of antitubercular treatment, since case fatality rates are high (WHO, 2003a). The WHO recommendations (revised in 2003) are outlined in Table 1.

Though clinical evidence supporting treatment recommendations are incomplete, the *first-line* ARV treatment regimen for tuberculosis-HIV co-infection is: (*Stavudine or Zidovudine*) + *Lamivudine* + *Efavirenz* (600 or 800 mg per day). Management of tuberculosis-HIV co-infection is complicated by the need to achieve patient acceptance of both the diagnoses, interaction of rifampicin with NNRTIs and PIs, pill burden, problems related to adherence, and drug toxicity (WHO, 2003a).

16.7 – ARV THERAPY FOR INJECTING DRUG USERS

HIV programmes need to ensure that this vulnerable subgroup in the HIV-infected population have access to life-saving ARV therapy and integrate treatment of drug dependence with HIV care. In such settings, directly observed

Table 1. Options for HIV-tuberculosis co-infection (WHO, 2003a)

CD4 Cell count	Recommended options and regimen
<200 per mm^3	• Start tuberculosis treatment. • Start antiretroviral therapy as soon as tuberculosis treatment is tolerated (between 2 weeks and 2 months) – recommend EFV-containing regimens.[a]
200–350 per mm^3	• Start tuberculosis treatment. • Start antiretroviral therapy after initial phase of tuberculosis treatment. Start earlier is severely compromised. • Consider antiretroviral therapy regimens – EFV-containing[a] (OR) NVP-containing.[b]
>350 per mm^3	• Start tuberculosis treatment and monitor CD4 counts. • Defer antiretroviral therapy.
CD4 cell counts – facility NOT available	• Start tuberculosis treatment. • Consider antiretroviral therapy.

[a] Efavirenz (EFV) is contraindicated in pregnant women and women of childbearing potential without effective contraception.
[b] Nevirapine (NVP) may be used in case of rifampicin-free continuation phase of tuberculosis therapy.

ARV therapy is a feasible option to ensure adherence. A number of ARV drugs are being explored for once daily use so that it can be considered for *directly observed therapy* (WHO, 2003a). The clinical and immunological criteria for starting ARV therapy in IDUs are the same as those in the general recommendations.

Problems in Treating Drug Users: Lifestyle instability in drug users affects adherence to ARV therapy. ARV drugs such as EFV, NVP, and RTV reduce plasma levels of methadone, leading to signs of opiate withdrawal. Patients receiving methadone along with ARV therapy should be monitored for signs of withdrawal and their methadone dose may have to be increased to alleviate withdrawal symptoms (WHO, 2003a).

16.8 – ADHERENCE TO ARV THERAPY

High levels of adherence to ARV therapy (more than 95 per cent) are desirable to maximise benefits such as improved virological and clinical outcomes, avoid drug resistance, and ensure durability of effect of ARV therapy (WHO, 2003a).

16.8.1 – Pretherapy Education

Peer counsellors can help in educating the patient on HIV and its manifestations, benefits and side effects of ARV medications, mode of taking the medications, and importance of not missing any dose.

16.8.2 – Ensuring Adherence during Therapy

Once treatment has begun, the keys to *successful adherence* to ARV therapy need to be remembered. These include minimum number of pills, such as fixed drug combinations, packaging of pills (such as co-blister packs), avoidance of food precautions, daily frequency of dosing not exceeding twice daily, fitting ARV therapy into the patient's lifestyle, and involvement of friends, relatives, and community members in support of the patient's adherence. Adherence should be assessed during visits to health centres. Home visits are useful. *Family-based care* is advised when more than one family member (especially mother and child) is HIV-infected. Support (from friends, relatives, community members) is essential for ongoing adherence since the duration of ARV therapy is lifelong duration (WHO, 2003a).

On 12 July 2006, the FDA of the United States announced the approval of Atripla, a FDC containing EFV, emtricitabine, and tenofovir (TDF) disoproxil fumarate for treatment of HIV-1 infection in adults. Atripla is the first one-pill, once-a-day ARV product that simplifies the treatment regimen for HIV-1 infected adults and has the potential to improve adherence of patients. Since May 2004, the FDA has approved seven co-packaged drugs or fixed drug combinations (FDA, 2006).

16.8.3 – Other Strategies for Ensuring Adherence

Though it has been suggested that cost sharing may assist adherence, studies from African countries indicate that cost sharing is detrimental to long-term adherence. Introduction of DOT, or its modifications, is a challenging task since DOT is resource-intensive and is difficult to sustain for the lifelong duration of ARV therapy. This approach may be useful for certain groups and for early patient training and has been successfully implemented in Haiti (John, 2004). Other strategies include using mobile vans to reach rural communities, ensuring regular and reliable supply of ARV drugs, and providing resources for culturally acceptable adherence programmes (WHO, 2003a).

16.8.4 – Community-Based Buyers' Club

The Thai Network of people living with HIV/AIDS (TNP+) established the Buyers' Club in October 2000 in partnership with local NGOs and Médecins Sans Frontières (MSF) with the objective of increasing access to a limited range of essential ARV drugs. Generic drugs were purchased directly from Thailand's Government Pharmaceutical Organisation. TNP+ has successfully negotiated a preferential price for EFV from the manufacturer Merck Sharp and Dohme. Members of Buyers' Club get selected ARV drugs along with suitable information on treatment, importance of adherence and follow-up of those who miss a prescription, and coping with side effects. The concept of Buyers' Club has spread to other countries such as Japan, Iran, Malaysia, and China (Kreudhutha *et al.*, 2005).

16.8.5 – Measuring Adherence

The numerous treatment challenges posed by ARV treatment include lifelong duration of treatment, pill burden, frequent dosing intervals, food restrictions, and adverse effects. Over 95 per cent adherence is required for viral suppression (Nwokike, 2005). Ensuring free medicines does not guarantee adherence. National ARV programmes need to develop adherence measurement and monitoring systems which are built into the national treatment protocols. Adherence measurement tools include:

1 *Pill count and/or estimation of volume of liquid medications during each visit* – a zero tolerance policy needs to be adopted viz. missing one dose in a four dose-per-day regimen translates to non-adherence for that day (Nwokike, 2005).
2 *Pill identification tests* – asking patients to identify their own pills from other pills that also include look-alike pills (Parienti *et al.*, 2001).
3 *Issuing monthly pill calendars.*
4 *Adherence partner* – having an adherence partner has been shown to be the most crucial factor that promoted adherence (Nwokike, 2005).
5 *Using a 7-day recall questionnaire* (Liu *et al.*, 2001) – However, this method may overestimate adherence.

16.8.6 – Adherence During Pregnancy and Post-Partum Period

Pregnancy-associated morning sickness, gastrointestinal upsets, and fears about potential effects on foetus may complicate ARV therapy. Post-partum physical changes and demands of caring for a newborn may compromise maternal drug adherence. Therefore, culturally appropriate adherence mechanisms need to be developed (WHO, 2003a).

16.8.7 – Adherence in Children

Adherence in children is complicated by disruption of the family unit by health or economic reasons. Keys to successful adherence to ARV therapy include family-based treatment programmes, wide availability of improved paediatric formulations, and matching of paediatric and adult regimens (i.e. frequency of dosing for mother and child should be similar) (WHO, 2003a).

16.8.8 – Innovative Methods for Ensuring Adherence

Innovative methods for ensuring adherence in antitubercular treatment may also be replicated in case of ARV therapy. Haiti has successfully experimented with administration of ARV drugs under direct observation or supervision – similar to directly observed treatment for tuberculosis (John, 2004). Since the duration of ARV therapy is lifelong, ensuring adherence may require many more innovative methods, as compared to that for antitubercular treatment.

Cape Town (South Africa) has 71 per cent cell phone usage. In a pilot project, Dr. David Green (www.compliance.za.net) used a freely available open source operating system, web server, and a database of patients receiving antitubercular treatment. Each day, the computer server sent half-hourly short messaging service (SMS) messages to remind patients to take their treatment. A variety of 800 other messages (jokes, lifestyle tips) were also on the database. Messages were changed daily to relieve boredom. The Cape Town Health Authority paid the equivalent of US$1.30 per patient per month to run the SMS reminder service. Out of more than 300 patients who were involved in the pilot project, there were only five treatment failures. WHO has singled out this novel scheme as an example of best practice (WHO, 2003b).

16.9 – SURVEILLANCE OF DRUG RESISTANCE

The WHO recommends the establishment of HIV drug resistance sentinel surveillance system to detect potential drug resistance at population level and to modify treatment regimens in view of the available information. To begin with, the prevalence rate of drug resistance is to be established in HIV-infected persons who have not received ARV therapy. After this, HIV-infected patients on ARV therapy (especially those diagnosed with treatment failure) should be monitored (WHO, 2003a). Drug susceptibility tests for ARV drugs is based on phenotypic and genotypic assays.

Phenotypic Assays: Phenotypic assays can only be used for cultivable viruses. These assays indicate whether a particular strain of a virus is sensitive or resistant to an ARV drug by determining the concentration of drug needed to inhibit growth of the virus in vitro. In case of HIV, plaque reduction assays may not be suitable since all HIV strains do not produce plaques in cell culture.

Genotypic Assays: Molecular techniques such as polymerase chain reaction and ligase chain reaction are used to assay mutations associated with drug resistant viruses. Being tedious, these assays are not suitable for routine diagnostic laboratories. Since HIV mutates rapidly, resistant strains may have merged in the early stages of infection.

16.10 – LIMITATIONS OF ARV THERAPY

16.10.1 – Limitations of Therapy

Due to the severity of the HIV epidemic, many ARV agents have been speedily licensed, often with little knowledge about their long-term safety (Wig, 2002). Emergence of drug-resistant strains of HIV is a widespread and growing problem (Durant *et al.*, 1999). The objective of ARV therapy is the long-term suppression of viral load. However, studies have revealed that even in successfully treated patients with extremely low (or undetectable) plasma HIV-1 RNA levels, HIV persists in sanctuaries where the drug cannot reach, or continues to exist in a *latent form* on which the drugs have no effect (Wong *et al.*, 1997). HIV has also been detected in the semen of patients who are on ARV therapy, without detectable HIV in the blood (Zhang *et al.*, 1998). Testes are a reservoir of HIV and the viral load in blood and semen is possibly different (Coombs *et al.*, 1998). Vasectomy in HIV-infected individuals does not reduce the quantity of HIV in the ejaculate, as most of the seminal fluid and cell-free HIV concentrate at a point proximal to the site of vasectomy (Zhang *et al.*, 1998). The persistence of latent HIV infection despite therapy for the prescribed duration suggests that *lifelong* treatment is necessary. The currently available drugs are expensive and difficult to tolerate for prolonged periods (Furtado *et al.*, 1999). In patients whose plasma HIV-1 RNA levels had been suppressed by ARV drugs to below undetectable levels, the plasma HIV-1 RNA levels invariably rebounded, within 3 weeks after the cessation of therapy (Harrigan *et al.*, 1999).

Problems in ARV treatment include intolerance of ARV drugs due to adverse reactions, poor adherence of patients and resulting emergence of drug resistance, use of non-standardised regimens, need for careful monitoring to evaluate response to treatment, high rate of HIV turnover and spontaneous mutation, and drug-induced selective pressure (Harries *et al.*, 2004; Potter *et al.*, 2004; NIAID, 2005).

16.10.2 – Limited Access to Antiretroviral Drugs

Research projects and programmes provide free ARV drugs only to HIV-infected individuals who meet inclusion criteria and live in a defined geographic area. In some countries, government programmes provide free ARV treatment at select institutions located in major cities. Even if treatment is free of cost, patients or their families have to bear costs of transport and possibly loss of daily wages of the accompanying family member. The dilemma faced by health care providers is how to tell impoverished patients that they must travel to specific government treatment centres in order to avail of free treatment (Whyte *et al.*, 2005).

In many resource-poor countries, prices of ARV drugs have fallen drastically, bringing fee-for-treatment within reach of more families. But for poor families, the decision to start ARV treatment involves painful prioritising. Supporting long-term ARV treatment for one family member would mean not being able to help another family member with money for education or some other important career goal. The situation gets worse when more than one family member is HIV-infected. When resources are scarce, the dilemma faced by family members is which family member is to be financially supported for ARV treatment. Even when a decision is made to initiate treatment, it is difficult to maintain the regimens for prolonged periods in the face of many other needs. The main reason for discontinuing treatment is economic. Families may have to make difficult choices that may influence adherence and hence these choices should be discussed during pretreatment counselling. Unequal access to ARV treatment poses questions of social justice particularly in poor nations where people cannot afford to buy drugs even at reduced prices and ultimately, unequal access may also influence adherence to treatment (Whyte *et al.*, 2005).

REFERENCES

ADRAC, 2003, Interactions with grape fruit juice – amendment. Austr Adv React Bull 22(2): 4.

Clinton W.J., 2003, Turning the tides on the AIDS pandemic. N Engl J Med 348: 1800–1802.

Collier A.C., Coombs R.W., *et al.*, 1996, Treatment of human immunodeficiency virus infection with saquinavir, zidovudine and zalcitabine. N Engl J Med 334: 1011–1017.

Connor E.M., Sperling R.S., Gelber R., *et al.*, 1994, Reduction of maternal-infant transmission of human immuno-deficiency virus type 1 with zidovudine treatment. N Engl J Med 331: 1173–1180.

Coombs R.W., Speck C.E., Hughes J.P., *et al.*, 1998, Association between culturable human immuno-deficiency virus type-I (HIV-1) in semen and HIV RNA levels in semen and blood – evidence for compartmentalization of HIV-1 in semen and blood. J Inf Dis 177: 320–330.

Durant J., Clevenbergh P., Halfon P., *et al.*, 1999, Drug resistance genotyping in HIV-1 therapy – the VIRADAPT randomized control trial. Lancet 353: 2195–2199.

Fischl M.A., Richman D.D., Griece M.H., *et al.*, 1987, The efficacy of azidothymidine (AZT) in the treatment of patients with AIDS and AIDS-related complex: a double-blind, placebo-controlled trial. N Engl J Med 317: 185–191.

Food and Drug Administration (FDA), 2006, FDA News. July 12. www.fda.gov/cder/drug/infopage/atripla/

Furtado M.R., Callaway D.S., Phair J.P., *et al.*, 1999, Persistence of HIV transcription in peripheral-blood mononuclear cells in patients receiving potent anti-retroviral therapy. N Engl J Med 340: 1614–1622.

Harries A., Maher D., and Graham S., 2004, TB/HIV – a clinical manual. 2nd edn. Geneva: WHO. pp. 137–154.

Harrigan P.R., Whaley M., and Montaner J.S., 1999, Rate of HIV-1 RNA rebound upon stopping anti-retroviral therapy. AIDS 13: F 59 – F 62.

John T.J., 2001, AIDS control and retrovirus drugs. Economic and Political Weekly July 7, pp 2489–2490.

John T.J., 2004, HAART in India: heartening prospects and disheartening problems (Editorial). Indian J Med Res 119: iii–vi.

Kelly M., 2002, The state of play: HIV treatment. HIV Australia 1: 13–15.

Kreudhutha N., *et al.*, 2005, Experience of a community-based antiretroviral Buyers' Club in Thailand. Essent Drugs Monit 34: 10–11.

Land S., McGavin C., Lucas R., *et al.*, 1992, Incidence of ZDV-resistant human immunodeficiency virus isolated from patients before, during and after therapy. J Infect Dis 166: 1139–1142.

Lewin S.R., Crowe S., Chambers D.E., and Cooper D.A., 1997, Antiretroviral therapies for HIV. In: Managing HIV (G.J. Stewart, ed.). North Sydney: Australasian Medical Publishing, pp 45–54.

Liu H., Golin C.E., Miller L.G., 2001, A comparison study of multiple measures of adherence to HIV protease inhibitors. Ann Int Med 134(10): 968–977.

Masquelier B., *et al.*, 2005, Prevalence of transmitted HIV-1 drug resistance and the role of resistance algorithms data from seroconverters in the CASCADE collaboration from 1987 to 2003. J AIDS 40: 505–511.

Mulder J.W., Cooper D.A., Mathiesen L., *et al.*, 1994, Zidovudine twice daily in asymptomatic subjects with HIV infection and a high risk of progression to AIDS: a randomised, double-blind placebo-controlled study. AIDS 8: 313–321.

NACO. National guidelines for clinical management of HIV/AIDS. New Delhi: Government of India.

National Institute of Allergy and Infectious Diseases (NIAID), 2005, HIV infection and AIDS: an overview, NIAID fact sheet. Bethesda: National Institutes of Health, March. www.niaid.nih.gov/

New Mexico AIDS Education and Training Center, 2005, Fact Sheet 430. Non-nucleoside reverse transcriptase inhibitors in development. University of New Mexico Health Sciences Center. www.aidsinfonet.org. Revised 19 November.

New Mexico AIDS Education and Training Center, 2006, Fact Sheet 157. Microbicides. University of New Mexico Health Sciences Center. www.aidsinfonet.org. Revised 9 March.

New Mexico AIDS Education and Training Center, 2006, Fact Sheet 401. Taking current anti-retroviral drugs. University of New Mexico Health Sciences Center. www.aidsinfonet.org. Revised 14 June.

New Mexico AIDS Education and Training Center, 2006, Fact Sheet 402. Anti-viral drug names. University of New Mexico Health Sciences Center. www.aidsinfonet.org. Revised 25 June.

New Mexico AIDS Education and Training Center, 2006, Fact Sheet 408. Salvage therapy. University of New Mexico Health Sciences Center. www.aidsinfonet.org. Revised 10 May.

New Mexico AIDS Education and Training Center, 2006, Fact Sheet 410. Nucleoside analog reverse transcriptase inhibitors. University of New Mexico Health Sciences Center. www.aidsinfonet.org. Revised 16 June.

New Mexico AIDS Education and Training Center, 2006, Fact Sheet 440. Protease inhibitors in development. University of New Mexico Health Sciences Center. www.aidsinfonet.org. Revised 30 June.

New Mexico AIDS Education and Training Center, 2006, Fact Sheet 449. Tipranavir. University of New Mexico Health Sciences Center. www.aidsinfonet.org. Revised 8 July.

New Mexico AIDS Education and Training Center, 2006, Fact Sheet 450. Darunavir. University of New Mexico Health Sciences Center. www.aidsinfonet.org. Revised 25 June.

New Mexico AIDS Education and Training Center, 2006, Fact Sheet 460. Attachment and fusion inhibitors in development. University of New Mexico Health Sciences Center. www.aidsinfonet.org. Revised 14 June.

New Mexico AIDS Education and Training Center, 2006, Fact Sheet 461. Enfuvirtide. University of New Mexico Health Sciences Center. www.aidsinfonet.org. Revised 9 March.

New Mexico AIDS Education and Training Center, 2006, Fact Sheet 470. Other anti-retroviral drugs in development. University of New Mexico Health Sciences Center. www.aidsinfonet.org. Revised 14 June.

New Mexico AIDS Education and Training Center, 2006, Fact Sheet 479. Hydroxyurea. University of New Mexico Health Sciences Center. www.aidsinfonet.org. Revised 20 July.

New Mexico AIDS Education and Training Center, 2006, Fact Sheet 480. Immune Therapies in development. University of New Mexico Health Sciences Center. www.aidsinfonet.org. Revised 30 June.

New Mexico AIDS Education and Training Center, 2006, Fact Sheet 481. Immune Restoration. University of New Mexico Health Sciences Center. www.aidsinfonet.org. Revised 20 June.

New Mexico AIDS Education and Training Center, 2006, Fact Sheet 724. DHEA. University of New Mexico Health Sciences Center. www.aidsinfonet.org. Revised 1 May.

Nwokike J.L., 2005, Baseline data and predictors of adherence in patients on anti-retroviral therapy in Maun General Hospital, Botswana. Essent Drugs Monit 34: 12–13.

Palella F.J. Jr., Deloria-Knoll M., Chmiel J.S., *et al.*, 2003, Survival benefit of initiating anti-retroviral therapy in HIV-infected persons in different CD4 + cell strata. Ann Intern Med 138(8): 620–626.

Parienti J.J., Verdon R., Bazin C., 2001, The pills identification test: a tool to assess adherence to antiretroviral therapy. JAMA 285(4): 412.

Potter S.J., Chew C.B., Steain M., *et al.*, 2004, Obstacles to successful anti-retroviral treatment of HIV-1 infections: problems and perspectives. Indian J Med Res 119: 217–237.

Sepkowitz K.A., 2006, One disease, two epidemics – AIDS at 25. N Engl J Med 354: 2411–2414.

WHO, 2003a, Scaling up anti-retroviral therapy in resource-limited settings: treatment guidelines for a public health approach. Geneva: WHO; 2003 Revision.

WHO, 2003b, South Africa – a novel approach to improving adherence to TB treatment. Essent Drugs Monit 33: 8.

Whyte S.R., *et al.*, 2005, Accessing retroviral drugs: dilemmas for families and health workers. Essent Drugs Monit 34: 14–15.

Wig N., 2002, Anti-retroviral therapy – are we aware of adverse effects? JAPI 50: 1163–1171.

Wong J.K., Hezareh M., Gunthard H.F., *et al.*, 1997, Recovery of replication-competent HIV despite prolonged suppression of plasma viremia. Science 278: 1291–1295.

Zhang H., Dornadula G., Beumount M., *et al.*, 1998, Human immuno-deficiency virus type-1 in the semen of men receiving highly active anti-retroviral therapy. N Engl J Med 339: 1803–1809.

APPENDIX 1

READY RECKONER

Table 2. Dosages for adults and adolescents (WHO, 2003a)

Group	Drug	Dose
Nucleoside reverse transcriptase inhibitors (NsRTIs)	Abacavir (ABC)	300 mg twice daily
	Didanosine (ddl)	400 mg once daily; (250 mg once daily if <60 kg); 250 mg once daily if administered with Tenofovir
	Lamivudine (3TC)	150 mg twice daily (or) 300 mg once daily
	Stavudine (d4T)	40 mg twice daily; (30 mg twice daily if <60 kg)
	Zidovudine (ZDV)	300 mg twice daily
NtRTI	Tenofovir disopril fumarate (TDF)	300 mg once daily*
NNRTIs	Efavirenz (EFV)	600 mg once daily
	Nevirapine (NVP)	200 mg once daily for 14 days, then 200 mg twice daily
Protease Inhibitors (PIs)	Indinavir/ritonavir (IDV/r)	800/100 mg twice daily
	Lopinavir/ritonavir (LPV/r)	400/100 mg twice daily (533/133 mg twice daily when combined with efavirenz or nevirapine)
	Nelfinavir (NFV)	1250 mg twice daily
	Saquinavir/ritonavir (SQV/r)	1,000/100 mg twice daily (or) 1,600/200 mg once daily

*TDF + ddl: drug interaction necessitates dose reduction of ddI.
Note: The dose of each drug may vary if anti-retroviral drugs are combined.

APPENDIX 2

NUCLEOSIDE ANALOGUE RTIs (NSRTIs)

RTIs (or "nukes") were the first ARV drugs. ZDV was approved for use in the United States in 1987 (Fact sheet 402, 2006). These drugs are analogues of nucleosides. ZDV is an analogue of thymidine; ddI is converted to an analogue of adenosine; and zalcitabine or di deoxy cytidine (ddC) is an analogue of cytidine. NsRTI are phosphorylated intracellularly to triphosphate, which inhibits synthesis of the DNA chain by reverse transcriptase enzyme (Lewin *et al.*, 1997). The dose of each drug may vary if ARV drugs are combined.

[A] ZIDOVUDINE (ZDV)

Synonym: Di deoxy thymidine or azidothymidine (AZT)

Pharmacology

The drug is a NsRTI and is active against both HIV-1 and HIV-2. It is indicated in adults infected with HIV. It prevents infection of uninfected cells and limits viral replication in infected cells. However, the drug does not eradicate established infection. Ribavarin and d4T antagonises the ARV action of ZDV (Harries *et al.*, 2004). Resistance to ZDV is seen mostly after at least 6 months of therapy and may decline on withdrawal of the drug (Land *et al.*, 1992; Mulder *et al.*, 1994). Resistance develops more rapidly in patients with lower CD4 cell counts. In HIV-infected pregnant women, ZDV treatment reduces the risk of transplacental transmission by up to 75 per cent (Connor *et al.*, 1994). However, combination therapy is superior to ZDV monotherapy (Lewin *et al.*, 1997).

Bioavailability on oral administration of the drug is 60 per cent. The serum and intracellular half-life is 1.1 hour and 3 hours, respectively. The drug is metabolised to ZDV glucuronide, which is excreted in urine. ZDV delays onset of opportunistic infections and prolongs survival in persons with advanced HIV infection (Fischl *et al.*, 1987). An increase in CD4 cell count and decline in plasma virus titre is seen within 4–8 weeks after commencement of therapy. By 6 months, these parameters recover to baseline levels. Clinical improvement is also seen in patients with HIV-related psoriasis, nail lesions, and arthritis (Lewin *et al.*, 1997).

Dosage Forms and Dosage

ZDV is available as 100 and 250 mg capsules, 300 mg tablets, and syrup containing 10 mg per mL. For infants less than 4 weeks old, the dose is 4 mg per kg body weight. Between 4 weeks and 13 years of age, 180 mg per m^2 of body surface area is recommended. Above 13 years of age, the dose is 300 mg twice daily. The drug can be administered along with food (Harries *et al.*, 2004). Doses of 1,000 mg per day or more have been used in patients with AIDS-related dementia. ZDV is given in the dose of 250 mg four times a day as prophylaxis for high-risk exposure (Lewin *et al.*, 1997). There are no food restrictions (Fact Sheet 401, 2006).

Storage Requirements

ZDV should be stored in amber-coloured glass jars and is light sensitive (Harries *et al.*, 2004).

Side Effects

Due to haematological toxicity, the bone marrow may be suppressed, resulting in anaemia and neutropenia. This toxicity is more severe in those with advanced disease. In symptom-free HIV-infected persons, the risk of anaemia is only about 2 per cent, after 18 months of continuous treatment. Mean corpuscular volume (MCV) is often elevated. ZDV can cause short-term decrease in platelet counts in patients with HIV-related thrombocytopenia (Lewin *et al.*, 1997). The drug should not be co-administered with d4T (Fact Sheet 401, 2006).

There may be subjective complaints of gastrointestinal intolerance, headache, insomnia, and asthenia. Nausea and headache are frequent complaints in the first 6 months of treatment and they decrease with continued treatment. Myopathy, associated with elevated levels of creatine phosphokinase (CPK), may be seen with long-term (more than 1 year) use. This condition is reversible within 8 weeks of stopping treatment. Rare toxic effects include fatty liver, lactic acidosis, mood disturbances, and bluish pigmentation of nails and mucosa (Lewin *et al.*, 1997).

Monitoring During Therapy

During treatment, complete blood count and liver function tests should be performed every month for the first 3 months and later, every 2 months. CPK levels should be monitored twice-monthly after 1 year of treatment or if symptoms of myopathy occur. The treatment should be stopped if aminotransferase levels increase rapidly or if progressive hepatomegaly occurs. ZDV should be cautiously used in first trimester of pregnancy (Lewin *et al.*, 1997).

[B] DIDANOSINE OR DI DEOXY INOSINE (DDL)

Pharmacology

Didanosine exhibits ARV activity against both HIV-1 and HIV-2, including strains resistant to ZDV. It has synergistic action with ZDV. During treatment, there is likelihood of emergence of HIV strains with reduced sensitivity to ddI. It is beneficial as initial treatment in children and in adults previously treated with ZDV. As compared to ZDV, it does not significantly cross the blood-brain barrier and is therefore ineffective in patients with AIDS-related dementia. The bioavailability on oral administration of the drug is 40 per cent. The serum and intracellular half-life is 1.6 hour and 12 hours, respectively. Half the ingested dose is excreted in urine (Lewin *et al.*, 1997).

Dosage Forms and Dosage

The drug is available as oral suspension or paediatric powder/water (10 mg/mL) and chewable tablets in the strengths of 25, 50, 100, 150, and 200 mg. Enteric-coated beadlets in capsules are also available in different strengths. These beadlets can be removed from the capsules and sprinkled on small amount of food before consumption (Harries *et al.*, 2004). The chewable tablets contain sufficient antacid since ddI is rapidly destroyed on exposure to acid (Lewin *et al.*, 1997).

The dose for infants less than 3 months old is 50 mg per m^2 of body surface area twice daily. Between 3 months and 13 years, 90 mg per m^2 twice daily (or 240 mg/m^2 once daily) is advised. For children older than 13 years or whose body weight is more than 60 kg, the dose is 200 mg twice daily or 400 mg once daily. The tablets should be taken on an *empty* stomach at least 30 minutes before, or 2 hours after eating (Harries *et al.*, 2004). The drug should not be administered within 1 hour of indinavir (IDV) or 2 hours of RTV (Fact Sheet 401, 2006).

Storage Requirements

Paediatric oral suspension should be refrigerated and shaken well before use (Harries *et al.*, 2004).

Drug Interactions

The drug should not be co-administered with d4T. The side effects include diarrhea, nausea, vomiting, pancreatitis, abdominal pain, and neuropathy (Fact Sheet 402, 2006). Antacids in ddI tablets can reduce the absorption of ketoconazole and dapsone. Therefore, these drugs should be taken with food, several hours after taking ddI. Oral administration of ganciclovir doubles the absorption of ddI, causing toxic effects. Conversely, ddI halves the serum levels of ganciclovir and can reduce the effectiveness of the latter against cytomegalovirus infection (Lewin *et al.*, 1997).

Side Effects

As compared to ZDV, ddI is less toxic to bone marrow. A major side effect is dose-related, predominantly sensory, symmetric peripheral neuropathy, which is reversible within weeks of stopping treatment. Persons with previous history of peripheral neuropathy are at high risk of recurrence. Mild to fatal dose-related pancreatitis may occur particularly in patients with previous history of pancreatitis or with history of risk factors for developing pancreatitis (drugs, alcohol). The patients should be warned about early signs of pancreatitis and peripheral neuropathy. Xerostomia may occur in 10 per cent of cases. Other toxic effects include hepatitis, electrolyte disturbances and potentially fatal rhabdomyolysis (Lewin *et al.*, 1997).

Monitoring During Therapy

During ddI therapy, it is essential to monitor peripheral nervous system and serum amylase levels. The dose must be reduced or the drug should be withdrawn if signs of peripheral neuropathy develop or if serum amylase level is more than twice normal (Lewin *et al.*, 1997).

[C] ZALCITABINE OR DI DEOXY CYTIDINE (DDC)

The production of this drug has been terminated in the United States in 2006 (Fact Sheet 401, 2006).

Dosage Forms and Dosage

The drug is available as 0.375 and 0.75 mg tablets. The initial oral dose is 0.75 mg three times a day, taken on an *empty* stomach. If side effects occur, the treatment should be stopped and resumption in a dose of 0.375 mg three times a day should be considered (Lewin *et al.*, 1997). The bioavailability on oral administration of the drug is 85 per cent. The serum and intracellular half-life is 1.2 and 3 hours, respectively. Almost 70 per cent of the ingested dose is excreted in urine.

Pharmacology

Antiviral activity is similar to that of ddI. The drug is also active against ZDV-resistant strains of HIV and has synergistic action with ZDV. Superior ARV response is seen when the drug is combined with ZDV or saquinavir. However, combination of ZDV and ddC does not retard emergence of ZDV resistance. A marginal decrease in CD4 cell counts is seen after 8–12 weeks of therapy. There is a dose-dependent reduction in levels of circulating p24 HIV antigen (Lewin *et al.*, 1997).

Side Effects

The patient should be warned about early signs of peripheral neuropathy. After about 8 weeks of treatment, dose-dependent, symmetric sensory peripheral neuropathy may occur. This initially involves the lower limbs and is reversible on stopping treatment. With the first few weeks of therapy, dose-dependent macular rash may develop mostly on the trunk and extremities. Stomatitis, including mouth ulcers, and fever may accompany the rash. Pancreatitis is rare (Lewin *et al.*, 1997).

Monitoring During Therapy

During ddC therapy, it is essential to monitor peripheral nervous system and serum amylase levels. The dose must be reduced or the drug should be withdrawn if signs of peripheral neuropathy develop or if serum amylase level is more than twice normal (Lewin *et al.*, 1997).

[D] LAMIVUDINE (3TC)

Dosage Forms and Dosage

It is available as 150 mg tablets and oral solution containing 10 mg per mL. FDC of 300 mg ZDV and 150 mg 3TC is available. In neonates (less than 1 month old), the dose is 2 mg per kg body weight twice daily. In patients older than 30 days and below 60 kg weight, the dose is 4 mg per kg body weight twice daily. The maximum dose for those weighing more than 60 kg is 150 mg twice daily (Harries *et al.*, 2004). In case of renal insufficiency, the dose should be reduced (Lewin *et al.*, 1997). The drug can be administered along with food (Harries *et al.*, 2004). There are no food restrictions (Fact Sheet 401, 2006).

Pharmacology

Bioavailability on oral administration is 86 per cent in adults and less in children. The serum half-life is 3–6 hours and intracellular half-life is 12 hours. Lamivudine is excreted unchanged in urine. 3TC is a NNRTI. The drug acts by terminating chain of reverse transcriptase and is a weak inhibitor of this HIV enzyme. Lamivudine is active against HIV-1 and ZDV-resistant strains. It has synergistic action with ZDV. Lamivudine-resistant strains remain sensitive to ZDV (Lewin *et al.*, 1997).

Storage Requirements

The oral solution should be stored at room temperature (up to 25°C or 77°) and used within 1 month of opening (Harries *et al.*, 2004).

Side Effects

In adults, 3TC is well tolerated (Harries *et al.*, 2004). The side effects include nausea, vomiting, fatigue, and headache (Fact Sheet 401, 2006). Macrocytosis and neutropenia are the commonest haematological side effects if the drug is used in combination with ZDV. Hair loss is rare. In children, past history and history of risk factors for pancreatitis should be elicited and their parents/guardians should be warned about early signs of pancreatitis. Serum amylase levels should be monitored in children (Lewin *et al.*, 1997). The drug should not be combined with abacavir (ABC) and TDF unless additional ARV drugs are used (Fact Sheet 401, 2006).

[E] STAVUDINE (D4T)

Pharmacology

Bioavailability on oral administration is 80 per cent. The serum half-life is 1 hour and intracellular half-life is 3–5 hours. Fifty per cent of the ingested dose is excreted in urine. Stavudine improves clinical and immunological parameters in HIV-infected persons with CD4 cell count less than 500 cells per µL, who had previously taken ZDV. Higher CSF levels are achieved as compared with ddI or ddC (Lewin *et al.*, 1997). Stavudine can be combined with 3TC or ddI, but *not* with ZDV, due to antagonistic ARV action (Harries *et al.*, 2004; Lewin *et al.*, 1997).

Dosage Forms and Dosage

The drug is available as oral solution containing 1 mg per mL and as 15, 20, 30, or 40 mg capsules. For persons weighing more than 30 kg, the dose is 30 mg twice a day. The maximum dose (more than 60 kg) is 40 mg twice daily (Harries *et al.*, 2004). The dosage should be reduced in persons with renal insufficiency (Lewin *et al.*, 1997). The capsules may be opened and mixed with small amounts of food. There are no food restrictions during treatment (Fact Sheet 401, 2006).

Storage Requirements

The oral solution should be stored under refrigeration and shaken well before use. The capsules should be stored in glass bottles (Harries *et al.*, 2004).

Side Effects

The drug should not be used with ZDV or ddI (Fact Sheet 401, 2006). The patient should be warned about early signs of peripheral neuropathy and those with a past history of this condition (particularly after taking ddI or ddC) have a high risk of recurrence. Peripheral neuropathy is reversible if treatment is stopped. Headache and nausea may occur and decrease with continued treatment. Though pancreatitis is rare, past history of pancreatitis, or risk factors for this condition (ganciclovir or pentamidine therapy) should be elicited. Anaemia and neutropenia are infrequent side effects (Lewin *et al.*, 1997). Chills, fever, and diarrhea have also been reported (Fact Sheet 401, 2006).

Monitoring During Therapy

During d4T therapy, it is necessary to monitor the peripheral nervous system, complete blood count, liver function tests, and serum amylase levels. The dose must be reduced or the drug should be withdrawn if signs of peripheral neuropathy develop or if serum amylase level is more than twice normal (Lewin *et al.*, 1997).

[F] ABACAVIR (ABC)

Dosage Forms and Dosage

The drug is available as oral solution (20 mg/mL) and 300 mg tablets. The dose is 8 mg per kg body weight for patients younger than 16 years or less than 37.5 kg. For those older than 16 years or weighing more than 37.5 kg, the dose is 300 mg twice daily. The drug may be administered along with food and there are no food restrictions. An FDC of *ZDV 300 mg + 3TC 150 mg + ABC 300 mg* is available as Trizavir. The maximum dose (those weighing more than 40 kg) of this FDC is one tablet twice daily. The tablet cannot be split and therefore, for children weighing less than 30 kg, Trizavir cannot be dosed accurately. At present liquid preparations of Trizavir are not available (Harries *et al.*, 2004).

Side Effects

Patients must be warned about possible hypersensitivity reaction, which occurs in about 8 per cent of patients (Fact Sheet 401, 2006). If hypersensitivity reaction occurs, ABC should be stopped permanently (Harries *et al.*, 2004). ABC should not be co-administered with 3TC or TDF (Fact Sheet 401, 2006).

[G] EMTRICITABINE (FTC)

The adult dose of the drug is 200 mg once daily. There are no food restrictions. The side effects include headache, nausea, vomiting, and skin rash (Fact Sheet 401, 2006). A FDC of EFV, emtricitabine, and TDF disopril fumarate called Atripla has been approved for use in HIV-1 infected adults by the FDA (USA) in July 2006. Atripla is the first once-pill, once-a-day ARV product (FDA, 2006).

[H] "NUKES" UNDER DEVELOPMENT

ELVUCITABINE (ACH-126, 443, beta-L-Fd4c) is an once-daily drug, which has shown activity against HIV that is resistant to several other "nukes" and is also effective against HBV.

MIV-210 (FLG) has shown activity against HIV that is resistant to several other "nukes".

RACIVIR has shown activity against HIV and HBV in laboratory studies. The drug exhibited anti-HIV activity that lasted more than 2 weeks after the drug was stopped. Probably, racivir can be used as a once-daily drug (Fact Sheet 410, 2006).

APRICITABINE (AVX 754, formerly SPD 754) has shown good activity against 3TC-resistant HIV and seems well tolerated.

AMDOXOVIR (MPD) is in Phase II studies and ocular problems detected in early studies are under investigation.

DIOXOLANE THYMIDINE (DOT) is being studied in the University of Georgia, USA.

KP1461 induces lethal mutations in HIV (Fact Sheet 410, 2006).

APPENDIX 3

NON-NUCLEOSIDE RTIs (NNRTIs)

These drugs (also called "non-nukes"), actually bind to reverse transcriptase and prevent its functioning. All "non-nukes" interact with many other drugs and therefore, the physician should elicit the medication history before starting treatment and advise the patient to seek advice before taking any medication. NOTE – The dose of each drug may vary if ARV drugs are combined.

[1] NEVIRAPINE (NVP)

Pharmacology

On oral administration, the bioavailability is 90 per cent. The serum half-life is 25–30 hours. NVP is metabolized by cytochrome P450. Almost 80 per cent of the ingested dose is excreted in urine as glucuronide, 5 per cent is excreted unchanged in urine, and 10 per cent is excreted in faeces. 12 months of triple drug therapy (NVP, ddI, and ZDV) causes significant decrease in viral load. Monotherapy or dual therapy with ZDV is associated with rapid development of drug resistance (Lewin *et al.*, 1997).

Dosage Forms and Dosage

NVP is available as oral suspension containing 10 mg per mL and as 200 mg tablets. The drug can be given along with food and there are no food restrictions (Fact Sheet 401, 2006). Since initiation of full-dose therapy is associated with development of skin rash in up to 50 per cent of patients, incremental doses are given as mentioned in Table 3.

Storage Requirements

Oral suspension can be stored in room temperature (up to 25°C or 77°F); must be shaken well before use (Harries *et al.*, 2004).

Side Effects and Interactions

Fever, headache, and nausea have been reported (Fact Sheet 401, 2006). Hepatitis may occur in 1 per cent of cases. Steven-Johnson syndrome is rare, but the risk is increased on ingestion of drugs containing clavulanic acid. The drug *induces* (stimulates) cytochrome P450 enzyme. Since rifampicin, rifabutin, antiepileptics, oral contraceptives, and other protease inhibitors *reduce* the plasma levels of NVP, these drugs should not be co-administered (Collier *et al.*, 1996).

Table 3. Incremental dosage for nevirapine (Harries et al., 2004)

Age	First 2 weeks	Next 2 weeks	Thereafter
15–30 days	5 mg per kg once daily	120 mg per m^2 twice daily	200 mg per m^2 twice daily
30 days to 13 years	120 mg per m^2 once daily	120–200 mg per m^2 twice daily	
>13 years	200 mg once daily	200 mg twice daily	

Monitoring during Therapy

Liver enzymes should be monitored during NVP therapy since abnormal liver function tests are common (Lewin *et al.*, 1997). Patients should be warned about skin rash. If mild or moderate rash occurs, the drug is withheld and dosing is restarted from the beginning of dose escalation. In case of severe rash, NVP is to be discontinued (Harries *et al.*, 2004).

[2] –DELAVIRIDINE (DLV)

Pharmacology

On oral administration, the bioavailability is 85 per cent. The serum half-life is 5–8 hours. The drug is metabolized partially by cytochrome P450. Of the ingested dose 51 per cent is excreted in urine as metabolites, 5 per cent is excreted unchanged in urine, and 44 per cent in faeces (Lewin *et al.*, 1997).

Dosage Forms and Dosage

This drug is available as 100 mg tablets. The adult dose is 100–300 mg four times a day. Antacids and ddI should be avoided within 1 hour of administering the drug, but there are no food restrictions (Fact Sheet 401, 2006). Delaviridine (DLV) should be used only in combination with other ARV drugs. Triple drug therapy (DLV, ddI, ZDV) is used (Lewin *et al.*, 1997).

Side Effects and Interactions

Diffuse maculopapular skin rash may develop in up to 30 per cent of patients. Mild headache, nausea, vomiting, diarrhea, and fatigue have been reported with its use (Fact Sheet 401, 2006). Liver enzymes may be elevated and hence, liver functions should be monitored (Lewin *et al.*, 1997). DLV inhibits cytochrome P450 enzyme and interacts with multiple drugs. Drugs that decrease levels of DLV, thus increasing risk of therapeutic failure of DLV include phenobarbital, phenytoin sodium, rifabutin, and rifampicin. On the other hand, DLV increases blood levels causing potential toxicity of the following drugs if co-administered – clarithromycin, dapsone, ergot alkaloids, IDV, quinidine, saquinavir, and warfarin (Lewin *et al.*, 1997). Terfenadine, astemizole, alprazolam, midazolam, and cisapride should also *not* be co-administered with DLV (Lewin *et al.*, 1997).

[3] EFAVIRENZ (EFV)

This NNRTI is recommended only for children aged over 3 years. The drug is available as syrup containing 30 mg per mL and as capsules in the strengths of 50, 100, and 200 mg (Table 4).

Table 4. Dosage of efavirenz (Harries et al., 2004)

Weight of child (kg)	Capsule (liquid) dose[*]
10–15	200 mg (270 mg = 9 mL) once daily
15–20	250 mg (300 mg = 10 mL) once daily
20–25	300 mg (360 mg = 12 mL) once daily
25–33	350 mg (450 mg = 15 mL) once daily
33–40	400 mg (510 mg = 17 mL) once daily
>40	600 mg once daily

[*]The syrup requires higher doses than capsules.

EFV is best given at bedtime, especially in the first 2 weeks, to avoid side effects pertaining to the central nervous system. The drug can be co-administered with food, but high-fat meals increase absorption by 50 per cent. The capsules may be opened and added to food and the very peppery taste of the drug can be disguised by mixing with sweet foods or jam (Harries *et al.*, 2004). Side effects include vivid dreams, insomnia, anxiety, dizziness, skin rash, nausea, diarrhoea, and headache (Fact Sheet 401, 2006).

[4] "NON-NUKES" UNDER DEVELOPMENT

(+)-CALANOLIDE A is derived from a rainforest plant. It can easily cross the blood-brain barrier and remain the blood stream for a long time.

GW5634 (also known as 695634) is a precursor that is broken down in the body to produce GW 8248 that has better bioavailability. It can cause skin rash and elevate levels of liver enzymes.

MIV-150 has shown activity against HIV strains that are resistant to other "non-nukes" in laboratory studies (Fact Sheet 430, 2005).

ETRAVIRINE (TMC 125) and TMC 278 have exhibited activity against some HIV strains that are resistant to other "non-nukes".

BILR 355 BS is being developed against wild-type and NNRTI-resistant HIV.

CAPRAVIRINE (AG 1549, formerly S-1153) has shown disappointing results in a recent Phase II trial (Fact Sheet 430, 2005).

APPENDIX 4

PROTEASE INHIBITORS

These ARV agents target the HIV enzyme protease, which cleaves the *gag* polyprotein that is required for production of mature, infectious virions. PIs also inhibit viral assembly. Nausea and diarrhoea are common with PIs. Bulking agents, loperamide, and dietary modifications can be helpful for diarrhoea (Wig, 2002). All PIs interact with many other drugs and therefore, the physician should elicit the medication history before starting treatment and advise the patient to seek advice before taking any medication. Drugs decreasing blood levels of PIs, causing their potential therapeutic failure include carbamazepine, NVP, phenobarbital, and rifampicin. PIs increase the blood levels of alprazolam, midazolam, triazolam, astemizole, terfenadine, cisapride, tricyclic antidepressants, felodepine, nifedipine, lovastatin, simvastatin, clarithromycin, cyclosporin A, and diazepam (Lewin *et al.*, 1997).

[A] NELFINAVIR (NFV)

The drug is available as 250 mg tablets and as powder for oral suspension, to be mixed with liquid before administration. Each 5 mL level teaspoonful of the suspension contains 200 mg. The powder is faintly bitter; gritty and hard to dissolve. The drug is to be taken with meals or a snack (Fact Sheet 401, 2006). Administration along with acidic food or juice increases its bitter taste (Table 5).

The tablets can be halved, or crushed and added to food or dissolved in water. Because of difficulties in dissolving the powder, crushed tablets are preferred even for infants if appropriate dose can be given. Powder and tablets can be stored at room temperature (Harries *et al.*, 2004). Side effects include nausea, diarrhoea, flatulence, abdominal pain, and weakness (Fact Sheet 401, 2006).

[B] INDINAVIR (IDV)

Pharmacology

This drug is indicated in HIV-infected adults. On oral administration, the bioavailability is 30 per cent. The serum half-life is 1.5–2 hours. IDV is effective when used alone and it exhibits synergistic ARV activity with ZDV, ddI, and NNRTIs. Undetectable viral loads have been achieved with triple therapy (IDV, ZDV, 3TC) in 88 per cent of patients (Lewin *et al.*, 1997).

Dosage Forms and Dosage

IDV is available as 200 and 400 mg tablets and should be stored in cool, dry place. The recommended standard dose is 800 mg *every 8 hours* (not three times a day), to be taken 1 hour before or after meals, or with low-fat snack or with lots of water on empty stomach (Fact Sheet 401, 2006). Patients should be advised to avoid consuming grape fruit and its juice altogether when taking certain medicines

Table 5. Dosage of nelfinavir (Harries et al., 2004)

Age	Dose
<1 year	40–50 mg per kg body weight three times daily (or) 75 mg per kg twice daily
1–13 years	55–65 mg per kg body weight twice daily
>13 years	1,250 mg twice daily

such as IDV because the interacting effect may persist for up to 3 days after ingestion (ADRAC, 2003). IDV should be started in full standard dose if liver functions are normal. Dosage must be reduced in cases of liver impairment. Liver
function tests should be monitored during IDV therapy (Lewin *et al.*, 1997).

Side Effects

Patients should be advised to consume at least 1.5 L of fluid daily to reduce risk
of nephrolithiasis, which may occur in 2–5 per cent of cases due to poor solubility of IDV in urine. Other toxic effects include headache, nausea, abdominal
pain, skin rash, dry skin, pharyngitis, and alteration of taste. Dose-dependent
hyperbilirubinemia and elevated levels of serum transaminases may occur and
these decrease with reduction in dosage. Oesophageal reflux is seen in 3 per cent
of cases. Since antacids prevent absorption of IDV, reflux should be treated with
H_2 blockers and proton-pump inhibitors (Wig, 2002).

Drug Interactions

IDV, which is metabolised by cytochrome P450 isoenzyme CTP 3A4, interacts
with drugs that *induce* or *inhibit* this enzyme. The drug should *not* be administered within 1 hour of ddI. Co-administration of the following drugs are to be
avoided – astemizole, terfenadine, cisapride, triaxolam, midazolam, ergot alkaloids, and itraconazole. With ketoconazole treatment, the dose of IDV should
be reduced to 600 mg three times a day. Use of rifampicin should be avoided for
treatment of mycobacterial infections. If rifabutin is used, the standard dose of
IDV should be halved (Lewin *et al.*, 1997).

[C] RITONAVIR (RTV)

Pharmacology

RTV is metabolised by cytochrome P450 isoenzymes 3A4, 2D6, 2C9, and 2C10.
Thus, it interacts with multiple drugs and a checklist (Table 6) should be used
when any drug is being prescribed to patients on RTV therapy. Serum half-life
is 3–5 hours (Lewin *et al.*, 1997). When RTV is used in small doses to boost the
action of other PIs, the name of the drug is suffixed with "/r". For example,
SQV/r means saquinavir boosted with RTV.

Dosage Forms and Dosage

This drug is available as 100 mg capsules. It is indicated in HIV-infected patients
older than 12 years of age. The recommended dosage is 600 mg twice a day, to
be taken with meals. If gastrointestinal intolerance develops, the dosage should
be halved and again increased to full dose within 3–4 days because subtherapeutic doses increase the risk of viral resistance (Lewin *et al.*, 1997). The drug
should be taken 2 hours apart from ddI (Fact Sheet 401, 2006).

Table 6. Checklist for drugs interacting with ritonavir (Lewin et al., 1997)

Group	Substrate
Analgesics, NSAIDS	Meperidine, piroxicam, propoxyphene
Antimicrobial agents	Clarithromycin, sulfamethoxazole
Anticoagulants	Warfarin
Antidepressants	Bupropion
Antihistaminics	Astemizole, terfenadine
Antimycobacterial drugs	Rifampicin, rifabutin
Antiretroviral drugs	Zidovudine, saquinavir
Barbiturates	Phenobarbital
Bronchodilators	Theophylline
Calcium blockers	Heptidil
Ergot alkaloids	Dehydro ergotamine
Gastrointestinal drugs	Cisapride
Neuroleptics	Phenytoin, Clozapine, pimozide
Opium derivatives	Codeine
Steroids	Dexamethasone, Ethinyl estradiol
Oral hypoglycemics	Tolbutaminde
Psychotropics	Chlorpromazine, alprazolam, estazolam, midazolam, triazolam, diazepam, flurazepam, clorazepate, zolpidem

Storage Requirements

RTV should be stored under refrigeration. But, single doses can be kept at room temperature (up to 25°C or 77°F) for 12 hours (Harries *et al.*, 2004).

Side Effects

Toxic effects include nausea, headache, vomiting, and diarrhoea, which may decrease with time. Alteration in taste has been reported. Circumoral paresthesia (numbness and burning) is a dose-dependent side effect.

Monitoring during Therapy

Elevated levels of serum triglycerides (more than twice normal), uric acid, CPK, and hepatic enzymes may be seen. These parameters should be monitored and RTV should be stopped if hepatic enzyme levels are more than three times normal (Lewin *et al.*, 1997).

[D] LOPINAVIR/RITONAVIR (LPV/R)

This combination is indicated in patients who are older than 6 months. LPV/r combination is available as oral solution (80 mg per mL LPV + 20 mg per mL RTV) and as capsules (133.3 mg LPV + 33.3 mg RTV). The oral solution has bitter taste. (Harries *et al.*, 2004). Liquid preparations are to be taken with food while there are no food restrictions when capsules are used (Fact Sheet 401, 2006).

Dosage

For patients aged more than 6 months to 13 years: *Surface area-based dosing* –
225 mg per m^2 LPV + 57.5 mg per m^2 RTV twice daily. *Weight-based dosing*:
7–15 kg – 12 mg per kg LPV + 3 mg per kg RTV twice daily; 15–40 kg – 10 mg
per kg LPV + 2–5 mg per kg RTV twice daily. The maximum dose for patients
weighing more than 40 kg is 400 mg LPV + 100 mg RTV (three capsules or 5 mL
oral solution) twice daily. Though the oral solution and capsules can be stored
at room temperature (up to 25°C or 77°F) for 2 months, they should be preferably
stored under refrigeration (Harries *et al.*, 2004).

[E] SAQUINAVIR (SQV)

Pharmacology

SQV is indicated in HIV-infected patients older than 12 years of age. Low
absorption may increase the risk of HIV-resistance to PIs. However, bioavailability is *improved* when SQV is taken with grapefruit juice or with RTV. Serum
half-life is 1–2 hours. SQV is metabolised in the liver by cytochrome P450 isoenzyme 3A4. Low bioavailability (about 4 per cent) on oral administration is due to
incomplete absorption and extensive first-pass metabolism (Lewin *et al.*, 1997).

Additive or synergistic effects are seen when dual or triple combinations (with
RTIs) are employed. Cross-resistance is not seen with RTIs but 20 per cent of
SQV-resistant strains exhibit cross-resistance with other PIs. When combined
with ZDV and ddC, saquinavir is highly effective in suppressing the viral load
and in increasing CD4 cell counts, even in patients with relatively advanced disease (Collier *et al.*, 1996). Saquinavir is available as 200 mg capsules. The recommended dosage is 600 mg twice a day, taken orally with, or within 2 hours of
a high-fat meal (Lewin *et al.*, 1997).

Drug Interactions

Drugs that induce hepatic enzymes (rifampicin, rifabutin, phenytoin, carbamazepine) further decrease bioavailability of saquinavir and should not be
co-administered with saquinavir. Like other PIs, it interacts with many drugs.
Rifampicin and rifabutin reduce the plasma levels of saquinavir by 80 and 40
per cent, respectively. Drugs that decrease levels of saquinavir, thus causing its
therapeutic failure include rifabutin, rifampicin, carbamazepine, NVP, phenobarbital, and terfenadine. Saquinavir increases the blood levels of the following
drugs, causing their toxicity if co-administered – alprazolam, astemizole, cime-
tidine, cisapride, clarithromycin, clindamycin, cyclosporin A, diazepam, felodepine, itraconazole, ketoconazole, lovastatin, midazolam, nifedipine, simvastatin,
terfenadine, triazolam, and tricyclic antidepressants (Lewin *et al.*, 1997). The
herb St. John's Wort lowers the blood levels of some PIs and therefore should
not be co-administered (Fact Sheet 449, 2006).

Side Effects

Triple therapy does not increase toxic effects (Fischl *et al.*, 1987). Toxic effects include gastrointestinal intolerance, abdominal discomfort, nausea, diarrhoea, oral ulceration, skin rash, headache, and elevated transaminases. It should be used cautiously in patients with hepatic impairment (Lewin *et al.*, 1997).

[F] OTHER PROTEASE INHIBITORS

TIPRANAVIR (Aptivus, PNU 140690) – This drug was approved by the US FDA in 2005 (Fact Sheet 449, 2006). Tipranavir boosted with RTV should not be used as part of an initial ARV regimen. The adult dose comprises two tablets (gel capsules) of 250 mg with one 100 mg tablet of RTV, to be taken with food, particularly high-fat meals, twice daily. The drug should be stored under refrigeration until the bottle is opened. After opening the bottle, the gel capsules may be stored for up to 60 days at room temperature. Its side effects include nausea, vomiting, diarrhoea, abdominal pain, fatigue, skin rash, headache, and aggravation of impaired liver function. Patients with hepatitis B or C should get their liver functions monitored periodically. Women taking oral contraceptives may get a skin rash and the efficacy of contraception may be reduced. The combination of tipranavir and RTV can cause increases in blood levels of cholesterol and triglycerides. The drug is best avoided in persons with bleeding disorders because few cases of internal bleeding were reported in 2006, some of them fatal. Tipranavir being a sulfa drug, history of sulfa allergy should be elicited before prescribing the drug. It interacts with multiple drugs (see under darunavir, below), lowers the blood levels of methadone, and causes excessive sedation if buprenorphine is co-administered (Fact Sheet 449, 2006).

Amprenavir (APV) – The drug may be used in combination with RTV. The adult dose of APV is eight tablets of 150 mg twice daily, to be taken with or without food. High-fat foods are to be *avoided*. The drug should not be taken within 1 hour of antacids. The reported side effects are nausea, vomiting, diarrhea, skin rash, circumoral paresthesia, and abdominal pain (Fact Sheet 401, 2006).

Atazanavir (ATV) – The adult dose is 200 mg once daily for patients new to ARV treatment or 150 mg with 100 mg of RTV once a day, to be taken with food. The adverse effects are bilirubinemia, nausea, vomiting, diarrhea, headache, skin rash, abdominal pain, tingling in hands or feet, depression, and cardiac arrhythmias (Fact Sheet 401, 2006).

Darunavir (TMC 114, Preztisa) – This drug was approved by the FDA (USA) in June 2006 (Fact Sheet 440, 2006) for use in HIV-infected persons who have already used other ARV drugs (Fact Sheet 450, 2006). The adult dose is 600 mg (two tablets of darunavir) along with 100 mg (one tablet) RTV, twice daily, to be taken with food. The drug should be stored at room temperature. Its side effects are nausea, vomiting, diarrhea, and rhinorrhoea. Skin rash may be serious in rare cases. The combination of darunavir with RTV can increase blood levels of cholesterol and triglycerides. Since darunavir belongs to the sulfa group of drugs, history of allergy to sulfa drugs should be elicited from patients. Darunavir may interact with antitubercular drugs, antifungals, antiarrhythmics, several antihistaminics, sedatives, anticholesterol drugs, and drugs used for treatment of erectile dysfunction and migraine (Fact Sheet 450, 2006).

Fosamprenavir (FPV) – The adult dose is 700 mg twice daily (or) 700 mg twice daily with 100 mg RTV once daily (or) 700 mg once daily with 100 mg RTV twice daily. There are no food restrictions. The drug is reported to cause nausea, vomiting, diarrhoea, skin rash, circumoral paresthesia, and abdominal pain (Fact Sheet 401, 2006).

[G] PROTEASE INHIBITORS UNDER DEVELOPMENT

BRECANAVIR (GW640385, VX-385) has shown activity against wild strains of HIV and strains that are already resistant to current PIs. If used in a relatively low dose that would reduce side effects, it will be boosted with RTV. The side effects are mild to moderate; the most common is skin rash (Fact Sheet 440, 2006).

APPENDIX 5

ATTACHMENT AND FUSION INHIBITORS

Attachment inhibitors prevent HIV from attaching to the host cell. *Fusion inhibitors* block the ability of HIV to enter host cell by blocking the merging (fusion) of the virus with the host cell membrane. This prevents the HIV from entering and infecting human immune cells (NIAID, 2005). These drugs may prevent infection of a host cell by free virus in the blood stream or by contact with an infected cell. Most of these drugs are administered by injections or intravenous infusions because they are destroyed by the action of digestive acids (Fact Sheet 460, 2006).

Enfuvirtide (ENF, T-20)

The first fusion inhibitor enfuvirtide was approved for use as an ARV drug in the United States by the FDA in 2003. When HIV infects a human cell, it attaches itself to the cell and fuses with the cell membrane. Enfuvirtide stops this process of fusion and thus prevents infection of human cells by HIV. This drug has been studied in adults and children over 6 months of age. It is recommended only when the other ARV drugs are not effective. The drug does not exhibit cross-resistance with any other ARV drug (Fact Sheet 461, 2006). It is expensive and requires parenteral administration. It is therefore, not suitable for use in situations with resource limitations (WHO, 2003a).

Dose – The adult dose is 90 mg per injection given subcutaneously, twice daily. For children, the dose is based on their body weight. The drug cannot be administered orally because it is destroyed by digestive acids (Fact Sheet 461, 2006).

Side Effects – Almost everyone develops skin reactions at the site of injection. These reactions may manifest as slight erythema, itching, swelling, pain, hardened skin, or hard lumps. Each reaction might last up to 1 week (Fact Sheet 461, 2006). Other side effects include headache, dyspnoea, pneumonia, chills and fever, skin rash, haematuria, vomiting, pain and numbness in feet or legs, dizziness, insomnia, and hypotension (Fact Sheets 401 & 461, 2006).

Drug Interactions – Enfuvirtide may increase the blood levels of tipranavir and RTV. There are very few known interactions with other ARV drugs. Enfuvirtide has not been studied for interactions with all drugs or dietary supplements. Health care personnel should elicit history of use of medications or dietary supplements from patients (Fact Sheet 461, 2006).

ATTACHMENT AND FUSION INHIBITORS UNDER DEVELOPMENT

AMD 070 blocks the CXCR4 receptor on CD4 T-lymphocytes to inhibit HIV fusion.

AK 602 is being developed by Kumamoto University, Japan and is in early human trials. It blocks the CCR5 receptor on CD4 T-lymphocytes to inhibit HIV fusion.

BMS-378806 is an attachment inhibitor that attaches to gp120 and prevents attachment of HIV.

MARAVIROC (MVC, UK-427–857) blocks the CCR5 receptor on CD4 cells to inhibit HIV fusion (Fact Sheet 461, 2006).

VICRIVIROC (SCH-417690, formerly called Schering-D) blocks the CCR5 receptor on CD4 cells. Phase III trials were discontinued due to poor virologic control in patients who had never received ARV treatment. The study is likely to be repeated at a higher dose of the drug (Fact Sheet 461, 2006).

TNA-355 is a genetically engineered monoclonal antibody that blocks the CD4 receptor. It may be administered by intravenous infusion or as a twice-monthly injection.

INCB 9471 is in Phase I trials in healthy volunteers.

PRO 140 and PRO 542 block fusion by binding to a receptor protein on the surface of CD4 cells.

SIFURVITIDE is being developed in China and human trials are to begin soon (Fact Sheet 461, 2006).

APPENDIX 6

MISCELLANEOUS DRUGS

NUCLEOTIDE ANALOGUE RTI (NTRTI)

TENOFOVIR or tenofovir disoproxil fumarate (TDF) – The adult dose is 300 mg orally, once daily. There are no food restrictions, but the drug is to be taken 2 hours before, or 1 hour after ddI. The side effects are usually mild and include nausea, vomiting, and loss of appetite. The drug should not be co-administered with 3TC and ABC unless additional ARV drugs are included in the regimen (Fact Sheet 401, 2006). The dose of the drug may vary if ARV drugs are combined. TDF, with or without emtricitabine, reduces infection rate in high-risk groups but its prophylactic use may accelerate emergence of drug resistance. TDF has reportedly been misused as a "party" pill by uninfected individuals who planned to engage in high-risk activity (Sepkowitz, 2006). The prevalence of resistance to TDF has increased from about 5 per cent before 1996 to at least 15 per cent between 1996 and 2003 (Masquelier *et al.*, 2005).

INTEGRASE INHIBITORS

Once the reverse transcriptase enzyme changes the HIV's genetic code from a single strand to a double strand, the double strand gets inserted into the genetic code of the host cell with the help of integrase enzyme. Integrase inhibitors block the action of integrase enzyme. *Gilead 9137* (JTK-303), discovered by Japan Tobacco and now licensed to Gilead Sciences, has shown promising results in early studies. *MK-0518* is moving to advanced stage of human trials (Fact Sheets 402 & 470, 2006).

ANTISENSE DRUGS

These are "mirror images" of part of the genetic code of HIV that lock on to the virus to prevent it from functioning. *HGTV43* an "antisense" therapy that aims to produce CD4 T-lymphocytes that resist infection by HIV. *VRX496* is another "antisense" drug. Both are under trial (Fact Sheets 470 & 480, 2006).

HYDROXYUREA

Hydroxyurea does not act on HIV directly, but enhances the action of ddI and d4T. When taken in the dose of 300 mg twice daily, hydoxyurea showed the best results in terms of tolerability and reduction of viral load. Since the drug blocks an enzyme produced by human cells, HIV cannot develop resistance to it. Hydroxyurea can slow down mutations in HIV so that it takes much longer for resistance to develop to other ARV drugs. It has been approved as an antimalignancy drug by the FDA (USA), but not as an ARV drug. Hydroxyurea is also effective in sickle cell anemia. The most serious side effect of hydroxyurea is pancreatitis that caused some deaths. Other side effects include nausea, vomiting, diarrhoea, weight gain, hair loss, peripheral neuropathy, and changes in skin colour. Since the drug can damage the bone marrow causing anaemia and neutropenia, hydroxyurea should not be co-administered with ZDV because both drugs can damage the bone marrow. Hydroxyurea can increase the side effects of ddI, cause serious birth defects, and reduce gains in CD4 cell counts (Fact Sheet 479, 2006).

MATURATION INHIBITORS

This group of drugs prevents maturation of internal structures in new virions. *BEVIRIMAT* (PA457), the first maturation inhibitor, has shown ARV activity and will probably be a once-a-day drug (Fact Sheet 470, 2006).

ZINC FINGER INHIBITORS (OR ZINC EJECTORS)

The inner core of HIV (the nucleocapsid) is held together by structures called "zinc fingers". Zinc finger inhibitors (or zinc ejectors) are drugs that can break apart these structures and prevent HIV from functioning. Since it is believed that the nucleocapsid core cannot mutate very easily, a drug that works against the zinc fingers might be effective for a long time. However, zinc fingers are not exclusive to HIV and zinc finger inhibitors could have serious side effects. *AZODICARBONAMIDE* (ADA) has been tested but there are no recent reports on its development (Fact Sheet 470, 2006).

MICROBICIDES

Technically microbicides are not intended to be used as part of ARV treatment. Microbicides are anti-HIV substances that could reduce the risk of HIV infection during vaginal or anal intercourse. Ideally, microbicides should be used in addition to condoms. Currently, the only available effective tools for HIV prevention are male and female condoms that need cooperation of the male partner. But in situations where male partners object to use of condoms, microbicides could reduce the risk of HIV transmission. Use of these products can be controlled by women without the need for cooperation from the male partner. Women might be able to use some products without their partners' knowledge. It is estimated that microbicides have the potential to prevent about 2.5 million HIV infections within 3 years, if they worked only 60 per cent of the time and used by only 20 per cent of women in 73 low-income countries. In addition, microbicides may prevent some other STIs besides HIV. However, success of microbicides depends on people remembering to use them correctly and consistently during each act of coitus. Microbicides can be incorporated in gels, foams, creams, thin films, or vaginal pessaries. The possible mechanisms of action include immobilising the virus, creating a barrier between the virus and epithelial cells, and preventing HIV from reproducing and establishing infection after it has entered the body. So far, no anti-HIV microbicide has been approved. Microbicides closest to approval are Carraguard, cellulose sulphate gel (also being studied as a contraceptive), PRO 2000 Gel, BufferGel, and Savvy (Fact Sheet 157, 2006).

IMMUNE THERAPIES IN DEVELOPMENT

Concept of Immune Restoration

"Immune restoration" is repairing the damage done to the immune system. A healthy immune system has a full range of CD4 T-lymphocytes that can fight different pathogens, including opportunistic organisms. The first CD4 cells that HIV attacks are the ones that specifically fight HIV. It is believed that increases in CD4 cell counts after ARV therapy is indicative of immune restoration. But, the new CD4 cells are probably copies of the existing types of CD4 cells. If some types of CD4 cells were lost, they do not reappear immediately. This could leave some gaps in the body's immune defences. Immune restoration explores ways to fill these gaps (Fact Sheet 481, 2006).

When HIV enters the blood stream through any one of the routes of transmission, it is attracted by lymphocytes that have matured in thymus and bear CD4 receptors on their surface. It was believed that the thymus shrinks and stops maturation of lymphocytes to CD4 cells by the age of 20 years. Recent research reveals that thymus continues working, may be up to the age of 50 years. If the viral load is under control for a few years as a result of effective ARV therapy, the thymus might make new CD4 cells that could fill these gaps and restore the immune system. Older persons with HIV infection might need hormonal stimulation of thymus or thymus transplantation (Fact Sheet 481, 2006).

Approaches for Immune Restoration

Cell Expansion: The non-infected cells from the HIV-infected patient are multiplied in vitro and then infused back into the body (Fact Sheet 481, 2006).

Cell Transfer: Transfusing immune cells from patient's HIV-negative twin or relative (Fact Sheet 481, 2006).

Cytokines: Cytokines are the body's chemical messengers that support the immune response (Fact Sheet 481, 2006). Some immunomodulators use cytokines to increase the strength of the immune response to HIV (Fact Sheet 402, 2006). Cytokines under trials include – Ampligen, a form of interferon; Interleukin-2 (IL-2, Aldesleukin, Proleukin); Multikine, a mixture of several cytokines; and Bay 50–4798, a modified recombinant form of IL-2. Interleukin-7 is being developed as a general immune system booster. Tumour necrosis factor-alpha is an immune system protein that is overproduced in immune disorders. A TNF-alpha blocker (that blocks this protein) is under trial (Fact Sheet 480, 2006).

Gene Therapy: Gene therapy attempts to make the bone marrow cells immune to HIV infection. These genetically changed cells would then travel to the thymus and mature to form CD4 cells (Fact Sheet 481, 2006). Gene therapies under study include – M 870 that makes T-lymphocytes resist infection by HIV; genetically modified CD4 and CD8 cells that block attachment of HIV; RRz2, a ribozyme that attacks *tat* gene of HIV; and VRX 496, a genetic factor that infects T-lymphocytes and attacks genetic code of HIV (Fact Sheet 480, 2006). Also being studied are Mifepristone (VGX410, RU486) that interferes with the viral protein vpr; and BI-201, an antibody designed to block HIV's *tat* gene (Fact Sheet 470, 2006).

Immunotherapy: HRG 214, a genetically engineered group of antibodies to HIV, is being studied for use as a passive immunotherapeutic agent. Dermavir, a novel immunotherapeutic agent that can be applied to the skin, is under study (Fact Sheet 481, 2006).

Other Immune Modulators: Immunitin (HE 2000), an immune regulating hormone has showed promising results in strengthening the humoral immune response. AVR 118 has showed promising results against AIDS-related wasting and anorexia. TH 9507, a growth hormone inducer, is being studied for treatment of visceral fat accumulation in lipodystrophy. Murabitide, that uses fragments of bacteria to stimulate the overall immune response, is being studied by Dr. Georges Bahr in France. Reservatrol is a chemical found in several plants and skin of red grapes. It protects plants against pathogens. It is being studied for immune boosting properties in HIV-infected persons. Reticulase, a nucleic acid, has shown increases in CD4 and CD8 cell counts, fewer opportunistic infections and weight gain in placebo-controlled studies. No toxic side effects have been reported so far. Reticulase is administered by subcutaneous injection. Zenapax (Dachzumab, anti-CD25) is being studied by National Institutes of Health, USA to reduce the viral load beyond what ARV therapy can achieve (Fact Sheet 480, 2006).

Dehydroepiandrosterone (DHEA): This is the steroid hormone produced by the adrenal glands. DHEA can be transformed in the body to testosterone, oestrogen, or other steroids. In normal adults, levels of the hormone decrease steadily after peaking at the age of about 20 years. HIV patients with lipodystrophy have very low levels of DHEA. A clinical trial is studying the effects of DHEA supplementation in HIV-infected individuals (Fact Sheet 724, 2006). Technically DHEA is not part of ARV treatment.

CHAPTER 17

TRADITIONAL ETHNOMEDICINAL SYSTEMS AND ALTERNATIVE THERAPIES

Abstract

Alternative therapies are tried by many HIV-infected persons on the assumption that these may enhance immune function, prevent, or treat various conditions. Alternative therapies usually presuppose that each individual possess an innate healing capacity and aim to restore strength and balance to the weakened system by using different modalities. Formal systems of traditional medicine are regulated by health authorities in many countries. But, unconventional non-formal therapies are not subjected to controls and patients may be vulnerable to exploitation by unscrupulous practitioners. Patients may not disclose their use of alternative therapies since they expect their doctors to taunt them for using such treatments. This leads to loss of opportunity to monitor benefits, side effects, and drug interactions. Health providers should acquire basic knowledge of alternative therapies, set aside prejudices, and facilitate open discussion since ridiculing these treatments will not prevent patients from trying them, but may stop patients from informing them. Combination of modern medicine with traditional or alternative therapies need not be discouraged unless they are expensive or interfere with prescribed treatment.

Key Words

Acupuncture, Acupressure, Alternative medicine, *Ayurveda*, Complementary medicine, Herbal medicine, Holistic treatments, Homoeopathy, *Jamu*, *Koryo*, Moxibustion, Serotonin syndrome, *Siddha* medicine, Traditional Chinese medicine, *Unani-Tibb*, *Yoga*.

17.1 – INTRODUCTION

Many different forms of healing have emerged in different parts of the world. Though the term "Western medicine" is applied to allopathy, which emerged in Western Europe during the 19th century, different medical systems, such as chiropractic, homoeopathy, and naturopathy have emerged in Europe and the United States (Tan, 2000a).

According to allopathic concepts, infectious diseases can be treated with medicines that act as "bullets", targeting the infectious organisms and killing them. Traditional ethnomedical systems are very complicated, often overlapping with religion and philosophy, and embedded in people's daily activities. The basic concepts of traditional ethnomedical systems are very different from those of Western medicine (allopathy) or Western naturopathy. Often, the human body is considered to be a microcosm of nature, with attributes similar to those of our environment. Individuals may be classified as "hot" or "cold" in terms of body type or personality. Illnesses are considered to be consequence of imbalances and healing techniques are means to restore the original balance. There is much overlap among traditional medical systems. Massage and use of natural products (derived from plants, animals, and minerals) is found in these systems. Dietary restrictions and fasting are also employed with the aim of "detoxifying" the body (Tan, 2000a).

HIV/AIDS care continues to be dominated by allopathic medicine, which centres on the germ theory. For people living with HIV/AIDS, traditional medical systems offer a bewildering variety of products and healing techniques that can be immediately tapped for treating symptomatic ailments such as pain, diarrhoea, nausea, and cough. For example, use of ginger is a remedy for nausea. Nausea can also be controlled by applying pressure on a Chinese acupuncture point called *nei-guan* on the wrist. These medical systems are potentially useful in strengthening the immune system and dealing with stress. Very little is being done about tapping traditional medicine for HIV/AIDS. It is unfortunate that many of those involved in HIV/AIDS care are sceptic about the role of traditional medicine. Such attitudes may lead to complete rejection of alternative forms of health care (Tan, 2000a).

Prior to the advent of the HIV epidemic, few people were familiar with the immune system. Now, the general public is curious about methods of bolstering immune system. Many HIV-infected persons use other systems of medicine or try alternative therapies in the belief that these may enhance immune function, prevent or treat various conditions. Alternative medicine has been variously called "natural", "complementary", "holistic", and by numerous other appellations, which refer to particular modality or tradition (www.lifepositive.com). These therapies may be considered "complementary" when used with conventional medicine, and "alternative" when taken instead of conventional medicine (McKnight & Scott, 1997). Alternative therapies usually assume that each individual possess an innate healing capacity and the objective is to restore strength and balance to weakened system by using a variety of modalities, such as detoxification, foods, and herbs, which are tailor-made for the individual's constitution and condition (www.lifepositive.com). The results of clinical trials in Western allopathic medicine reveal that simple lifestyle changes related to food and exercise can boost the immune system. Such discoveries are not new in traditional medical systems (Tan, 2000a).

17.2 – TYPES OF "ALTERNATIVE" THERAPIES

Formal Systems: These include *Ayurveda, Siddha, Unani-Tibb*, traditional Chinese medicine, and Homoeopathy. Traditional ethnomedical systems are "holistic" by nature – they aim to treat the whole individual rather than a specific disease or symptom. They address not only the physical aspect of the patient but also the mind and the spirit (www.lifepositive.com). *Jamu*, the traditional system of medicine practiced for centuries in Indonesia, is being developed by the Government of Indonesia into a safe and effective option (Chaudhury & Rafei, 2003). *Koryo* is the traditional system of medicine in People's Republic of Korea, where health care is provided *free of cost*. *Koryo* is combined with modern medicine in University medical curricula (Chaudhury & Rafei, 2003).

Holistic Treatments: These rely on the individual's interaction with the environment. Holistic treatments include yoga, herbal therapies, aromatherapy, reiki, and meditation.

Combination Therapy: Many practitioners have developed protocols for HIV-infected patients, using a mixture of herbal medicines, homoeopathy, acupuncture, nutrition, exercise, and massage (McKnight & Scott, 1997). Combination of allopathic drugs and traditional Chinese medicine are a common practice in China. However, the potential for drug–herb interactions remains to be investigated (Li *et al.*, 2003).

17.3 – INTRICACIES

Frequency of Use: A survey in South Australia found that 52 per cent of adults had used at least one complementary medicine in the previous year, and 57 per cent had not told their doctor about their use of these products (ADRAC, 2005). Considering the frequency of use of other systems of medicine, it is necessary to conduct formal clinical investigations of these therapies with an open view (Golleridge & Riley, 1996). Though the medical profession has not ratified their use, formal systems of traditional ethnomedicine have enjoyed centuries of acceptance and credibility in the countries of their origin (McKnight & Scott, 1997).

Scientific Studies: Though a review of clinical studies on antimicrobial effects of several "natural" therapies has been published (Golleridge & Riley, 1996), very little information is available on interaction of other alternative treatments with modern orthodox medicine used in HIV-infected persons (McKnight & Scott, 1997). Some alternative therapies treat the whole person, not an illness. Moreover, the treatment is not standardised for a particular disease or condition. For a particular disease (as diagnosed by allopathic medicine), the prescribed treatment in traditional medicine would be based on the concepts of "imbalance in humours", "disturbances in energy flow", and "temperament" of the patients. It is difficult to evaluate the mechanism of action of traditional medicines because their classification systems and concepts, such as "hot" and "cold" have no equivalents in Western allopathic medicine (Tan, 2000a). Practitioners of traditional medicines usually dispense extracts that may contain multiple active ingredients and certain other agents that modify or alter action of the active ingredient or reduce its adverse effects. In clinical trials of these medicines, purified active ingredients may be administered by a different route and therefore results obtained may not be comparable. Since most alternative therapies are inexpensive and cannot be patented, it is difficult to find a sponsor to pay for the clinical trials (Fact Sheet 700, 2006).

Controls: Formal systems of traditional medicine, such as *Ayurveda, Unani-Tibb*, Chinese medicine, and Homoeopathy are open to scrutiny by the health authorities and their respective Councils. On the other hand, unconventional non-formal therapies such as aromatherapy, reiki, and meditation are not subjected to controls and patients may be vulnerable to exploitation by unscrupulous practitioners, who may have a vested interest in sustaining the credibility of their occupation. The popularity of these non-formal therapies can change rapidly (McKnight & Scott, 1997). Unregulated complementary medicines, including those obtained through the Internet may be contaminated with other drugs such as steroids, or with toxic heavy metals such as lead, mercury, or arsenic (ADRAC, 2005).

17.4 – ROLE OF THE PHYSICIAN

Many HIV-infected patients are unwilling to disclose their use of alternative therapies since they expect their doctors to mock or ridicule the use of such treatments. Thus, the opportunity to monitor benefits, side effects, and drug interactions is lost. Hence, the treating family physician or specialist should set aside such prejudices so that patients can freely disclose whether they use alternative therapies (McKnight & Scott, 1997). Basic knowledge of these therapies is essential for providers since ignoring or rejecting these treatments will not prevent patients from trying them, but may stop patients from informing their providers. Open discussion about realistic treatment and choices should be promoted. Combination of modern medicine with alternative therapies need not be discouraged unless they are expensive or interfere with prescribed treatment (Scott & Irvine, 1997).

17.4.1 – Dos and Don'ts for Health Care Providers

- Elicit history of use of alternative medicines from patients.
- Collect information about the therapy from persons administering the therapy and from persons who have undergone the treatment recently and in the past.
- Enquire about merits, demerits, side effects, costs, benefits experienced, and duration of time for benefits to accrue.
- Inform patients that currently, health insurance does not reimburse the cost of alternative therapy.
- Enquire about formal training and expertise of alternative medicine practitioner (www.lifepositive.com).

17.5 – *AYURVEDA*

Ayurveda (Sanskrit: *ayu* = life; *veda* = knowledge) originates from the Vedic times. It is believed that this system of medicine is being practised since 12th century BC (Chaudhury & Rafei, 2003). *Atharvaveda* has 114 hymns on treatment of diseases (Bannerman *et al.*, 1983). *Ayurveda* is based on the theory of three humours (*tridosha*) theory (Fact sheet 702, 2006). The *doshas* (humours) are *vata* (wind), *pitta* (gall), and *kapha* (mucus). Balance between these three *doshas* meant good health, while disharmony or imbalance between them resulted in disease. About 8,000 recipes have been described in Ayurvedic pharmacy (www.lifepositive.com).

A disease with symptoms identical to that of AIDS has been termed *Rajayakshma* (the king of diseases) and has been described by the Ayurvedic physician Vaghbhata in *Ashtanga Hridayam* (*Chikitsitam* section), and its supplementary text *Ashtanga Sangraham*. The ancient Indian physician Charaka's treatise "Charaka Samhita" ("Nidanam" section) also describes this disease. The symptoms of *Rajayakshma* are drastic loss of body weight, fatigue and lethargy, vulnerability to allergies and contagious diseases, skin irritations, bronchial disorders often leading to tuberculosis of the lungs, damage to intestinal flora resulting in diarrhoea and dysentery, and wide fluctuations in body temperature. *Rajayakshma* is thought to be caused by anal intercourse, indiscriminate sexual intercourse with multiple partners, not cleaning the genitals after coitus, washing the body with contaminated or dirty water, bestiality, and contaminated blood. It is debatable as to whether *Rajayakshma* and AIDS are one and the same disease. Some practitioners believe that the treatment used for *Rajayakshma* can be used effectively against HIV infection (www.lifepositive.com).

The treatment is started with *Rasayanams* (tonics and rejuvenators) to boost immunity. Following this, select medications are administered to counter the virus. In order to stimulate appetite and strengthen the system, *Ajamamsa Rasayanam*, a preparation containing extract of goat's meat, cow's milk, and clarified butter (or *ghee*) and *Indukantham Ghritham* are given. Later, *Rasasindoor* (purified mercury preparation) is applied along with other medications (www.lifepositive.com).

If the patient shows signs of recovery after 6 months of this regimen, "shodana" (elimination) techniques, such as enemas, purgation, and emesis are used to expel the toxins from the body. As the general immunity increases, the blood is purified using cooling medications. After blood purification, a non-vegetarian diet along with *ghee* (clarified butter) preparations and soups is advocated. Acidic, oily, and spicy foods are to be avoided. A small quantity of alcohol is advised to assist the process of digestion and to remove blockages in the blood vessels (www.lifepositive.com).

After bathing the patient is bathed twice or thrice a day with cold water, sandal wood paste is applied on the body. The baths are said to cool the body and blood by penetrating the follicles. If the patient has weakness, steaming "swedanam" is advised. The results of the treatment depend on the severity of the case. It is believed that HIV may continue to be present in the body, but is unable to cause further damage due to bolstered immunity.

17.6 – *SIDDHA* MEDICINE

The principles and doctrines are similar to that of *Ayurveda*. The Government of India officially recognizes this system of medicine (Bannerman *et al.*, 1983). *Siddha* system of medicine originated in ancient Tamil Nadu, thousands of years ago. Tamil *Siddhars* were 18 enlightened men and women who documented the causes of 4,448 different diseases and prescribed herbs, roots, salts, metals, and mineral compounds in the form of *churnam* (powdered formulation), *chenthuram*, and *leyham* (thick batter-like formulation). *Siddha* medicines have total rejuvenating effect on the body and are not only effective against a particular disorder. Nearly all preparations are compound formulations. Due to their synergistic action, toxicity is decreased and bioavailability is increased. The pharmacodynamics of this system is distinctly different from that of other systems of medicine (www.lifepositive.com).

AIDS was already known to *Siddha* system of medicine and was called *Vettai Noi*. *Vetta Noi* was caused by defect in the three humors (*tridoshas*), and primarily due to *Azhai Kurtrum* (*Pittam* or bile, acidic nature), exhibited in the blood stream. *Vetta Noi* was subclassified into 21 types, most of which were caused by depletion of *prana* and/or *ojas* through excessive indulgence and abuse of the body, rendering the immune system weak and vulnerable to pathogens. Drugs that could be used for HIV-infected patients include: (a) herbal preparations: *Serankottai Nei, Mahavallathy Lehyam, Parangi Rasayanam*; (b) herbo-mineral preparations: *Gandhak Parpam, Gandhaka Rasayanam*; (c) herbo-mercuric preparations: *Idivallathy Mezhugu, Poona* Chandrodayam; and (d) herbo-mercuric-arsenical preparations – *Rasagandhi Mezhugu, NandhiMezhugu, Sivandar Amritham, Kshayakulanthan Chenduram* (www.lifepositive.com).

Three formulations *(rasagandhi mezhugu, amukkara churnam,* and *nellikai lehyam)* have been found to be effective for HIV-infected patients who do not have overt neural involvement. Since 1992, all three formulations have been tested on over 35,000 HIV-infected patients at the Government Hospital of Thoracic Medicine, Tambaram Sanatorium, Chennai, India. They are apparently without side effects. They are said to reduce viral load, improve CD4 and CD8 cell counts, control symptoms and increase body weight. So far, these drugs are unable to cure HIV infection, although prolonged viral suppression has occurred in a few patients (www.lifepositive.com). The following *Siddha* medicines are used in supportive therapy to control opportunistic infections: (a) for purifying blood: *Kanthaga Rasanayam, Paranki Pattai Churnam, Palakaria Parpam*; (b) for reducing fever: *Linga Chenduram, Gowri Chintamani, Thirikadugu Churnam, Rama Banam, Vadha, Piththa, Kaba Sura Kudineer*; (c) for persistent diarrhoea: *Thair Sundi Churnam, Kavika Churnam, Amaiodu Parpam*; (d) immune system revitalizers and rejuvenators: *Orilai Thamarai Karpam, Serankottai Eagam, Thertan Kottai Lehyam, Amukkara* Churnam; (e) antiviral agents: *Rasagandhi Mezhugu, Murukkanvithutu, Masikai, Edi Vallathathy Mezhugu*; and (f) for restoring the disturbed mind: *Vallarai* (www.lifepositive.com).

17.7 – *UNANI-TIBB*

This medical system is of Greco-Arab origin (www.lifepositive.com). It was introduced in India by Muslim rulers around 10th century AD and enjoyed State patronage till the advent of British rule. Prescriptions, usually written in Persian, begin with the legend *Howash Shafi* (meaning, "God is the Healer"). Governments of India and Pakistan officially recognize *Unani-Tibb* medicine (Bannerman *et al.*, 1983). Arab physicians who contributed to development of modern medicine were Zakaria Al Razi, or Rhazes (850–923 AD) and Abu Ali ibn Sana, or Avicenna (980–1037 AD) who wrote *Al-Qannan-fil-Tibb* (Canon of Medicine) in five volumes. It is believed that all life forms have originated from the sea. *Unani* medicine believes that using clean and fresh water, breathing clean air and consuming fresh food can prevent diseases. A balance between the mind and body helps facilitate metabolic processes and to evacuate body wastes easily. The human body is believed to have seven components (called *Umoor-e-Tabaiyah*), which are responsible for maintaining health (www.lifepositive.com) – *arkan* (elements), *mizaj* (temperament), *akhlaat* (the four humours), *aaza* (organs), *arwah* (vital forces), *quwa* (faculties), and *afaal* (functions). The loss of any one of these basic components or alteration in their physical state could lead to disease or death. Each individual is believed to have a special hidden defence mechanism called *Tabiyat-e-Muddabare*. *Tabiyat* is the sum total of structural function and psychological character of the human being (www.lifepositive.com). Temperament (*mizaj*) has an important role in treatment (Bannerman *et al.*, 1983).

   Unani medicine is distinct from other systems of medicine as it uses medications that are natural in their sources and forms. The emphasis is on retaining the natural compounds that belong to the human body and hence, only natural remedies are prescribed. Herbs are used in various forms like food, medicine, cosmetics or perfumes. There are no reports about *Unani's* efforts in developing a cure for HIV infection. *Unani* medical researchers in Florida (USA) are conducting trials of *Nigella sativa*, an herb which is said to be effective in boosting function of the immune system (www.lifepositive.com).

17.8 – CHINESE MEDICINE

Chinese medicine is probably the world's first organized body of medical knowledge that dates back to 2700 BC. Traditional Chinese medicine involves the use of 6,000–8,000 substances derived from plants, animal parts and minerals. The first work on traditional Chinese medicine is Shen Nong's Herbal (2nd century AD), which describes 365 substances – 252 botanicals, 67 zoologicals, and 46 minerals. The first official Pharmacopoeia – *Xin Xiu Ben Cao* (Newly Revised Materia Medica) – was published by the Chinese Government in 659 AD (Chaudhury & Rafei, 2003).

   Traditional Chinese medicine has developed from knowledge accumulated through clinical observations and treatment over several millennia. This system of medicine has grown rapidly in Western countries where concerns have been raised about its quality and safety. Many preclinical and clinical studies carried out in China have been published in Chinese literature but the results are not readily available to Western communities. The *Pharmacopoeia of the People's Republic of China*, published by the State Pharmacopoeia Commission, Beijing, contains standards of materia medica and patent preparations. The quality assurance process comprises the correct identification of a plant with reference to its scientific (botanical) name and ensuring quality in cultivating, harvesting, processing, and manufacturing practices (Li *et al.*, 2003).

Chinese medicine is based on a holistic view of supporting the innate ability of the mind and body to maintain health and to heal itself should illness occur. The system is also intertwined with a philosophy of life and is the result of thousands of years of accumulated experience. The concept of *Qi* (pronounced "chee") is central to the philosophy of Chinese Medicine. *Qi* is the vital energy of the universe, of which all things are made. *Qi* is believed to vitalize the body by moving through pathways called "meridians". The "meridian theory" of Chinese medicine is not accepted in Western medicine, because they have not been identified anatomically. Points along the meridians have been used successfully as acupuncture sites for thousands of years. This provides circumstantial evidence for existence of meridians. *Qi* patterns fluctuate between the polarities of *yang* (active masculine principle) and *yin* (negative feminine principle). Balance between these two opposing forces meant good health. Illness can be viewed as excesses or deficiencies in either the *yang* or *yin* components of *Qi* (www.lifepositive.com). Illness is not defined by symptoms or by the name of a disease, as in Western medicine (Fact sheet 704, 2006).

In the *Yellow Emperor's Classic of Medicine*, "yin", representing the sunny side of a hill is even and well while "yang" representing the shady side of a hill is firm. Together with the theory of yin and yang, the Theory of Five Elements (wood, fire, earth, metal, and water) constitutes the basis of traditional Chinese medicine. Each element checks the other, e.g. water checks fire. A typical prescription in traditional Chinese medicine comprises – principal herbs, associate herbs, assistants and envoys – often described as emperors, ministers, assistants and envoys, respectively (Chaudhury & Rafei, 2003). *Raw herbs* (the most potent form but difficult to prepare) are dispensed as dried pieces of root, bark, leaves, or seeds. Alcohol-based extracts of dried herbs are available as "tinctures". *Powdered herbs* are to be mixed with water and consumed or ingested in the form of a capsule. Patent medicines contain combinations of herbs and are available as tablets, capsules, or creams. Usually, these medicines are labelled in the Chinese language (Fact sheet 704, 2006).

The approach in Chinese medicine is to support the body's innate ability to heal itself, align the functions of organs as a whole, and promote balance of energy. It involves acupuncture, dietary advice, massage, and *QiGong* – a form of physical or psychosomatic exercise (Bannerman *et al.*, 1983; Chaudhury & Rafei, 2003).

Chinese medicine is non-toxic when used by licensed practitioner. Self-medication is not recommended (McKnight & Scott, 1997). Since the communist revolution of 1949, traditional medicine is fully integrated with modern ("Western") medicine and almost all hospitals of allopathic medicine in China have departments of traditional medicine (Bannerman *et al.*, 1983).

In San Francisco, the American College of Traditional Chinese Medicine has treated many symptomatic HIV-infected patients. Seven HIV-related conditions appear to be most responsive to Chinese medicine: weight loss, diarrhoea/loose motions, abdominal pain, nausea, headache, enlarged lymph nodes, and neuropathy. Some studies are being conducted to study efficacy of Chinese medicine in HIV-infected individuals (www.lifepositive.com).

17.8.1 – Acupuncture

Acupuncture (Latin: *acus* = pin; *puncture* = prick), moxibustion and acupressure are used in Chinese medicine. Energy channels containing vital force (*Qi*) are thought to permeate the human body and 360 acupuncture points have been identified (Bannerman *et al.*, 1983).

Acupuncture involves the insertion of extremely thin needles into the skin at specific points along the energy channels (meridians). The procedure is relatively painless. It is believed that the insertion of needles at specific points removes the blocks in flow of energy and symptoms can be relieved. Just as energy can be depleted or diminished, it can also be rechannelled and replenished (Fact sheet 703, 2006). Thus, the acupuncture needles may stimulate the body's own energy reserves or they may transmit energy from the environment into the body. The treatment is individualised because each individual has a unique character, and a unique interplay of energies, organs, and elements. It is important to find an acupuncturist who has the expertise in treating HIV-infected persons. The practitioner uses several traditional diagnostic methods to ascertain whether treatment should be directed at boosting or dispersing energy. Needles are then inserted at definite points along the suitable meridian. Acupuncture is used to bolster the immune system and to relieve some HIV-related symptoms like neuropathy, fatigue, and pain (www.lifepositive.com).

After the first four or five sittings, most HIV-infected patients begin to experience a decrease in abnormal sweating, diarrhoea, and skin rashes. Some have reported higher energy levels and many patients have gained weight appreciably. Somerville Acupuncture Centre (Boston), AIDS Alternative Health Project (Chicago), and Quan Yin Herbal Support Program (San Francisco) have reported similar symptomatic relief and overall improvement in HIV-infected persons. Though acupuncture may not cure HIV infection or increase CD4 cell counts, it increases endorphins and possibly reduces stress, pain, spasms in gastrointestinal conditions, drug-induced nausea, and some neurological problems. There is no claim that acupuncture has direct antiviral effect on HIV. However, the validity of acupuncture and traditional Chinese medicine continues to be controversial in Western culture (www.lifepositive.com).

Acupressure: Acupressure involves the application of deep finger pressure at specific points along the energy channels (meridians). Reflexologists concentrate on the hands and feet (Tan, 2000a).

Moxibustion: Acupuncture is sometimes accompanied by the application of heat from burning herbs. This method may be useful for patients with chronic pain, nausea, and vomiting. This technique is being scientifically studied for treating patients with addictions and stroke (Tan, 2000a).

17.9 – HOMOEOPATHY

Dr. Samuel Hahnemann (1755–1843) of Germany developed this system of medicine, which is based on the principle: *Similia similibus curentur* (Latin for "like cures like"). This principle is similar to that of Newton's third law of motion (Bannerman *et al.*, 1983). Homoeopathy is a low-cost system that uses relatively non-toxic drugs. It is based on law of direction of cure, law of single remedy, and law of minimum dose. A substance, which causes a specific set of symptoms in a healthy person, will cure similar symptoms in a sick person, when the substance is administered in infinitesimally small doses (Bannerman *et al.*, 1983). Homoeopathic medicines include natural substances like minerals, vitamins, and animal products (www.lifepositive.com). Homoeopathy gained foothold in India in early 19th century. India has the largest number of homoeopathic practitioners in the world (Bannerman *et al.*, 1983). India and Mexico have officially recognized homoeopathic medical colleges.

Patients with HIV infection may have unusual symptoms and syndromes that cannot be diagnosed immediately. However, a homoeopath can prescribe a remedy, based on the symptoms, before a definitive diagnosis is made using conventional methods. Since curative remedies are prescribed on the basis of a patient's unique pattern of symptoms, a conventional diagnosis is not necessary. Homoeopathic medicines provide promising results for HIV-infected persons and those with early onset of AIDS. Significant improvement has not observed in persons with advanced AIDS, though there have been exceptions. Research studies suggesting immune modulatory effects of homoeopathic medicines and beneficial response seen in HIV-infected persons in clinical studies has not drawn the attention of physicians and public health authorities. Homoeopathy is still ignored and is not yet considered as a viable component of a comprehensive programme for care of HIV-infected persons (www.lifepositive.com).

17.10 – HERBAL THERAPIES

Astralagus: Active substances from the root of the non-toxic plant *Astralagus membranaceous* are used in China for enhancing immune function and to prevent chemotherapy-related nausea and bone marrow suppression and were also used in the former USSR and Japan for treating myocardial infarction and strokes. This plant should not be mistaken for *A. lentignosus*, which is toxic (McKnight & Scott, 1997).

Bitter Lemon: Bitter lemon (*Momordica chantia*) is a climbing plant related to Chinese cucumber (*Tricosanthes kirilowii*). Extracts of fruit, seeds, and plant are used for inducing abortion and for treating diabetes, gastrointestinal disorders, some cancers, and viral infections. Extracted active proteins (MAP-30, alpha- and beta-momorcharins) are believed to inhibit HIV replication (McKnight & Scott, 1997).

Cannabis: *Cannabis indica* (also known as *C. sativa* or Indian hemp) grows in many parts of the world and all parts of the plant are poisonous. The active principle is not an alkaloid but oleoresins called cannabinols, the best known being tetrahydrocannabinol. Though not legal, cannabis is widely used in four forms in India: (a) *bhang* (also known as *siddhi, patti,* or *sabji*) consists of dried leaves and fruits shoots that are used as an infusion and is the least potent of all cannabis preparations; (b) *majun*: a confection made with *bhang*, flour, milk, and butter; (c) ganja: consists of dried flowering tops of the female plant, specially growing so that there is a large amount of resinous exudate and as little leaf as possible. The rusty green coloured resin is smoked with tobacco in a clay pipe chiefly by *sadhus, fakirs,* and temple priests to get into a religious mood; (d) *charas* (known as "hashish" in Egypt): is the resinous exudate from the leaves and stems of the plant, which is the most potent of all cannabis preparations. It is smoked with tobacco in a pipe or "hookah" (similar to hubble-bubble). In America, cannabis is called "marihuana" or "marijuana" and is also known by various slang words such as pot, weed, grass, tea, Mary, and Jane. "Marihuana" is a Mexican word meaning "pleasurable feeling". The plant is chiefly grown in Mexico. In America and many parts of the western world, marijuana is chiefly used in a form similar to *ganja* in India. Marijuana cigarettes are known as "reefers" or "weed". Marijuana is eaten alone, or as part of confection, or consumed with beverages such as beer. Many scientific studies have documented marijuana's ability to (a) treat glaucoma by reducing intraocular tension, (b) increase appetite, (c) reduce nausea and vomiting in patients receiving cancer chemotherapy, (d) reduce muscle spasms in patients with multiple sclerosis, and (e) help in relieving pain in peripheral neuropathy. This led to the development of the drug dronabinol, a synthetic version of tetrahydrocannabinol. Many patients with HIV infection have low appetite, nausea, and peripheral neuropathy. In May 2001, the US Supreme Court ruled that medicinal use of marijuana is illegal under federal law. Eleven states in the USA have passed "medical marijuana" laws that permit limited use for health reasons. In 2005, the US Supreme Court ruled that federal officials can take action against users of medical marijuana or "buyers' clubs" even in states with medical marijuana laws (Fact Sheet 731, 2006).

Cat's Claw: Cat's claw (*U. tomentosa*) is a vine that grows in Peru, South America. The plant has pairs of long curved thorns that grow along the vine. Its Spanish name is *Uña de Gato*, meaning "cat's claw". The inner bark and root of the vine are used to make tea. Since many centuries, Peruvian natives have used cat's claw to treat inflammatory conditions such as arthritis, dysentery, and women's hormone imbalances. *U. tomentosa* is not to be confused with *Acacia gregii*, a plant that grows in northern Mexico and southern Texas and is also known as "cat's claw". This plant has no known health benefits and its bark may be poisonous. Though Klaus Keplinger patented oxindole alkaloids present in *U. tomentosa*, the herb or its purified extracts have not been approved by government agencies. Studies sponsored by manufacturers of purified versions of the herb have reported rapid healing in genital herpes and herpes zoster. However, studies on humans, including people with HIV/AIDS by independent researchers have produced inconclusive results (Fact Sheet 722, 2006).

Chinese Cucumber: Compound Q (GLQ223) is derived from Chinese cucumber (*T. kirilowii*). It is used in China to induce abortions and to treat cancers. Clinical trials have not demonstrated beneficial effects in HIV-infected persons (McKnight & Scott, 1997; Sepkowitz, 2001).

Echinacea: *Echinacea*, a flowering plant, also called "purple coneflower", grows mainly in Europe and North America. Red Indians (Native Americans) living in the Great Plains region have used leaves and roots of *Echinacea augustifolia* or *E. purpurea* for treating a wide range of pains and illnesses. In vitro and animal studies have reported broad immunostimulatory effects. Closely related species (*purpurea, augustifolia,* and *pallida*) of the plant have different medicinal properties. *E. purpurea* and *E. augustifolia* appear to be more effective in vitro and in vivo, respectively. The German Government has approved the use of *E. pallida* root and *E. prupurea* leaf for short-term treatment (not more than 1–2 weeks) of colds, influenza, and chronic respiratory, or urinary infections. It has also been used for skin wounds, psoriasis, and eczema. Echinacea is available in the form of capsules (containing a powder of the dried plant or root) or tincture (alcohol-based extract).The dosage depends on the species and part of the plant used. Use of Echinacea in HIV-infected persons is controversial. An animal study has reported that Echinacea increased levels of tumour necrosis factor-alpha, which has been linked to progression of HIV disease. Though Echinacea is believed to increase immune function, the German Government has advised against use of this herb in persons with diseases of the immune system such as HIV, tuberculosis, or multiple sclerosis. These warnings are based on laboratory studies in animals. So far, there is no published research on people with HIV (Fact Sheet 726, 2006).

Essiac: In 1922, Rene Caisse, a Canadian nurse, developed a herbal tea based on a formula from a Red Indian (Ojibwa tribe) medicine man and named it "Essiac" (her last name spelt backwards). The original ingredients are:
(a) Burdock root (*Articum lappa*) that is used in ethnic medicine as an aid to digestion, diuretic, and laxative
(b) Indian or Chinese rhubarb (*Rheum palmatum*), a strong laxative
(c) Sheep sorrel (*Rumex acetosella*), claimed to be effective against cancer
(d) The inner bark of slippery elm (*Ulmus fulva* or *U. rubra*), safely used for many centuries for alleviating sore throats.
Some versions have additional ingredients (commonly watercress and *Pau d'Arco*: bark from a South American tree) that supposedly improve the product's

efficacy and taste. Yellow dock or curly dock may be substituted for sheep's sorrel. According to advocates of Essiac, the only correct method is to consume freshly brewed tea one to three times a day on an empty stomach. They do not advocate use of tea bags, pills or capsules containing the ingredients. Proponents of Essiac claim that the preparation shrinks tumours, prolongs the life of cancer patients, strengthens the immune system, improves appetite, relieves pain, and improves the overall quality of life. Rene Caisse probably used different variations of the herbal tea to treat different types of cancer. In 1938, the Canadian legislature enacted a law that required her to disclose the formula for Essiac, which she refused. In 1977, she sold the formula to Resperin Corporation of Canada. In 1982, the Canadian Government authorised availability of Essiac for "compassionate use" partly based on lack of reports on adverse effects in 78 patients who used it between 1978 and 1982, but it does not consider Essiac to be an effective anticancer treatment. Dr. Gary Glum, a chiropractor, has claimed beneficial effects for HIV-infected persons but there are no research publications supporting his claim (Fact Sheet 727, 2006).

Garlic: Allicin, a sulfur-containing compound present in garlic (*Allium sativum*), is responsible for its suggested medicinal properties (Golleridge & Riley, 1996). Garlic also has antibacterial (Foster, 1990) and antifungal (Johnson & Vaughn, 1969) properties and may help in preventing oral candidiasis. However, garlic may interfere with protease inhibitors (Fact Sheet 501, 2006).

Shiitake Mushrooms (*Lentinus edodes*): These are used for making *L. edodes* mycelia (LEM). An in vitro study has demonstrated that ethanol-precipitated LEM prevents interaction of HIV with CD4 receptors (Carson & Riley, 1994; Tochikura *et al.*, 1988).

Silymarin (*Silybum marianum*): Silymarin is the name given to the extract from the seeds of the plant *Silybum marianum*, also called "milk thistle". Flavonoids (silybin, silydianin, and silychristin) are the active ingredients. Milk thistle has been used for over 2,000 years and during the Middle Ages, the seed of the milk thistle was used to treat liver diseases. Its antioxidant action protects the liver, promotes the growth of new hepatocytes, and aids fat digestion. Silymarin can help prevent or reverse liver damage caused by alcohol, recreational drugs, pesticides, poisoning by certain types of mushrooms, or hepatitis. There is no evidence that silymarin has ARV action but it can be used to prevent or treat liver damage and indigestion caused by several ARV drugs. A standardised extract of the seeds of the milk thistle plant (silymarin) should contain 80 per cent of silymarin. The shelf life is only about 3 months. The usual dose is 300–600 mg daily. Milk thistle does not dissolve easily in water. Some people get an upset stomach, diarrhoea, bloating, or flatulence when they start using silymarin. In such cases, the dosage is reduced and then progressively increased. Few people may develop allergic reactions. There are no clearly documented serious adverse effects (even in high doses) and interactions with ARV drugs. Most scientific studies on silymarin have been published in Europe (Fact Sheet 735, 2006).

St. John's Wort (*Hypericum perforatum*): This flowering plant grows in many parts of the world. "Wort" is an old English word for an herb or plant. All the above-ground parts of the plant are collected while the plant is flowering and has been used since many centuries to treat burns, bruises, mild depression, or anxiety. The herb contains many substances that work together and a major compound is *Hypericin*. Activity against cytomegalovirus, human papilloma, hepatitis B, and herpes viruses has been demonstrated in high doses in in vitro and animal studies. But there are no scientific studies to show that the herb can reduce HIV load in HIV-infected persons. The herb's antiviral effect is probably by oxidation and is stronger when exposed to light. In a 1991 study on persons infected with HIV, purified doses of hypericin were given intravenously in doses that were much higher than for treating depression. The study was stopped when every white-skinned participant in the trial developed skin rash and photosensitive reactions. A lone black-skinned participant did not develop photosensitivity. Many HIV-infected persons use St. John's Wort for mild to moderate depression and anxiety. The mechanism of action in depression is unclear. St. John's Wort is available as capsules (containing a powder of the dried plant) and as oil-based preparations for topical application. The optimum dosage is not known. Adverse affects (skin rash and photosensitivity) have been reported with purified extracts or at extremely high doses. It may have a negative effect on fertility in both males and females. Its safety in pregnancy is not established. St. John's Wort should *never* be combined with allopathic medications for treating depression (Fact Sheet 729, 2006). This herb may reduce plasma concentrations of a number of drugs including antidepressants, ARV drugs such as indinavir, cyclosporine, and oral contraceptives (ADRAC, 2005). It is also serotonergic (Hall, 2003; ADRAC, 2005). For the above-mentioned reasons, health care providers should be informed if St. John's Wort is being used (Fact Sheet 729, 2006).

Tea Tree Oil: Aboriginal people of Australia use oil of tea tree (*Melaleuca alternifolia*), which has antibacterial and antifungal properties and is non-toxic, when used in low concentrations.

17.11 – YOGA

Yoga is an ancient system based on Indian philosophy and it aims at helping the individual to balance the body's energy centres (called *chakras*). Yoga comprises breathing exercises, postures, stretching exercises, and meditations. It is gaining in popularity as an alternative therapy because of its adaptability and its physiological and psychological benefits. Yoga practitioners believe that yoga can promote detoxification, improve stamina, and mitigate chronic fatigue. Stress experienced by HIV-infected individuals produces biological changes that can worsen the damage to the immune system. Since it is difficult for a person under stress to practise yoga, a group of like-minded HIV-infected persons attend yoga classes. *Yoga nidra* is a state of psychic sleep, which is reached by systematically inducing complete physical, mental, and emotional relaxation. A group of four yoga poses – headstand (*Sirsasana*), shoulder stand (*Sarvangasana*), bridge pose (*Setu Banda Sarvangasana*), and plough pose (*Halasana*) – is believed to promote strength and flexibility, relieve pressure on abdominal organs, and increase blood circulation. Certain poses assist in relieving fatigue, diarrhoea, anxiety, and depression. Some poses require a bench or chair for support. Head, back, or buttocks may be supported using blankets, pillows or bolsters. In the Iyengar system of yoga, the sequencing of postures is important (www.lifepositive.com).

17.11.1 – Primary Sequence of Yoga Poses

The *primary sequence* of yoga poses for HIV-infected persons are – *ardha mukha vakrasana* (headstand), *pinca mayurasana* (peacock pose – optional), *ardha mukha svanasana* ("dog pose" with head supported by a block), *sirsasana* (headstand – contraindicated for persons with neck problems), *viparita dandasana* (supported inverted staff pose), *setu banda sarvangasana* (supported bridge pose), *sukhasana* (cross leg pose, leaning forward, head supported), *salamba sarvangasana* (supported shoulder stand), *ardha halasana* (supported plough pose), *viparita karani* (legs rested on wall with buttocks supported), *supta badhda konasana* (supine angle pose with back and head supported), and *savasana* (relaxation pose with head and back supported) (www.lifepositive.com).

17.11.2 – Alternative Sequence of Yoga Poses

Individuals who cannot attain the full yoga pose may try the *alternative sequence* of poses. This comprises – *ardha mukha svanasana* ("dog pose" with head supported by a block), *janu sirsasana* (head/knee forward bend with head supported), *triang mukhaipada paschimottanasana* (three part forward bend with head supported), seated forward bend or *halasana* (using support), *ardha baddha padmma paschimottanasana* (half-bound lotus forward bend with head supported), *paschimottanasana* (stretch with head supported), *ardha salamba sarvangasana* (supported shoulder stand), *setu banda sarvangasana* (supported bridge pose), *viparita karani* (legs rested on wall with buttocks supported), *supta badhda konasana* (supine angle pose with back and head supported), and *savasana* (relaxation pose with head and back supported) (www.lifepositive.com).

17.11.3 – Benefits of Yoga

Though yoga cannot replace professional counselling, yoga techniques are known to help in reducing excessive fear, feeling of loneliness, depression, anxiety, and in learning stress-coping skills. Meditation helps in self-awareness and in building inner strength through relaxation (www.lifepositive.com).

17.12 – OTHER ALTERNATIVE THERAPIES

Aromatherapy: Essential oils extracted from flowers, herbs, or minerals are inhaled to relieve stress and pain (McKnight & Scott, 1997).

Chiropractic: Also called "osteopathic medicine" or "spinal manipulation", this system originated in the western countries and involves the manipulation of joints. It has been found to be effective for many types of pain (Tan, 2000a).

Coenzyme Q (CoQ10): This is found in high concentrations in mammalian heart, liver, kidney, and muscle. In persons with immune dysfunction, HIV infection, congestive cardiac failure, and muscular dystrophy, its levels are abnormally low. Toxicity has not been reported with oral administration of CoQ10 for inhibiting replication of HIV in infected cells (McKnight & Scott, 1997).

Dinitro Chlorobenzene (DNCB): This potentially hazardous chemical was first used as a treatment for warts (Happle, 1985), and is also used in colour photography. When applied to skin, DNCB causes delayed-type hypersensitivity. The first topical application of DNCB may produce a chemical burn in addition to the normal pruritic rash. Advocates of DNCB claim that it can clear up all manifestations of HIV disease except for *Pneumocystis* pneumonia. They believe that ARV drugs, long-term use of acupuncture, more than short-term use of most herbs, and high doses of vitamins reduce the action of DNCB. Though this chemical appears to restore cutaneous immune responses that are lost when HIV disease progresses, these skin responses may not be a good indicator of overall immune health. There is very little research (none recent) to support the benefits of DNCB and its long-term side effects are not known. Since DNCB is inexpensive and cannot be patented, it is difficult to find a sponsor to pay for the clinical trials (Fact Sheet 725, 2006).

Massage: Many traditional medical systems use a wide variety of massage techniques that involve systematic application of pressure to soft tissues of the body, often with the help of oils. These techniques promote circulation of blood and lymph and may thus help in the healing process. Massage therapies have been found to be useful in relieving stress-related ailments (Tan, 2000b).

Meditation: There are many forms of meditation that have similar principles and involve the deliberate suspension of the stream of consciousness. Studies have found meditation to be useful for lowering blood pressure and relieving chronic pain. Biofeedback involves use of monitoring equipment to measure degree of relaxation in a person (Tan, 2000b).

Mind–Body Techniques: Visualization is a process in which persons are guided to imagine themselves in a state of vibrant health, in the belief that through such mental images, the immune system may be directed to fight HIV. Guided imagery is a process where the subject is asked to focus mentally on selected images, such as relief from pain. Other mind–body techniques include hypnosis, humour therapy, biofeedback training, and listening to inspirational or relaxation-inducing audiotapes (www.lifepositive.com).

Music Therapy: Music therapists believe that music (preferably instrumental) should have a frequency of 70–80 beats per minute, corresponding to human heart beat and should have a low pitch. Music therapy has been used to help overcome anxiety, insomnia, and phobias and to relieve stress (Tan, 2000b).

Native American Healing: Most Red Indian (Native American) tribes have healing traditions based on the tribe's beliefs about how individuals fit in the web of life that includes the tribe, all humanity, the Earth, and the universe. Healing is believed to occur when the person is restored to harmony and connected to universal powers. Healing techniques do not follow written guidelines; they differ from tribe to tribe and from patient to patient. Healers use herbs, "sweat lodges", ceremonial smoking of tobacco, talking circles, shamans, animal spirits, or "vision quests" (Fact Sheet 708, 2006).

Reiki: It is believed that when the reiki master lays his or her hand on a subject, there is a natural flow of healing energy that percolates to sites in the body, where it is needed (McKnight & Scott, 1997).

Vitamins and Minerals: High-dose vitamin C (ascorbic acid) is most commonly used in the belief that it will enhance immune function (McKnight & Scott, 1997).

17.13 – SIDE EFFECTS AND INTERACTIONS

There is a widespread belief among the public that complementary medicines are safe, because they are "natural". Many of these products have not been subjected to evaluation of efficacy and adverse effects. Lack of systematic data, together with a misplaced perception of safety, and frequent non-disclosure of use to health care personnel may lead to occurrence of unrecognized side effects due to use of these products (ADRAC, 2005).

Side Effects: *Aristolochia* species (renal failure), bee products (anaphylaxis), black cohosh or *Cimicifuga racemosa* (reportedly causes liver failure, requiring its transplantation), and *Echinacea* species (allergic reactions). Caffeine content of guarana (*Paullina cupana*) may cause overdose (ADRAC, 2005).

Drug Interactions: A large number of substances including glucosamine and cranberry juice (*Vaccinium* species) increase the activity of warfarin, causing bleeding. Herbs having similar interaction with warfarin include garlic (*A. sativum*), Korean ginseng (*Panax ginseng*), and *Gingko biloba* (ADRAC, 2005). St. John's Wort (*H. perforatum*) may reduce plasma concentrations of a number of drugs and is also serotonergic. (Hall, 2003; ADRAC, 2005).

Serotonin Syndrome: Serotonin syndrome is a toxic state caused mainly by excess serotonin within the central nervous system. Most cases are self-limiting, but the condition can range in severity from mild to life-threatening and result in a variety of mental, autonomic and neuromuscular changes. This condition is not an idiosyncratic reaction but a dose-related range of toxic symptoms. Drugs implicated in severe serotonin syndrome include L-tryptophan (precursor of serotonin), tricyclic antidepressants, monoamine oxidase inhibitors (MAOIs), pethidine, tramadol, buspirone, amphetamines, anorectics, lithium, and LSD. If two serotonergic drugs are co-administered, the manifestations of the syndrome are severe. Management involves withdrawal of the serotonergic drugs, supportive care, and use of serotonin antagonists like cyproheptadine (Hall, 2003).

Grape Fruit Juice: Health care providers should be aware that there are several groups of drugs that interact with grape fruit juice (Bailey *et al.*, 1998). Bitter Seville oranges also interact with drugs (ADRAC, 2002). Grape fruit juice can inhibit the amount of drug available for absorption, which may increase its pharmacological or toxic effects. Significant interactions have been found to occur with some of the HMG-CoA reductase inhibitors (statins) and saquinavir, among many other drugs (McNeece, 2002). Grape fruit juice inhibits a cytochrome P450 enzyme (CYP3A4) in the intestinal cells, but not in the hepatocytes (ADRAC, 2002). Patients should be advised to avoid consuming grape fruit and its juice altogether when taking certain medicines because the interacting effect may persist for up to 3 days after ingestion (ADRAC, 2003).

17.14 – APPROACH TO ALTERNATIVE THERAPIES

17.14.1 – Guidelines for Patients

Many individuals harbour the notion that medicines that have been used for several centuries must be safe. However, many traditional remedies have not been evaluated for their side effects especially if used for prolonged periods. Traditional medicines are generally more affordable as compared to Western allopathic medicine. However, in recent times, many traditional and alternative medicines have become expensive because of the fad that has grown around natural medicines (Tan, 2000b). Even if the unproven treatments are affordable, patients must know that health insurance will not defray costs and that use of unproven treatments may delay use of proven treatments (Fact Sheet 206, 2006). The *PDR Family Guide to Natural Medicines and Healing Therapies* published by Ballantine Books, New York (www.randomhouse.com/BB/) contains information on many alternative therapies and herbal remedies along with a review of scientific literature for each of these therapies. The book has a "Western" perspective and may be useful for readers seeking scientific information. The Ayurvedic Institute (www.ayurveda.com/) established in 1984 in Albuquerque, New Mexico (USA) is one of the leading Ayurvedic schools and Ayurvedic health spa outside of India.

17.14.2 – Dos and Don'ts

When choosing to use traditional or alternative therapies, individuals must generally avoid medicines with multiple ingredients because they increase the risk of harmful interactions. Individuals must be wary of products and their promotional literature that carry:

(a) Testimonies and anecdotes, which are not adequate proof of a therapy's efficacy – placebos may work for some persons because of the element of faith or belief
(b) Claims of "approval" by government agencies
(c) Offers of instant cures for a wide range of diseases or claiming to be effective for incurable diseases, such as AIDS or cancers
(d) Claims of being free of side effects: medicines, or therapeutic techniques cannot be completely free of side effects particularly when they have to be used for prolonged periods
(e) Claims of being "ancient cures" or cures from a far-away land
(f) No mention of their ingredients – if ingredients are listed, it is better to check out their safety and effectiveness.

Claims of "detoxifying" or "purifying" the body (Tan, 2000b).

Products being promoted by discrediting competing products or other medical systems, promotional materials that use pseudoscientific language, and products being promoted as "completely natural" are better avoided. The latter ride on the "back to nature" fad and may be adulterated (Tan, 2000b). Asian, particularly Chinese and Indian herbal medicines have been found to be contaminated with toxic heavy metals and adulterated with undeclared allopathic drugs, such as steroids, antibiotics, and analgesics that could have long-term harmful effects (Ernst, 2002; Tan, 2000b). One should be suspicious if words such as "amazing breakthrough" or "miraculous cure" are used in promotional material. To avoid government regulations, the product may be available privately for a short period of time or only from one source (Fact Sheet 206, 2006).

17.14.3 – The Indian Approach

Practitioners of indigenous systems of medicine, such as *Ayurveda, Unani, Siddha* and Homeopathy (known by the acronym – AYUSH in India), provide the bulk of health care to rural Indians. Of the estimated 400,000 *Ayurvedic* physicians in India, about 90 per cent serve the rural areas (DGHS, 2004). Most of them are local residents and are close to the community socially and culturally. In addition, many of the *Ayurvedic* dispensaries are run by the State Governments. The Government of India has established a National Institute of *Ayurveda* in Jaipur and a National Institute of Homeopathy in Kolkata. A Central Council of Indian Medicine was established in 1971 to prescribe minimum standards of education in Indian Medicine.

REFERENCES

ADRAC, 2002, Interactions with grape fruit juice. Austr Adv Drug Reactions Bull 21(4): 2.
ADRAC, 2003, Interactions with grape fruit juice – Amendment. Austr Adv Drug Reactions Bull 22 (2): 4.
ADRAC, 2005, Adverse reactions to complementary medicines. Austr Adv Drug Reactions Bull 24(1): 2.

Bailey D.G., Malcolm J., Arnold O., and Spence D., 1998, Grape fruit juice interaction. Br J Clin Pharmacol 46: 101–110.

Bannerman R.H., Burton J., and Chieh C.W. (eds.), 1983, Traditional medicine and health care coverage. Geneva: World Health Organization.

Carson C.F. and Riley T.V., 1994, The antimicrobial activity of tea-tree oil. Med J Austr 160: 236.

Chaudhury R.R. and Rafei U.M. (eds.), 2003, Traditional medicine in Asia. New Delhi: WHO-SEARO.

Directorate General of Health Services (DGHS), 2004, Health statistics of India. New Delhi: Government of India.

Ernst E., 2002, Toxic heavy metals and undeclared drugs in Asian herbal medicines. Trends Pharmacol Sci 23: 136–139.

Foster S., 1990, Echinacea: nature's immune enhancer. Rochester, NY: Healing Arts Press.

Golleridge C.L. and Riley T.V., 1996, 'Natural' therapy for infectious diseases. Med Jour Austr 164: 94–95.

Hall M., 2003, Serotonin syndrome. Austr Prescr 26(3): 62–63.

Happle R., 1985, The potential hazards of dinitrochlorobenzene. Arch Dermatol 121: 330–332.

Johnson M.G. and Vaughn R.H., 1969, Death of *Salmonella typhimurium* and *Escherichia coli* in the presence of freshly reconstituted garlic and onion. Appl Microbiol 17: 903–905.

Li G.Q., Duke C.C., and Roufogalis B.D., 2003, The quality and safety of traditional Chinese medicine. Austr Prescr 26(6): 128–130.

McKnight I. and Scott M., 1997, HIV and complementary medicine. In: Managing HIV (G.J. Stewart, ed.). North Sydney: Australasian Medical Publishing.

McNeece J., 2002, Grape fruit juice interactions. Austr Prescr 25: 37.

New Mexico AIDS Education and Training Center, 2006, Fact Sheet 206. How to spot HIV/AIDS fraud. University of New Mexico Health Sciences Center. www.aidsinfonet.org. Revised 10 August.

New Mexico AIDS Education and Training Center, 2006, Fact Sheet 501. Candidiasis. University of New Mexico Health Sciences Center. www.aidsinfonet.org. Revised 16 March.

New Mexico AIDS Education and Training Center, 2006, Fact Sheet 700. Alternative and Complementary therapies. University of New Mexico Health Sciences Center. www.aidsinfonet.org. Revised 16 March.

New Mexico AIDS Education and Training Center, 2006, Fact Sheet 702. Ayurvedic medicine. University of New Mexico Health Sciences Center. www.aidsinfonet.org. Revised 5 September.

New Mexico AIDS Education and Training Center, 2006, Fact Sheet 703. Chinese acupuncture. University of New Mexico Health Sciences Center. www.aidsinfonet.org. Revised 23 May.

New Mexico AIDS Education and Training Center, 2006, Fact Sheet 704. Chinese herbalism. University of New Mexico Health Sciences Center. www.aidsinfonet.org. Revised 23 May.

New Mexico AIDS Education and Training Center, 2006, Fact Sheet 708. Native American traditional healing. University of New Mexico Health Sciences Center. www.aidsinfonet.org. Revised 23 May.

New Mexico AIDS Education and Training Center, 2006, Fact Sheet 722. Cat's claw (*Uña de Gato*). University of New Mexico Health Sciences Center. www.aidsinfonet.org. Revised 2 May.

New Mexico AIDS Education and Training Center, 2006, Fact Sheet 725. DNCB (Dinitrochlorobenzene). University of New Mexico Health Sciences Center. www.aidsinfonet.org. Revised 9 March.

New Mexico AIDS Education and Training Center, 2006, Fact Sheet 726. Echinacea. University of New Mexico Health Sciences Center. www.aidsinfonet.org. Revised 19 April.

New Mexico AIDS Education and Training Center, 2006, Fact Sheet 727. Essiac. University of New Mexico Health Sciences Center. www.aidsinfonet.org. Revised 22 June.

New Mexico AIDS Education and Training Center, 2006, Fact Sheet 729. St. John's Wort (Hypericin). University of New Mexico Health Sciences Center. www.aidsinfonet.org. Revised 18 July.

New Mexico AIDS Education and Training Center, 2006, Fact Sheet 731. Marijuana. University of New Mexico Health Sciences Center. www.aidsinfonet.org. Revised 5 September.

New Mexico AIDS Education and Training Center, 2006, Fact Sheet 735. Silymarin (milk thistle). University of New Mexico Health Sciences Center. www.aidsinfonet.org. Revised 6 April.

Scott M. and Irvine S.S., 1997, What do people with HIV, their carers and families want of their medical carers? In: Managing HIV (G.J. Stewart, ed.). North Sydney: Australasian Medical Publishing.

Sepkowitz K.A., 2001, AIDS – the first 20 years. N Engl J Med 344: 1764–1772.

Tan M.L., 2000a, Traditional and alternative medicine. AIDS Action. Asia-Pacific edition. 49: 1–2.

Tan M.L., 2000b, Making the right choice in traditional and alternative medicine. AIDS Action. Asia-Pacific Edition. 49: 2–3.

Tochikura T.S., Nakashima S., Ohashi Y., and Yamamoto N., 1988, Inhibition (*in vitro*) of replication of the cytopathic effects of human immunodeficiency virus by an extract of the culture medium of *Lentinus edodes* mycelia. Med Microbiol Immunol 177: 235–244.

Website: www.lifepositive.com/Body/body-holistic/AIDS/hiv-treatment.asp. Accessed on 15 August 2006.

SECTION FOUR

PREVENTION AND CONTROL

CHAPTER 18

STRATEGIES FOR PREVENTION AND CONTROL

Abstract

At present, information, education and communication (IEC), contact tracing (partner management), cluster testing, and condom promotion are the available measures for effectively preventing the spread of HIV epidemic. Education of various subgroups in the population helps making life-saving choices. "Safer sex" refers to any sexual act without direct contact with body fluids of the sexual partner.

India has adopted a six-pronged approach to ensure safety of blood and blood products, which includes mandatory licensing of all blood banks and testing of every unit of blood for blood-borne pathogens, including HIV. ARV drugs are used for prolonging the life of HIV-infected individuals, preventing transmission from mother to child, and decreasing the complications of immune suppression. Other strategies for controlling HIV epidemic include early diagnosis and treatment of STIs, integration of HIV-related activities with primary health care, and targeted interventions. Specific prophylactic measures are universal biosafety precautions, concurrent disinfection and decontamination, and prevention of opportunistic infections.

Key Words

Antiretroviral drugs, Cluster testing, Condom promotion, Contact tracing, Harm reduction, Information, education, and communication, Needle exchange programmes, Out-of-school youth, Partner management, President's Emergency Plan for AIDS Relief, Safer sex, Sex workers, Targeted interventions, Truck drivers, Universal biosafety precautions

18.1 – INFORMATION, EDUCATION AND COMMUNICATION

Information, education and communication (IEC) is a "pre-planned, concerted effort, with specific objectives, focussed towards specific programme goals, in order to reach specific target audiences" (NIHFW, 1996). Knowledge, attitude and practice (KAP) studies act as a barometer of the level of knowledge, the prevailing attitudes, and the type of behaviour. KAP studies should be undertaken *before* and *after* IEC programmes, in order to know the impact of these programmes (NIHFW, 1996). The objective of each of the components of IEC is clear and specific, and is focused towards specific programme goals. The *information component* strives to generate awareness in the target audience and improve their knowledge by dissemination of information, either in individual, group, or mass settings. *Education component* deals with the development of

275

favourable attitudes. *Communication component* aims at bringing about desired behaviour changes, through motivation and persuasion.

18.1.1 – Approaches

Mass Approach: This is used to sensitise the community, and other *large* groups, about the programme. Various media used for mass approach include film and puppet shows, exhibitions, television and radio programmes, print media, and drama, or films with participation of popular artistes.

Group Approach: This is used to create interest among *smaller* target groups to accept specific health-related programmes. This approach allows for discussion with intended beneficiaries in order to clarifying their doubts about a particular programme, or intervention. The methods comprise discussion with members of women's organisations, and youth clubs, training camps for village leaders or *panchayat* (village council) members and other opinion leaders in the community, and targeting members of fan clubs of celebrities such as film stars and sports persons (NIHFW, 1996).

Individual Approach: This involves meeting decision-makers in the family namely mothers-in-law, husbands, grandparents, or meeting peer groups. Flash cards, charts, and posters may be used in this approach (NIHFW, 1996).

18.1.2 – IEC for Control of HIV Epidemic

Currently, the only available approach for effectively preventing the spread of HIV epidemic is *education* of various subgroups in the population, to enable making life-saving choices. The topics and modes of presentation will vary according to target group.

General Population: Among the *general population*, particularly adolescents, education programmes need to emphasise the importance of sexual discretion (avoiding multiple sexual partners and indiscriminate sexual intercourse with strangers, and use of condoms). Piercing of nose, ear, or circumcision is to be done only in health facilities.

High-Risk Groups: Persons with high-risk behaviour should be advised to refrain from donating blood, body organs, semen, or other tissue. HIV-infected women and women at high risk need to be advised about the possible risk of transmission to unborn or newborn baby.

18.1.3 – Ethical and Socio-Cultural Issues

Since the apprehensions about HIV infection and epidemic among the lay persons are not based on scientific facts, public education would be the best antidote. It is said that *fear is bred in the unknown*. Educators and persons involved in IEC activities have an ethical obligation to provide correct scientific information. In their zeal to promote "desirable" behaviour, they must not create fear and panic or target groups known for high-risk behaviour for selective discrimination. While it is necessary to promote correct scientific information, there may be vested interests and self-appointed "guardians of public morals" in communities, who may oppose frank discussions on sex and sexuality. IEC activities ought to sensitise the public regarding nondiscrimination, acceptance, and respect for human dignity while dealing with HIV-infected individuals.

18.1.4 – HIV-Related IEC Activities

From 2006, students in 21,000 secondary and upper secondary schools in Bangladesh will be taught about HIV/AIDS as part of a "life skills" curriculum (UNDP, 2006). In India, NACO has launched a nationwide multimedia campaign. NGOs have been provided financial support to undertake awareness and intervention programmes for vulnerable groups such as sex workers, truckers, IDUs, street children, and migrant labourers. A National Counselling Training Programme has been launched to train grassroots-level counsellors. National AIDS Helpline has been set up with a toll-free telephone number "1097" so that the caller can avail of counselling services in an atmosphere that maintains privacy and confidentiality of the caller. School AIDS Education Programme provides lifestyle education and HIV-related information to high school students. A programme called "University talk AIDS" is targeting higher secondary and university students. School health curriculum and many training modules have been prepared, including one on counselling. The mass media have a role in inculcating moral values in society, and in educating the public about the link between alcohol use, substance abuse, and promiscuity.

18.2 – PREVENTING SEXUAL TRANSMISSION

Sexual intercourse is the most common mode of transmission of HIV. Promotion of safer sexual behaviour holds the key to preventing sexual transmission. *Safer sex* refers to any sexual act in which, there is no direct contact with body fluids of the sexual partner. Correct use of quality condoms effectively prevents contact with cervicovaginal secretions and semen. Studies in homosexual men (Detels *et al.*, 1989) and heterosexual couples (Vincenzi, 1994) have shown that consistent use of condoms prevents transmission of HIV.

18.2.1 – Preventive Advice

Target groups for preventive advice include all patients with STIs, MSM, persons with multiple sex partners, sex workers, IDUs, persons seeking contraceptive advice, and overseas travellers (Bradford *et al.*, 1997). Techniques for preventive advice include:

(a) Taking a detailed sexual history to assess the risk of acquiring HIV infection
(b) Using explicit and clear messages that individuals can understand
(c) Dispelling myths and erroneous beliefs
(d) Providing tailor-made preventive advice to suit the individual's needs

18.2.2 – Partner Counselling

Synonyms: Contact tracing and partner management. Based on the information provided by the "source client" (the STI- or HIV-infected individuals), his or her sexual and/or drug-using partners are identified, located, investigated, and treated. It requires the cooperation of the source client. Partner counselling has been used as a public health response in case of STIs such as syphilis and gonorrhoea (UNAIDS & WHO, 2000). In the context of control of HIV epidemic, the objective is to encourage the partners to come in for HIV counselling and testing. Where possible, confidentiality of the source client is maintained and partner counselling is done with the source client's consent. In many countries, the legal position regarding "partner notification" is not clear. If the HIV-infected person refuses to cooperate, health care providers have a moral and ethical responsibility to see that the partner is informed. The doctor may take the help of the public health authorities to trace the contacts (Bradford *et al.*, 1997).

As per directives of the Supreme Court of India, "partner counselling" has been included as a component of India's National HIV Policy. All HIV-infected individuals need to be encouraged to disclose their HIV status to their spouse/ sexual partners. However, HIV status should be disclosed to the spouse/sexual partners, only after proper counselling (NACO, Training Manual for Doctors). Due to intensive use of contact tracing, persons in high-risk groups have become wary of the health care system and have become reluctant to volunteer for HIV testing. Some HIV-infected persons may require professional assistance or support for notifying their partners (Bradford *et al.*, 1997).

18.2.3 – Cluster Testing

In this method of case detection, the *source client* is asked to name other persons, of either sex, who move in the same socio-sexual environment. These persons are screened, using blood tests. This technique almost doubles the number of detected cases.

18.2.4 – Condom Promotion

Unprotected multipartner sexual activity is the major cause of HIV transmission in India. The disadvantages of *free distribution* of condoms are difficulty in ensuring supplies and doubts about their actual use. Condom promotion works best when targeted at those with high-risk behaviour. Promoting condom use among married couples has so far, met with little success. Health care personnel need to overcome their discomfort and reluctance to provide condoms to unmarried young males. Advertising through the mass media is directly related to demand for both socially marketed and public sector-distributed condoms. In communities where condom use is identified with promiscuity, strategies need to be devised to counter stigma of condom use.

Perception of risk of HIV infection and social acceptance are among the factors that influence condom use. The reported use of condoms is higher among unmarried persons as compared with the married, and with casual partners as compared with steady (or regular) partners. Even when aware of the risk of HIV infection, youth frequently do not use condoms and tend to establish the trustworthiness of their partners with criteria other than sexual history (Longfield *et al.*, 2002). A four-country intervention programme in sub-Saharan Africa found that interventions increased risk perception of pregnancy only among females in two countries, but did not change the perceived risk of HIV infection (Agha, 2002).

Adolescent males are uncomfortable getting condoms where they might be recognised. In urban Botswana, youth were reluctant to get condoms from public sector clinics, even when available free of cost because health workers asked them questions about the use of condoms and their age. However, they were willing to get condoms from a public hospital where boxes of condoms were kept at a window with no questions asked (Meekers *et al.*, 2001). Studies suggest that females can negotiate condom use more easily for prevention of pregnancy than for prevention of STIs. More studies are needed to identify the reasons youth prefer pharmacies, bars, or kiosks as outlets for condoms (Finger & Pribila, 2003).

18.2.5 – Dos and Don'ts Regarding Condom Promotion and Use

Potential clients ought to be informed that condoms prevent deposition of semen and urethral discharge, exposure to penile lesions, and unwanted pregnancies. Condoms should be used each and every time an individual has sexual contact (vaginal, anal, and oral) and with every partner (known or unknown). A condom should never be reused. Dates of manufacture and expiry should be checked before using the condom. The correct method of using condoms is to be demonstrated. Potential clients need to be told about common causes of condom failure like

(a) Damage to latex caused by exposure to heat, moisture, sunlight, and prolonged storage
(b) Failure to expel air in the "receptacle" or "teat" while wearing condom, which causes tearing
(c) Incorrect use, under influence of alcohol and/or drugs

Most condoms are pre-lubricated to increase sexual stimulation. Oil-based lubricants such as vaseline, oils, or creams can damage male and condoms. If the condoms are not lubricated, *water-based lubricants* should be used. Use of alcohol or drugs before or during coitus increases the risk of ignoring safer sex guidelines or faulty use of condoms (Fact Sheet 151, 2006).

18.3 – PREVENTING INJECTING DRUG USE

Harm Reduction: Harm reduction programmes encourage IDUs to abstain from using drugs. The drug users are provided with psychological assistance to abstain from drugs and substitution treatment with methadone. However, in some countries, substitution treatment with methadone is illegal. Although giving up drugs altogether is an ideal objective, harm reduction programmes recognise that total abstinence is very difficult or impossible in some cases. Harm reduction includes needle exchange programmes wherein drug users exchange dirty (used) needles for clean ones. If a supply of clean needles is available, drug users would be less likely to share needles and expose themselves to the risk of HIV infection. Some harm reduction clinics have encountered resistance from potential beneficiaries due to fear of government agencies (Kirby, 2006). The first comprehensive needle exchange programme for IDUs was established in Tacoma, Washington, DC, USA in 1988 (Kaiser Network, 2006). Nepal was the first developing country to establish a harm reduction programme with needle exchange (UNDP, 2006).

One of the strategies to contain the HIV epidemic is to impart information on risk of HIV transmission to IDUs, general public, policy makers, health care providers, and law enforcement personnel (Wodak & Dolan, 1997). However, in case of IDUs, the results of HIV education are not encouraging and studies have not revealed beneficial effects like changes in attitudes, behaviour or HIV seroprevalence (Des Jarlais *et al.*, 1992). In countries where outbreaks of HIV infection have occurred among IDUs, the infection subsequently spread to rest of the population (Wodak & Dolan, 1997).

In view of their disappointing experiences in trying to prevent injecting drug use, some countries have launched needle exchange programmes where IDUs are provided sterile injecting equipment in exchange for used ones (Wodak & Dolan, 1997). In some states of Australia, imprisoned IDUs are supplied with bleach (hypochlorite-containing compounds) to decontaminate syringes and needles (Dolan *et al.*, 1995). The effectiveness of bleach against HIV or hepatitis B/C viruses, when used under prison conditions is not known (Wodak & Dolan, 1997).

18.4 – PREVENTING BLOOD-BORNE TRANSMISSION

In 1982, the first case of possible HIV transmission through transfusion of blood or blood products was reported from California, USA (CDC, 1982).

Safety of Blood and Blood Products: India has adopted a multipronged approach for ensuring blood safety. The approaches include:
(a) Mandatory licensing of all blood banks and mandatory testing of every unit of blood for blood-borne infections such as HIV-1, HIV-2, hepatitis B/C, malaria, and syphilis

(b) Establishing zonal blood testing centres (to provide linkage with other blood banks, and test blood samples from blood banks in the zone), and blood component separation facilities (to reduce the wastage of whole blood transfusion)
(c) Training blood bank staff and modernising blood banks in public and voluntary sectors
(d) Promoting voluntary blood donation
Blood donation by professional blood donors has been banned in India since 1 January 1998.

Other Measures: People in high-risk groups need to refrain from donating blood, semen, body organs, or other tissues. Injectable medications are to be avoided, unless absolutely essential. As far as possible, *disposable* gamma-sterilised needles and syringes should be used. Blood transfusion is to be given only when strictly indicated. Heat-treated Factors VIII and IX should be given to haemophiliacs, instead of coagulation concentrates. All health care providers ought to be trained in *universal biosafety precautions* and these should be strictly enforced in all health care facilities. Voluntary confidential testing, counselling, and referral services need to be provided at STI clinics, antenatal clinics, family planning centres, and at places where people belonging to high-risk groups gather for rest, or recreation such as resting points on highways for *hijras* (transgendered persons) and sex workers, or "gay clubs". In some countries, only blood products from regular, voluntary donors are used since these donors have lower titres of markers for infectious diseases, including HIV antibodies (Wylie & Dodd, 1997). Other risk-reducing measures include autologous blood donation, directed, or designated donation such as parental donors for neonatal blood transfusion (Pink *et al.*, 1994).

Limitations of Screening Procedures: The available screening procedures do not ensure zero-risk and the screening tests do not detect all the HIV variants (Wylie & Dodd, 1997). In the United States, estimates of risk of HIV transmission through blood or blood products vary between 1 in 450,000 donations to 1 in 660,000 donations (LaKritz *et al.*, 1995). Blood from new donors is quarantined for a specific period. This is because the window period is currently at least 25 days (Busch *et al.*, 1995). However, this policy may reduce the supply of blood to unacceptably low levels (Wylie & Dodd, 1997).

Possible Risk-Reducing Measures in Future: In the developed countries, due to screening of donated blood and heat-treatment techniques to destroy HIV in blood products, the risk of blood-borne transmission of HIV is extremely small (NIAID, 2005). Irrespective of the screening procedure employed, it is unlikely that zero-risk blood supply can be obtained. In spite of this, researchers are exploring new techniques for reducing the risk of HIV transmission through donated blood and blood products (Wylie & Dodd, 1997).

Reducing the "window period" – Before the formation of detectable anti-HIV antibodies, there is a brief period when HIV antigens are present in the blood

(Wylie & Dodd, 1997). The test for p24 antigen, which is believed to have the potential to reduce the window period, was introduced in the United States in 1996. Introduction of this additional screening test costs US$ 20–50 million per year, but still does not meet the goal of zero-risk blood transfusion (Alter *et al.*, 1990). However, in Thailand, where the prevalence of HIV infection in the donor population is as high as 4 per cent, p24 antigen screening has identified one additional infected donor for every 250 seropositive donors that were detected by existing screening tests (Mundee *et al.*, 1994).

Molecular diagnostic techniques – Techniques such as PCR have the potential to reduce the window period for HIV by about 6–10 days. But, the existing molecular diagnostic techniques such as PCR, ligase chain reaction, and branched chain DNA do not permit large-scale, cost-effective screening of donated blood. Moreover, contamination of specimens can lead to false positive results. Currently, PCR will have to be used as a supplement and not as a substitute for HIV antibody screening tests.

Inactivating HIV – It is theoretically possible to inactivate HIV in plasma and blood cells by using solvent detergents, treating with heat, and by photoinactivation with methylene blue. But, in the immediate future, it is unlikely that virus-inactivated plasma or blood cells would be available.

Blood substitutes – Synthetic oxygen-carrying products would be safer than transfusing human blood. Possible oxygen-carrying products under research include polymerised haemoglobin derivatives and fluorocarbons (Wylie & Dodd, 1997).

18.5 – ANTIRETROVIRAL THERAPY

Though no vaccine or cure is available so far, ARV drugs have been useful in *prolonging the life* of HIV-infected individuals, preventing MTCT in HIV-positive pregnant women, and for decreasing the complications of immune suppression. *Postexposure prophylaxis* (PEP) with ARV drugs should be also considered after accidental exposure to blood or body fluids for health care providers and for victims of rape. MTCT of HIV infection is negligible in the United States due to appropriate ARV treatment (NIAID, 2005). Certain ARV drugs have been approved for use in other countries under the President's Emergency Plan for AIDS Relief (PEPFAR), a US$15 billion initiative announced by US President George Bush in 2003 to fight the HIV/AIDS pandemic. The objective of PEP-FAR is to prevent 7 million new HIV infections, treat at least 2 million HIV-infected persons and care for 10 million HIV-infected individuals, AIDS orphans, and vulnerable children worldwide (FDA, 2006).

18.6 – SPECIFIC PROPHYLAXIS

Universal biosafety precautions and management of biomedical waste are to be enforced in health care facilities. Concurrent disinfection and decontamination involves the disinfection of all materials and equipment contaminated with blood and body fluids, irrespective of the patient's HIV serostatus.

Prevention of Opportunistic Infections: In the absence of a preventive vaccine, specific prophylaxis will be currently directed at preventing opportunistic infections in patients with CD4 count below 200 cells per μL. For preventing *Pneumocystis* pneumonia, the drug of choice is cotrimoxazole and alternative drugs are aerosolised pentamidine or dapsone. Rifabutin is used, after excluding presence of tuberculosis, to prevent *M. avium intracellulare* infection. Similar prophylaxis is feasible for other opportunistic infections such as cytomegalovirus retinitis. For preventing *M. tuberculosis* infection isoniazid may be given daily in the dose of 300 mg for 9–12 months, to all HIV-infected persons, who test Mantoux positive (taken as induration of more than 5 mm). But, NACO Technical Resource Group on Chemoprophylaxis has deferred isoniazid chemoprophylaxis till more scientific data is available in the Indian setting (Pathni & Chauhan, 2003).

18.7 – SEXUALLY TRANSMITTED INFECTIONS: DIAGNOSIS AND TREATMENT

(See Chapters 10 and 15 for details.)

18.8 – PRIMARY HEALTH CARE

HIV-related programmes should be integrated with all aspects of primary health care, including maternal and child health (MCH), family planning, and health education. On 1 February 1987, the WHO launched the Global Programme on AIDS (GPA) to support the development of National AIDS Control Programmes. The Global AIDS Program (GAP) of the Centers for Disease Control and Prevention supports HIV-related activities such as training, information exchange, programme for IDUs, programme for MSM, and cross-border programmes in five Asian countries – Cambodia, China, India, Lao PDR, Thailand, and Vietnam (CDC, 2006).

18.9 – TARGETED INTERVENTIONS

Pilot projects in India for targeted interventions include those for sex workers in Kolkata, MSM in Chennai, IDUs in Manipur, Nagaland, and Assam, and truck drivers in Rajasthan (NACO, Training Manual for Doctors). In Mumbai, the AIDS Workplace Awareness Campaign targets truck drivers at the regional transport authority, where the drivers get their licences renewed annually. Some projects include petrol pump employees and owners (Fredriksson-Bass & Kanabus, 2006).

18.9.1 – Sex Workers

Sex workers constitute a cost-effective target population for reducing new infections. In communities where commercial sexual activity is a delicate issue, the support of community leaders is essential. Peer educators are valuable partners in this strategy since sex workers tend to be wary of unfamiliar persons. This group utilises reproductive health services, if offered with primary health care. For promoting use of condoms, the condoms should be accessible and afford-able and the sex workers should be trained in negotiating condom use with their clients (UNFPA, 1998).

Targeted intervention programmes that empower sex workers have demon-strated that HIV transmission can be curbed. Condom use among sex workers in Kolkata's Sonagachi red-light area was about 85 per cent and HIV prevalence declined to less than 4 per cent in 2004, having exceeded 11 per cent in 2001. By contrast, in Mumbai, HIV prevalence among female sex workers has not declined below 52 per cent since 2000 (UNAIDS/WHO, 2005). This is probably because brothel-based sex workers in Mumbai are controlled by "madams", pimps, and moneylenders making HIV prevention activities more difficult (Fredriksson-Bass & Kanabus, 2006).

18.9.2 – Out-of-School Youth

Targeted interventions work better if obstacles to free and frank discussion on sexual matters are overcome. Income generation schemes may be used as an entry point. Youth-friendly reproductive health services are to be linked to IEC activities. Promotion of sexual abstinence rarely works and peer student educa-tors are not best placed for reaching this group. Condom promotion is the cen-tral component of strategies to prevent infection in sexually active youth. Health care personnel need to overcome their discomfort and reluctance to provide condoms to unmarried young males. In communities where condom use is identified with promiscuity, strategies need to be devised to counter stigma of condom use (UNFPA, 1998).

REFERENCES

Agha S., 2002, A quasi-experimental study to assess the impact of four adolescent sexual health interventions in sub-Saharan Africa. Int Fam Plann Perspect 28(2): 67–70, 113–118.

Alter H.J., Epstein J.S., Swanson S.G., *et al.*, 1990, Prevalence of human immunodeficiency virus type 1 p24 antigen in US blood donors – an assessment of the efficacy of testing in donor screen-ing. N Engl J Med 323: 1312–1317.

Bradford D., Kippax S., and Baxter D., 1997, HIV prevention in the community: sexual transmis-sion. In: Managing HIV (G. J. Stewart ed.), North Sydney: Australasian Medical Publications.

Busch M.P., Lee L.L., Satten G.A., *et al.*, 1995, Time course of detection of viral and serological markers preceding human immunodeficiency virus type 1 seroconversion: implications for screen-ing blood and tissue donors. Transfusion 35: 91–97.

Centers for Disease Control and Prevention (CDC), 1982, Possible transfusion-associated acquired immunodeficiency syndrome (AIDS). Morb Mortal Wkly Rep 31: 652–654.

Centers for Disease Control and Prevention (CDC), 2006, Global AIDS Program. www.cdc.gov.27 February.

Des Jarlais D.C., Friedman S.R., Choopanya K., *et al.*, 1992, International epidemiology of HIV and AIDS among injecting drug users. AIDS 6: 1053–1068.

Detels R., English P., Visscher B.R., *et al.*, 1989, Serum conversion, sexual activity, and condom use among 2915 HIV seronegative men followed for up to two years. JAIDS 2: 77–83.

Dolan K., Wodak A., and Penny R., 1995, AIDS behind bars: preventing HIV spread among incarcerated drug injectors (Editorial). AIDS 9: 825–832.

Finger W. and Pribila M., 2003, Condoms and sexually active youth. Youth Lens on Reproductive Health and HIV/AIDS. Arlington, VA: YouthNet, March 2003.

Food and Drug Administration (FDA), 2006, FDA News. Washington, DC: US Department of Health and Human Services. www.fda.gov/cder/drug/infopage/atripla. 12 July.

Fredriksson-Bass J. and Kanabus A., 2006, HIV in India. www.avert.org. Last updated July 19.

Kaiser Network, 2006, Global HIV/AIDS Timeline. www.kff.org/hivaids/timeline

Kirby M., 2006, AIDS in Eastern Europe and Central Asia. www.avert.org. Last updated 10 July.

LaKritz G.M., Satten G.A., Alberle-Grass J., *et al.*, 1995, Estimated risk of transmission of human immunodeficiency virus by screened blood in the United States. N Engl J Med 333: 1721–1725.

Longfield K., Klein M., and Berman J., 2002, Criteria for trust and how trust affects sexual decision-making among youth. Working Paper No. 451. Washington, DC: Population Services International.

Meekers D., Ahmed G., and Molatihegi M.T., 2001, Understanding constraints to adolescent condom procurement: the case of urban Botswana. AIDS Care 13(3): 297–302.

Mundee Y., Kamtorn N., Chaiyaphruk S., *et al.*, 1994, Prevalence of HIV antibodies and p24 antigen among blood donors in Northern Thailand. Transfusion 34 (Suppl 635): Abstract.

NACO, Training manual for doctors. New Delhi: Government of India.

National Institute of Allergy and Infectious Diseases (NIAID), 2005, HIV infection and AIDS: an overview. NIAID Fact Sheet. Bethesda: National Institutes of Health. www.niaid.nih.gov/

National Institute of Health and Family Welfare (NIHFW), 1996, Inter-sectorial co-ordination and IEC Management. Module-5. New Delhi: NIHFW.

New Mexico AIDS Education and Training Center (NMAETC), 2006, Fact Sheet 151. Safer sex guidelines. University of New Mexico Health Sciences Center. www.aidsinfonet.org. Revised 18 July 2006.

Pathni A.K. and Chauhan L.S., 2003, HIV/TB in India – A public health challenge. JIMA 2003 101(3): 148–149.

Pink J., Thomson A., and Wylie B., 1994, Infectious disease markers in autologous and directed donations. Transfus Med 4: 135–138.

UNAIDS & WHO, 2000, Opening up the HIV/AIDS epidemic. Geneva: WHO.

UNAIDS/WHO, 2005, Aids epidemic update. Geneva: UNAIDS/WHO, pp 31–44.

United Nations Development Programme (UNDP), 2006, Asia-Pacific at a glance. www.youand aids.org

UNFPA, 1998, Desk study on HIV/AIDS interventions for commercial sex workers, out-of-school youth and condom promotion. Evaluation Findings. New York: United Nations Population Fund. Issue No. 9, September.

Vincenzi J.D., 1994, A longitudinal study of human immunodeficiency virus transmission by heterosexual partners. N Engl J Med 331: 341–346.

Wodak A. and Dolan K., 1997, HIV prevention in the community: injecting drug users. In: Managing HIV (G.J. Stewart ed.), North Sydney: Australasian Medical Publications.

Wylie B.R. and Dodd R.Y., 1997, Protecting the blood supply from HIV. In: Managing HIV (G.J. Stewart ed.), North Sydney: Australasian Medical Publications.

CHAPTER 19

SURVEILLANCE FOR HIV

Abstract

HIV surveillance in India involves annual cross-sectional survey of the same risk group in the same place over few years by unlinked anonymous testing by two ELISA, rapid, and simple (E/R/S) tests. Adequate numbers of representative samples are collected within the shortest possible time period. Patients attending STI clinics, IDUs, and MSM represent the *high-risk groups* in the population. The *low-risk* group comprises women attending antenatal clinics. Each sentinel site conducts an annual round of surveillance with 250 samples from *high-risk* groups and 400 samples from *low-risk* groups. From time to time, sentinel sites were increased to provide adequate representation to high- and low-risk groups in urban and rural populations. As compared to *point estimate*, the *range estimate* is more scientific, reflects actual situation in the field, and also helps planners to formulate specific interventions for HIV-affected persons. Upper limit of the range was set at 20 per cent higher than the lower limit in order to take care of unaccounted number of HIV-positive persons in high-risk groups and the other age groups.

In 2005, the number of adults (aged 15–49) living with HIV in India was estimated to be 5.21 million, of whom 39 per cent were women. The adult HIV prevalence in 2005 (0.91 per cent) is comparable with that in previous 2 years. The prevalence of HIV infection in rural and urban inhabitants was 58.7 per cent and 41.3 per cent, respectively. The number of children under 15 years of age who were newly infected in 2005 was 59,007. The *high-prevalence* states were Andhra Pradesh, Karnataka, Maharashtra, Manipur, Nagaland, and Tamil Nadu. Gujarat, Goa, and Pondicherry were categorised as *medium-prevalence* states, while all the remaining states and Union Territories were *low-prevalence* states. HIV prevalence was more than 1 per cent among antenatal clinic attendees in 95 districts, including nine districts in low-prevalence states. The prevalence in this category has dropped to less than 1 per cent during the last 4 years in the high-prevalence state of Tamil Nadu while it has steadily declined from 2.08 per cent in 2003 to 0.88 per cent in 2005 in the low-prevalence state of Mizoram.

Key Words

Behavioural surveillance, Epidemiological surveillance, Epidemic Projection Package, Linked testing, Named case reporting, Participation bias, Point estimate, Range estimate, Second-generation surveillance, Sentinel surveillance, Unlinked anonymous testing, Unnamed case reporting

19.1 – INTRODUCTION

Epidemiological surveillance has been defined as "ongoing systematic collection, analysis and interpretation of outcome-specific data" (Thacker & Berkelman, 1988). The WHO has defined it as "continuous scrutiny of the factors that determine the occurrence and distribution of disease and other conditions of ill health" (WHO, 2002). The objectives of surveillance are to provide timely information about changing trends in health status of a population so that appropriate interventions may be devised; and to provide feedback – this may be used for redefining objectives. HIV surveillance continues to be the backbone of surveillance systems since the WHO established the GPA in 1987. Data on incidence and prevalence of HIV *seropositivity* is more useful as compared with case reporting of AIDS for health planning and programme evaluation (WHO, 2002).

19.2 – TYPES OF HIV SURVEILLANCE

19.2.1 – Population-Based Surveillance

This is the best method of assessing HIV prevalence and includes *serosurveillance* by HIV testing of blood specimens from a random sample of the given population and *behaviour surveillance*. For serosurveillance, the health authorities should obtain consent of all persons whose blood is collected. However, this is an expensive method, and *participation bias* is likely (WHO/UNAIDS, 2000).

19.2.2 – Sentinel Surveillance

This involves systematic collection of data on incidence and prevalence of HIV/AIDS in a given population. It also involves collecting and testing blood samples from people living in a defined area or those belonging to a particular subgroup in the population such as high-risk groups. In general, sentinel surveillance involves two types of testing:

Linked Testing: The blood sample or report is traceable by code to an individual. Consent is required.

Unlinked Anonymous Testing: This is the most commonly used method for HIV surveillance. The blood sample and its report are anonymous and not traceable to an individual. Blood samples collected routinely for other purposes such as blood donation and antenatal check-up is used for HIV testing. There is no need for consent and participation bias is minimised (WHO/UNAIDS, 2000).

19.2.3 – Case-Based Surveillance

This form of surveillance is used in countries where notification of HIV/AIDS is legally required. Information is provided on a standard reporting form.

Named Case Reporting: The names of individuals with HIV/AIDS are provided to the health authorities to ensure appropriate interventions. This information is to be kept confidential. However, there is a risk of breach of confidentiality.

Unnamed Case Reporting: The health authorities are provided only with information (age, sex, risk factors, and occupation) that is required for disease surveillance and data analysis (WHO/UNAIDS, 1999). The names are replaced by *name code* made up of first two letters for surname, followed by first two - letters of first name (Kaldor & Crofts, 1997). In order to avoid duplication of reporting, additional information (that does not identify the patient) may be given. For example – date of birth, postal code of residence, AIDS-defining conditions, date of diagnosis of HIV infection, date of diagnosis of AIDS, diagnosing doctor and hospital, prior ARV treatment (Kaldor & Crofts, 1997).

19.2.4 – Behavioural Surveillance

The spread of HIV epidemic is primarily driven by individual behaviour, which puts other persons at risk of getting infected. Individual behaviour is mainly guided or influenced by socio-cultural and religious norms, and economic status. Very little information is available on patterns of sexual behaviour and sexual networking that determines the spread of HIV infection in different socio-cultural groups in the developing countries. Where possible, behavioural surveillance should be used along with surveillance of HIV seropositivity among young people in order to understand the recent trends in HIV incidence and risk behaviour. The changes are first detected in HIV prevalence figures for 15- to 19-year-old group.

19.2.5 – Surveillance of Young People

Percentage of infected young people is a useful indicator of recent trends in HIV infection in areas where the epidemic spreads by heterosexual transmission. However, this indicator is not reliable where the infection remains confined to high-risk groups.

19.2.6 – Second-Generation Surveillance

This was launched in order to combine biological and behavioural data to assess trends. Young adults have the highest rates among newly HIV-seropositive individuals. Hence, many heavily affected countries have aimed for a 25 per cent reduction in HIV prevalence among young people, by the year 2005. For monitoring progress towards this goal, surveillance data (grouped by 5-year intervals, e.g. 15–19 and 20–24) are obtained from antenatal clinics (WHO/UNAIDS, 2000; UNAIDS/WHO, 2000).

19.3 – EPIDEMIC PROJECTION PACKAGE

This software package has been developed by UNAIDS/WHO Reference Group on Estimates, Models, and Projections in order to enable standardised and systematic estimation of the burden of HIV and the course of the epidemic. It uses HIV prevalence time series derived from data on women attending antenatal clinics. There are separate models for urban and rural trends. The trend in HIV prevalence is estimated by fitting a simple epidemiological model to the surveillance data. The urban and rural estimates are used to provide a national estimate. The model uses four parameters:

1. Estimated year of commencement of the HIV/AIDS epidemic
2. Rate of spread of infection
3. Initial proportion of high risk (susceptible) adult population
4. Extent to which an individual dying from AIDS is replaced by a new susceptible

Epidemic projection package (EPP) outputs fit into a software package called *SPECTRUM* (www.futuresgroup.com and www.unaids.org), that can be used to calculate relevant indicators such as trends in mortality due to AIDS, life expectancy, and number of pregnant women needing ARV therapy for prevention of MTCT (WHO, 2002).

19.4 – CASE REPORTING

Reporting is the procedure through which health care providers systematically inform the health authorities about each individual case of HIV infection and AIDS. This is a legal requirement in countries where HIV infection is notifiable (UNAIDS/WHO, 2000; WHO/UNAIDS, 1999).

19.4.1 – Choice of Method of Case Reporting

Countries should choose a suitable method for HIV/AIDS surveillance, as per their needs and resources and the prevailing socio-political environment (Kaldor & Crofts, 1997). Resource-poor countries with inadequate infrastructure to maintain confidentiality should avoid HIV case reporting and opt for sentinel surveillance.

19.4.2 – HIV Case Reporting

Criteria for using this option include:
(a) Wide access to HIV testing and ARV therapy
(b) Presence of mechanisms to maintain confidentiality
(c) Availability of mechanisms for follow-up of clients
(d) Enforcement of legislation to protect the right to privacy and to prevent non-public health use of data

(e) Periodic scanning to data for detecting extent of over- or under-reporting and duplication of data (multiple reporting of same cases)

19.4.3 – AIDS Case Reporting

This is done on an yearly basis. Its limitations are:
- Completeness of AIDS reporting varies from *less than* 10 per cent (developing countries), to *more than* 90 per cent (developed countries). This high variability restricts the utility of data for estimating extent of AIDS epidemic.
- There is a wide variation in case definitions (i.e. CDC, WHO, and national) for AIDS between various countries.
- Increasing number of individuals receiving ARV treatment, also affects AIDS reporting (WHO/UNAIDS, 2000).
- This system records cases that became HIV seropositive many years ago. In order to provide early warning of transmission patterns, some countries have adopted HIV surveillance (Kaldor & Crofts, 1997).

19.4.4 – Uses of Case Reporting

In *high-income* countries, both HIV case reporting and AIDS case reporting are employed. It is possible for these countries to monitor access of seropositive individuals to appropriate services, including the availability of ARV therapy. But, in *low-income* countries, the data obtained by HIV/AIDS case reporting have not been beneficial for surveillance because

1. Less than 10 per cent of HIV-positive persons have been tested for HIV. Thus, most people are unaware of their HIV status.
2. Many people may not want to know their HIV status due to stigma, discrimination, and lack of access to services including ARV therapy.
3. In *named* case reporting, there is reluctance of health care providers to report cases by name for reasons of confidentiality and privacy. There is also a problem of under-diagnosis and under-reporting of HIV/AIDS cases.
4. Where *unnamed* case reporting is employed, multiple reporting of the same cases is a problem (WHO/UNAIDS, 2000).
5. Data accuracy depends on the context in which HIV testing was done, e.g. patients presenting with clinical manifestations suggestive of HIV infection, voluntary testing on individual's request, testing of donated blood (Kaldor & Crofts, 1997).
6. Insufficient capacity or infrastructure for maintaining confidentiality.
7. Case reporting requires *expensive* infrastructure and trained manpower.
8. By itself, case reporting does not improve access to supportive services.

Hence, in low-income countries, resources should be set aside for developing health and social support services, before they are utilised for case reporting.

19.5 – HIV SENTINEL SURVEILLANCE IN INDIA

19.5.1 – Chronology of Events

1. 1981 – AIDS was first recognised in the United States.
2. 1985 – India joined the ranks of the first few countries that initiated serosurveillance among high-risk groups. Screening of blood samples for HIV was started in 1985 in Christian Medical College, Vellore (Tamil Nadu) and National Institute of Virology, Pune (Maharashtra).
3. April 1986 – The first group of seropositive individuals (10 female sex workers) was detected in April 1986 in Madras (now Chennai) and subsequently among sex workers in Mumbai. Within a short period of 18 months, it became apparent that heterosexual promiscuity was the major mode of transmission in India and that the seropositivity was low (approximately 4 per 1,000). HIV seropositive pregnant women (first detected in 1986) and their infants were followed up and HIV-infected children were detected in 1987–1988 (Ramachandran, 1990).
4. 1986 – The first case of AIDS was detected in Mumbai.
5. 1990 – Maharashtra and Manipur became "medium-prevalence" states.
6. 1994 – Maharashtra and Manipur became "high-prevalence" states. Gujarat and Tamil Nadu came in as "medium-prevalence" states.
7. 1998 – "High-prevalence" states were Maharashtra, Manipur, Nagaland, Tamil Nadu, Karnataka, and Andhra Pradesh; while "medium-prevalence" states were Gujarat and Goa.
8. 2001 – 49 districts in eight states were identified as "high-prevalence" districts.
9. 2005 – 95 districts in India were identified as "high-prevalence" districts.

After the detection of HIV infection in the country in April 1986, the testing facilities for HIV were made available in different parts of India through 62 centres and nine reference centres. The results of HIV screening indicated that HIV infection had reached almost all parts of the country and that the major mode of its transmission was through the *heterosexual* route. Once these findings were known, the objectives of surveillance were redefined to monitor the trends in HIV infection and a *National HIV testing policy* was adopted. India's prevalence estimates are solely based on sentinel surveillance data. Since HIV/AIDS is not a notifiable disease in India, HIV testing information from the private sector is not compulsorily reported to the national information system (Fredricksson-Bass & Kanabus, 2006).

19.5.2 – Technique for Sentinel Surveillance

- The surveillance is best carried out by annual cross-sectional survey of the same risk group in the same place over few years by *unlinked anonymous testing* by two ELISA, rapid, and simple (E/R/S) tests.
- The number of samples to be screened must represent the risk groups under study. Adequate numbers of samples are collected within the shortest possible time period.

- Clinic-based approach has many advantages, including the procedure for collecting samples, which need to be repeated every year. This approach avoids selection and participation bias.
- Each sentinel site conducts an annual round of surveillance with 250 samples from *high-risk groups* and 400 samples from *low-risk groups*.
- Patients attending STI clinics, IDUs, and MSM represent the *high-risk groups* in the population. The *low-risk* group comprises women attending antenatal clinics.
- If the sample size is not adequate, the number of samples collected "up to 3 months" is taken as "adequate" sample size. After each round of surveillance, the collected data are compiled and analysed.

19.5.3 – Number of Sentinel Sites

- Based on the National Testing Policy for HIV, the sentinel surveillance system was initiated in 55 sites in 22 states and Union Territories in the year 1994.
- It was found that these 55 sites did not produce adequate data for estimating the degree of spread of HIV/AIDS in the country. Thus, the sentinel sites for HIV surveillance were increased from time to time to provide adequate representation to high-risk and low-risk groups in urban, as well as rural population.
- In 2005, the number of sentinel sites was increased to 703.

19.5.4 – Prevalence of HIV Infection

The prevalence of HIV infection has been worked out for each year, based on data from annual rounds of HIV surveillance using *consistent methodology* (i.e. number of samples for different risk groups, inclusion and exclusion criteria, number of sites from urban and rural areas).

In 2001, the Government of India appointed a core group of experts to estimate the HIV prevalence. It was decided to present a *range estimate* instead of a *point estimate*, since the former is more scientific and reflects the actual situation in the field. The range estimate also helps planners to formulate specific interventions for HIV-affected persons. The upper limit of the range was 20 per cent higher than the lower limit in order to take care of unaccounted number of HIV-positive persons in high-risk groups and the other age groups.

19.6 – HIV SENTINEL SURVEILLANCE – 2005

HIV sentinel surveillance was conducted in 703 sites across the country. Using the same methodologies and assumptions an in previous years, the number of adults (aged 15–49 years) living with HIV was estimated at 5.21 million. The adult HIV prevalence in 2005 (0.91 per cent) is comparable with that in 2004 (0.92 per cent) and 2003 (0.93 per cent). Women comprises 39 per cent of the adults (15–49 years) living with HIV. The prevalence of HIV infection in rural and urban inhabitants was 58.7 per cent and 41.3 per cent, respectively. The number of children under 15 years of age who were newly infected in 2005 was 59,007. The *high-prevalence* states were Andhra Pradesh, Karnataka, Maharashtra, Manipur, Nagaland, and Tamil Nadu. Gujarat, Goa, and Pondicherry were categorised as *medium-prevalence* states, while all the remaining states and Union Territories were *low-prevalence* states (NACO, 2006).

19.6.1 – Prevalence Among Antenatal Clinic Attendees

Prevalence of HIV infection among pregnant women is indicative of HIV prevalence in the general population. HIV prevalence was more than 1 per cent among antenatal clinic attendees in 95 districts, including nine districts in low-prevalence states. The median prevalence in this subgroup has remained more than 1 per cent in all the high-prevalence states with the exception of Tamil Nadu, where the prevalence has dropped to less than 1 per cent during the last 4 years. In the low-prevalence state of Mizoram, HIV prevalence among antenatal clinic attendees has steadily declined from 2.08 per cent in 2003 to 0.88 per cent in 2005 (NACO, 2006).

19.6.2 – Trends in High-Risk Groups

The median prevalence among STI clinic attendees increased significantly, as compared with that in 2004, in the low-prevalence states of Delhi, Rajasthan, and Orissa. Among STI clinic attendees, the prevalence was more than 10 per cent at 34 sites across the country indicating multiple heterogeneous epidemics. During 2003–2005, HIV prevalence among STI clinic attendees has remained between 5 and 6 per cent in high- and moderate-prevalence states with the exception of Nagaland, Pondicherry, and Gujarat (NACO, 2006).

Assam, Chandigarh, Delhi, Kerala, and West Bengal revealed an increase in HIV prevalence among IDUs between 2002 and 2005. Frequency of HIV infection increased between 2003 and 2005 among female sex workers in Bihar, Nagaland, Rajasthan, and West Bengal. HIV seropositivity among MSM was higher in 2005, relative to that in 2004, in Delhi, Goa, Gujarat, and Kerala (NACO, 2006).

REFERENCES

Fredricksson-Bass J. and Kanabus A., 2006, HIV/AIDS in India. www.avert.org. Last updated 19 July.

Kaldor J.M. and Crofts N., 1997, Epidemiological surveillance for HIV/AIDS. In: Managing HIV (G.J. Stewart, ed.). North Sydney: Australasian Medical Publishing.

National AIDS Control Organisation (NACO), 2006, HIV/AIDS epidemiological surveillance and estimation report for the year 2005. New Delhi: Government of India, pp 1–11.

Ramachandran P., 1990, HIV infection in women. ICMR Bulletin 20 (11&12): 111–119.

Thacker S.B. and Berkelman R.L., 1988, Public health surveillance in the United States. Epidemiol Rev 10: 164–190.

UNAIDS/WHO, 2000, Opening up the HIV/AIDS epidemic. Geneva: UNAIDS/WHO.

WHO, 2002, HIV surveillance and global estimates. Wkly Epidemiol Rec 77(50): 425–430.

WHO/UNAIDS, 1999, Questions and Answers on Reporting, Partner notification and Disclosure of HIV and/or AIDS sero-status. Public Health and Human Rights implications. Geneva: WHO/UNAIDS.

WHO/UNAIDS, 2000, Guidelines for Second Generation HIV Surveillance. Geneva: WHO/UNAIDS.

CHAPTER 20

COUNSELLING

Abstract

Counselling is indicated in all conditions where client is likely to be under psychological stress. Any person who has the necessary aptitude, values, attitude, knowledge, and skills can undertake this challenging task. The purpose is to provide the means to overcome the stress-inducing condition and to take personal decisions that are not affected by moods, emotions or sentiments. In HIV medicine, both *counselling* and *medical intervention* are accorded equal importance. HIV-related counselling is somewhat unique because it starts before HIV testing (pre-test counselling) and continues for the spouse/partner and family members even after the death of the HIV-infected person. A counsellor should have a non-judgemental approach and help clients to reach for sources of support. Respect for confidentiality protects the privacy of clients. The process of counselling should continue for persons living with HIV infection. Special situations for counselling include that for the infected client's partner(s), accidentally exposed health care providers, victims of partner violence, victims of rape, and IDUs. Recording of counselling sessions will permit continuity even if there are different counsellors for subsequent counselling sessions and prevent the client from having to repeat the same information from the previous session. The quality of counselling can be evaluated in the areas of interpersonal relationships, gathering information, giving information, and ability to handle special situations.

Key Words

Confidentiality, Contact tracing, Counselling in special situations, Crisis counselling, Empathy, Evaluation of counselling, Food safety, Home care, International travel, Nutrition, Partner counselling, Personal hygiene, Pre-test counselling, Post-test counselling, Recording counselling, Referral, Safer sex, Shared confidentiality, Source client

20.1 – INTRODUCTION

Counselling is a "confidential dialogue between the client and the health care provider, which is aimed at enabling the client to cope with stress and to take realistic personal decisions" (UNAIDS/WHO, 2000). Counselling is indicated in all conditions where client is likely to be under psychological stress. The purpose is to provide the means to overcome the stress-inducing condition and to take personal decisions that are not affected by moods, emotions, or sentiments. Any person (professionally trained counsellor, doctor, nurse, paramedical worker, or volunteer) who has the necessary aptitude, values, attitude, knowledge, and skills can undertake counselling. This is a challenging task that requires patience, dedication, and commitment.

20.1.1 – Need for Counselling

In the past, traditional systems were in place for counselling apparently healthy persons, patients, and their families for stress-related problems. Usually, respectable elderly persons who had the necessary aptitude and know-how undertook this challenging task within their own communities. Social support systems were also community-oriented and well known to members of the community. A classic example of counselling is depicted in the Hindu scripture Bhagavad Gita, where Lord Krishna advises the hesitant Pandava Prince Arjuna and helps him to overcome his doubts, anxiety, and predicament about fighting his own kith and kin on the battlefield. In recent times, the need for counselling has increased due to breakdown of traditional support systems and values, and stressful situations caused by "modern" lifestyle. Counsellors have to develop a wide range of skills in order to provide services to individuals from varying social backgrounds.

20.2 – FUNDAMENTALS OF COUNSELLING

20.2.1 – Dos and Don'ts

1. Ensure privacy so that the client can ventilate his or her personal feelings without being inhibited.
2. Use a form of address that is appropriate to the client's age.
3. Give undivided attention to the client and make eye contact.
4. Ideally, the same counsellor should provide *both* pre- and post-test counselling.
5. At first, speak to the client alone. Later on, others (spouse/sexual partner and/or family members) may be involved, with the prior consent of the client.
6. Client's personal particulars including marital status and lifestyle are to be recorded.
7. Find out the emotional, social, medical support systems of the client.
8. Be familiar with local terms for sexuality.
9. Encourage the client to ask questions and to come back, as soon as possible.
10. *Never* give advice, but give information. Likewise, *never* frighten a client.
11. Use a language, which is understood by the client.
12. Repeat the client's message clearly and summarise to show that you have understood.
13. Encourage the client to speak, by showing that you are listening attentively.
14. Use humour or other "ice-breaking" techniques to reduce tension and stress.
15. Use gestures occasionally.
16. Keep suitable conversational distance.

17. Use questions that are centred on the concerns of the client (NACO, Handbook for Counsellors; CDC, 1995).

20.3 – HIV-RELATED COUNSELLING

In HIV medicine, both *counselling* and *medical intervention* are accorded equal importance. HIV-related counselling is somewhat unique because it starts before HIV testing (pre-test counselling) and continues for the spouse/partner and family members even after the death of the HIV-infected person. Clients understand the social, ethical, and legal implications of HIV testing only when given correct information. In "crisis counselling", clients are provided with psychosocial support at times of crises. Basically, HIV counselling is of three types: pre- and post-test counselling and partner counselling.

HIV-related counselling aims to achieve sustained behavioural changes that are necessary to prevent the transmission of HIV infection, which persists for life. Revealing the diagnosis of HIV infection can cause tremendous psychosocial stress. It is essential to address psychological concerns, which include:

1. Feeling of guilt about one's high-risk behaviour or having spread HIV infection to others
2. Fear of physical isolation and loss of relationships
3. Fear regarding dissemination of HIV infection in family and community, financial problems, and possible loss of housing, education, employment
4. Feeling of anger, loneliness, depression, and vulnerability to psychosocial problems

Counselling enables individuals to make decisions that facilitate "coping" and change high-risk behaviour, helps clients in identifying their immediate needs and possible sources of support, and helps in evaluating personal risk of HIV transmission that facilitates prevention.

20.3.1 – Values and Attitudes

Maintaining Confidentiality: The client should be assured that the disclosed personal, intimate feelings, or events, would be kept confidential in order to build trust and rapport. The result of the HIV test is revealed only to the client. If the client reveals a desire for *shared confidentiality* (i.e. sharing the test result with spouse, partner, a family member, or a close friend), the counsellor should offer guidance.

Positive Approach: The client's sexual preferences, behaviour, and lifestyle should not be judged or criticised. Moralising or preaching may add to the guilt (or "self-blame") in clients.

Acceptance: The counsellor should accept the emotions and reactions of clients (including hostility) when they realise that they are HIV-infected. This acceptance of clients should not be affected by the counsellor's subjective feelings about their high-risk behaviour, lifestyle, sexual preference, or social background.

Empathy: Empathy is "trying to place oneself in another's situation". Display empathy by making the counselling process more culturally acceptable and understandable, and by using culturally acceptable gestures (non-verbal communication).

Self-Determination: A frightened client may look to others for support and decision-making. Try to counter this dependence and support the autonomy of the client(s) by giving information and guidance.

Understanding Grief: The counsellor should be familiar with various culturally acceptable ways of expressing grief and help clients to reach for spiritual sources of comfort and support.

Resources: The counsellor should be well informed about the availability and location of various resources such as medical facilities, legal aid, peer support organisations, and social support services in order to help clients who may need them (NACO, Handbook for Counsellors).

20.3.2 – Target Groups

Individuals, couples, families, and groups are the intended beneficiaries for counselling. The persons who should be involved in the counselling process will vary in each case. The client should choose the person(s) who should be involved (UNAIDS/WHO, 2000). Types of clients are:
(a) *General clients*: Men with multiple sex partners, health care providers with risk of occupational transmission, and pregnant women.
(b) *Special groups*: MSM, sexually abused children; sexually exploited persons (rape or sodomy); child prostitutes and commercial sex workers (CSWs); street children; and IDUs.
(c) *Referred clients*: These are referred cases for routine HIV testing. The policy of NACO is to discourage HIV testing before surgical interventions. There is no public health rationale for mandatory HIV testing.
(d) *Voluntary* or *self-reported clients*: Clients with high-risk behaviour who have obtained information on HIV/AIDS through the mass media, individuals who already know their HIV status and desire retesting, and potential blood donors who are aware of the disease (NACO, Handbook for Counsellors).

20.3.3 – Written Protocol for Counselling

The counsellors should have a written protocol or manual, which identifies specific procedures for pre- and post-test counselling (CDC, 1995). A sample of this written protocol is given below:
1. Allot adequate time for counselling sessions.
2. Cover content of counselling session as appropriate.

3. During pre-test counselling, obtain consent for HIV testing (written or verbal) and record it in the client file.
4. During post-test counselling, review test result and record it in client file.
5. Ensure that the client file is stored in a locked cabinet (CDC, 1995).

20.4 – PRE-TEST COUNSELLING

This dialogue between the client and the health care provider or counsellor is aimed at discussing the HIV test and the possible implication of knowing one's HIV status. The quality and content of pre-test counselling ought to be ensured because clients should be able to make an informed decision on whether or not to take the HIV test. If they decide to take the test, they should be well prepared for the result. Pre-test counselling provides an opportunity to help clients assess their personal risk and to know how to reduce that risk, even if they decide not to take the HIV test (CDC, 1995).

20.4.1 – Procedure for Pre-Test Counselling

1. Discuss the reason for attending the counselling session and assure confidentiality.
2. Assess the client's knowledge of STD/HIV and transmission of these diseases. Correct myths and misconceptions, if any.
3. Provide information on HIV test (process, possible outcomes of HIV testing, and window period), and discuss the meaning of positive and negative HIV test results and possible implications of each.
4. Allow time for client to think over the issues, ask questions, and get clarifications.
5. Help the client to assess personal risk and discuss personal *risk reduction plan*.
6. Provide information about referral services appropriate for the client's needs.
7. Discuss whether or not the client will take the HIV test, and follow up arrangements (NACO, Handbook for Counsellors; CDC, 1995; UNAIDS, 2000).

20.5 – POST-TEST COUNSELLING

This dialogue between the client and the health care provider, or counsellor intends to discuss the HIV test result. It aims to provide appropriate information, support and referral, and information on behaviour changes that reduces the risk of becoming infected, even if the client is not infected. It also aims to reduce the risk of transmission of HIV to others, if the client is infected (WHO, 1992).

20.5.1 – Post-Test Counselling for Seronegative Clients

1. Give the test results simply and clearly. Check that the client has understood the test result.

2. Discuss the meaning of the test result. Explain about the "window period", which may last 3–6 months. During this period the HIV test result may be negative, even if the client is infected. Assess the client's risk and encourage repeat testing, after considering the "window period".
3. Discuss the benefits of sharing test results with sexual partner ("shared confidentiality") and encouraging partner to take the HIV test.
4. Discuss personal risk reduction plan and give information on precautions to be taken by the client to prevent HIV infection in future. Discuss needs for referrals, sources of support and follow-up plans (NACO, Handbook for Counsellors; CDC, 1995; UNAIDS, 2000).

20.5.2 – Post-Test Counselling for Seropositive Clients

It is difficult to tell clients that they are infected with HIV. Use clear language and check whether the client has understood the result. Clients should be given time to express their feelings. The procedure is as follows:
1. Give test results simply and clearly and allow time for the results to sink in.
2. Check whether the client has understood the result and discuss the meaning of the result.
3. Deal with immediate emotional reactions – grief, hostility, fear, denial, anger, or other feelings.
4. Discuss personal, family, and social implications including benefits of sharing test results with sexual partner and encourage partner to take the HIV test.
5. Discuss personal plan for risk reduction, check for availability of sources of immediate support, and identify options and resources for support.
6. Review follow-up care and support (ongoing counselling, counselling of other family members, social support, legal advice, referral for STIs and family planning, and medical referral).
7. Discuss follow-up plans and referrals (NACO, Handbook for Counsellors; CDC, 1995; UNAIDS, 2000).

20.5.3 – Ongoing Counselling for Seropositive Clients

Ongoing counselling is necessary during *each visit* and can be undertaken by any trained health care provider. Repetition reinforces messages.

"Do not fear to repeat what has already been said. Men need the truth dinned into their ears many times and from all sides. The first rumour makes them prick up their ears, the second registers, and the third enters" – René Théophile Hyacinthe Laennec (1781–1826), Regius Professor of Medicine, College de France (cited in: Gottlieb, 2001).

Physical Exercises: Regular physical exercises contribute to a feeling of well being; lead to better health and stamina; help in reducing stress, anxiety, and depression; increase muscle mass, strength and endurance; increase bone strength; decrease abdominal fat; and improve appetite and sleep. Cardiovascular exercises such as brisk walking, jogging, bicycling, or swimming improve heart and lung endurance. Patients taking certain ARV drugs face an elevated risk for cardiovascular disease and diabetes. A moderate exercise programme can help decrease blood levels of cholesterol, triglycerides, and sugar. Patients with heart disease or other risk factors ought to seek medical opinion so that they can exercise safely. Drinking adequate quantity of fluids prevents exercise-induced dehydration. With increased activity, one may need to consume extra calories to avoid losing weight. Meals should be consumed at least 2 hours *before* or half an hour *after* an exercise session. A regular schedule of exercise for about 30 minutes a day should be adequate for most individuals. Weight training increases lean body mass that may be lost due to HIV disease or ageing and also prevents osteoporosis. Exercising too much can cause injuries and loss of lean body mass. Should any injuries occur during exercising, the client should be advised to cover all open wounds immediately (Fact Sheet 802, 2006).

Nutrition: When the body fights any infection, it needs more energy (calories) and proteins, as compared with the healthy state. HIV-infected persons may tend to consume less food due to loss of appetite, gastric upset caused by ARV drugs, or opportunistic infections affecting the mouth, throat, or oesophagus. Loss of body weight is a common manifestation in HIV infection. Each meal should be balanced in relation to proteins, carbohydrates, fats, vitamins, and minerals. *Protective foods* such as fruits and vegetables are rich in vitamins, minerals, micronutrients, and dietary fibre. Deficiency of nutrients can cause disturbances in the immune system. Some ARV drugs cause diarrhoea, which causes loss of nutrients and fluids. Drinking extra water or other liquids (fruit juice and soups) also prevents dry mouth and constipation. Consumption of caffeine-containing beverages such as coffee, tea, colas, and/or alcohol can aggravate dehydration.

Nutritional Supplements: Many medications cause deficiencies of essential nutrients. The recommended dietary allowances (RDA) are the minimum amount of nutrients required to prevent deficiencies in healthy persons and are therefore not applicable to HIV-infected persons. Some molecules (called "free radicals") are produced in the body as a product of normal metabolism. These free radicals react easily with other molecules and can damage cells. HIV infection leads to higher levels of free radicals. Antioxidants that prevent or limit the damage caused by free radicals, are naturally present in many fruits and vegetables. If an HIV-infected person is unable to consume a balanced diet with plenty of fruits and vegetables, the health care provider may prescribe suitable nutritional supplements. Patients should take nutritional supplements only after consulting their health care providers because many nutrients interact with each other or with medications. *Lactobacillus acidophilus* is a commensal bacterium in the small intestine, which is destroyed by many antibiotics. Not much information is available on specific nutrients and HIV disease (Fact Sheet 801, 2006). It has been suggested that vitamin A may have immune stimulatory properties and a role in maintaining the integrity of vaginal mucosa or placenta, thus limiting MTCT of HIV. The possible role of micronutrients like zinc and selenium has also been implied (UNAIDS/WHO, 1999). Women with vitamin A levels below 1.4 μM/L were found to have a 4.4-fold higher risk of transmitting HIV infection to their offspring (Semba *et al.*, 1994).

Food Safety: HIV-infected individuals, particularly those living in or travelling to developing countries, should protect themselves against food- and water-borne infections. Since public water supply may not be safe in all situations, it is better to consume boiled and cooled water or packaged bottled water. Cleanliness in the kitchen is mandatory. Raw vegetables and fruits should be washed carefully in clean water. Leftovers ought to be immediately refrigerated and consumed within 3 days. All packaged foods should be checked for date of expiry and date-expired products should be discarded (Fact Sheet 800, 2006).

Personal Hygiene: The frequency of daily bath should be according to the climate. Teeth should be brushed twice daily: after waking up in the morning and before bedtime. Regular haircut and cutting of nails of fingers/toes is a must. Emollient creams can be helpful for patients with dry skin and scalp, particularly in winter. The hands must be washed before every meal and after defecation. After each outdoor trip, hands, feet, and face should be washed. Clients must be advised to disinfect their homes periodically (NACO, Handbook for Counsellors). Sharing of towels may result in non-sexual contact transmission of STIs such as gonorrhoea, particularly in women due to anatomical differences. While washing after defecation, some individuals tend to wash from anal region towards the genitalia. This causes transfer of commensal bacteria such as *Escherichia coli* and *Streptococcus faecalis* to the external genitalia, where they act as pathogens. Due to anatomical differences, women are more vulnerable to this type of infection.

Occupation and Recreation: The HIV-infected client should continue working, if possible and remain occupied in productive or meaningful activities. Socialising with friends and family members is helpful. The client should have a free and frank discussion about his or her diagnosis with friends and family members.

Safer Sex: *Safer sex* refers to any sexual act in which, there is no direct contact with body fluids of the sexual partner. Consistent and correct use of male or female condom during each act of sexual intercourse even with steady partners and avoiding sexual activity with *multiple* (known or unknown) or *casual* partners can prevent reinfection and transmission of infection to others. Condoms should be used each and every time an individual has sexual contact (vaginal, anal, and oral) and with every partner. Most condoms are pre-lubricated to increase sexual stimulation. Oil-based lubricants such as vaseline, oils, or creams can damage male and condoms. If the condoms are not lubricated, *water-based lubricants* should be used. To be safe, one should *assume* that one's partner is infected with HIV. In situations where both partners are already infected, safer sex practices help in preventing reinfection with HIV, or infection with a different strain of HIV and/or other STDs. Use of alcohol or drugs before or during coitus increases the risk of ignoring safer sex guidelines or faulty use of condoms (Fact Sheet 151, 2006).

Medical Aid: The client should identify and avoid potential and actual stress factors and scrupulously follow advice for preventive care. He or she should be advised to seek medical attention for health problems and cover all open wounds with bandage or plaster. Ready-to-use wound dressings are available under brand names such as Bandaid and Handyplast (NACO, Handbook for Counsellors).

Child Immunisation: If an HIV-positive baby has *no symptoms* of HIV-related diseases, all the vaccines (live and killed) are to be given as per the national schedule. If the baby is *symptomatic*, all the vaccines *except* live vaccines should be given as per the national schedule (NACO, Training Manual for Doctors).

Vaccinations for Adults: With progression of HIV infection, the strength and duration of immune response to vaccines may be adversely affected. There is not much published research on vaccination of persons with HIV. The health care provider should consider the risk of possible adverse effects vaccination and the likelihood of infection with vaccine-preventable diseases. In general, HIV seropositive persons should *not* receive live vaccines and avoid close contact with persons who have been immunised with live vaccines in the last 2–3 weeks. Tetanus toxoid may be safely administered. HIV-positive persons at risk of hepatitis B or those exposed should receive a complete course of vaccination because HBV could cause serious infection in immune deficiency states (Fact Sheet 207, 2006).

International Travel: Countries have varying immunisation requirements for entry. Inactivated (killed) versions of typhoid or polio vaccines may be safely administered to HIV-positive travellers. Most countries may accept a letter from a health care provider explaining that the traveller has a medical reason not to be immunised with a live vaccine (Fact Sheet 207, 2006). However, travellers arriving in India from yellow fever endemic countries of Africa and South America should possess a valid certificate of immunisation using 17D live attenuated yellow fever vaccine, or else be quarantined. Travellers should avoid unprotected sexual activity with *multiple* (known or unknown) or *casual* partners because of the risk of infection with different serotypes of HIV.

20.6 – PARTNER COUNSELLING

Partner counselling (also called "partner management" and "contact tracing") is the process of contacting the sexual and/or drug-injecting partners of an HIV-positive individual, also called the "source client". WHO/UNAIDS recommends that the previously used term "partner notification" was associated with coercion and pressure and hence the word "partner counselling" should be used instead (UNAIDS/WHO, 2000). The purpose is to encourage the partners to come in for HIV counselling and testing. Where possible, confidentiality of the source client is maintained and partner counselling is done with the source client's consent. Partner counselling has been used as a public health response in case of STIs like syphilis and gonorrhoea. As per decision of the Supreme Court of India, partner counselling has been included as a component of the National HIV Policy. The National AIDS Committee has a policy of encouraging HIV-positive individuals to disclose their HIV status to their sexual partner (NACO, Handbook for Counsellors). All HIV-positive individuals should be encouraged to disclose their HIV status to their spouse/sexual partners. However, the attending physician should disclose the HIV status to spouse/sexual partners, only after proper counselling.

Procedure: After obtaining consent of the client, the spouse/sexual partner is counselled. Only then the HIV status of the client is disclosed.

Precautions: It is the counsellor's duty to conceal the identity of the client, and to ensure support, to prevent family disruption, and violence (WHO/UNAIDS, 1999). Female clients have very valid reasons such as possibility of abandonment and physical violence, for fearing to disclose their HIV status to their husbands/partners (United Nations, 1998).

20.6.1 – Types of Partner Counselling

Mandatory Partner Counselling: The source client is legally bound to reveal the names of his or her sexual, or drug-injecting partners. This approach has many disadvantages. People will be deterred from using voluntary counselling and testing (VCT) services because of fear of disclosure. It is impracticable to implement such a scheme on a large scale due to the costs of training and deploying a large number of personnel to trace and counsel all the partners whose names are given. Many HIV-positive individuals may not remember the names of their sexual partners and it is not possible to force them to reveal such names (UNAIDS/WHO, 2000).

Voluntary Partner Counselling: Since confidentiality is maintained, voluntary partner counselling can create a climate of trust (UNAIDS/WHO, 2000).

Third Party Counselling: The counsellor may be authorised to notify an identifiable third party (whose identity is known to the counsellor), if the HIV-positive client does not wish to reveal the name of his or her partners, in spite of repeated efforts by the counsellor and there is a danger of the third party being affected. Third party counselling is indicated only when client has been thoroughly counselled about the need for partner counselling, but the client has failed to achieve the desired behavioural changes and risk of HIV transmission to an identifiable third party. The counsellor should give prior notice of his or her intention to counsel the third party (UNAIDS/WHO, 2000).

20.6.2 – Encouraging Partner Counselling

1. Legislation – Should protect the principles of confidentiality and consent, and define the circumstances under which partner counselling may take place, without the consent of the source client (United Nations, 1998).

2. Establishing codes of Professional Conduct – For medical and social service professionals, along with provision of penalties for unethical conduct. Health care institutions should also constitute mechanisms for ensuring accountability in relation to ethics (UNAIDS/WHO, 2000).
3. Training – Of health care providers and counsellors in techniques for voluntary partner counselling (UNAIDS/WHO, 2000).
4. Support Mechanisms for Women – Effective legal safeguards and social support mechanisms for women who are more vulnerable to stigma, physical violence, and abandonment (UNAIDS/WHO, 2000; United Nations, 1998; Gielen *et al.*, 1997).

20.7 – COUNSELLING IN SPECIAL SITUATIONS

Counselling other Family Members: Involvement of family members and friends of HIV-infected persons is essential in advanced stages of the disease or if cerebral involvement develops. They need access to accurate information, social support, and referral services. In case of persons belonging to socially marginalised groups, their "family" of choice may be different from their family of origin. Health care providers should recognise significance of these relationships (UNAIDS/WHO, 2000; NACO, Handbook for Counsellors). Other family members of the client may be counselled after obtaining consent of the client to
(a) Reduce stress, anxiety and provide moral support.
(b) Inform family members about available support systems in the community.
(c) Prepare family members for *home care* of the patient – by informing them about client's risk of developing various opportunistic infections and precautions for protecting themselves, and by training them in safe disposal of body fluids, secretions, and excretions of the patient.
Precautions – Some family members may seek details, which the HIV-infected person may not like to be revealed. In such situations, health care providers should to strike a balance, in the best interests of the patient.
HIV-Positive Tuberculosis Patients: In India, it is estimated that about 50 per cent of the entire adult population harbours *Mycobacterium tuberculosis*, the causative organism. When compared to HIV-negative persons, the spread of the disease is faster when an HIV-positive individual is newly infected by the organism. The infection rapidly advances to clinically active disease, which shortens the patient's lifespan. The counsellor should ensure that every person reporting to the VCT centre with symptoms of tuberculosis is referred to the designated microscopy centre for three sputum examinations, as per protocol of RNTCP. If the client is diagnosed to be suffering from tuberculosis, the counsellor should emphasise that tuberculosis is *curable* when regular and complete treatment is taken under supervision and help patients in identifying a convenient location for DOTS. Patients should be told that the diagnosis and treatment for the disease is available *free of cost* at health centres run by the government, municipalities, and certain NGOs. The counsellor should also advise sputum-positive patients about screening their contacts.

HIV-Positive Pregnant Women: Preferably involve the spouse (or sexual partner), with the consent of the client, so that the couple can support one another in decisions regarding pregnancy and safe sex (NACO, Handbook for Counsellors). Counselling aims to enable HIV-infected pregnant woman to make informed choices to decide whether to continue the pregnancy, or to terminate it; discuss interventions such as ARV treatment and infant feeding options to prevent MTCT of HIV if the woman decides to continue the pregnancy; and decide about her sexual behaviour and future fertility (NACO, Handbook for Counsellors). The counsellor should inform the pregnant woman about

(a) Process of child birth
(b) Implications of a positive HIV test and the risk of the baby getting infected during pregnancy, childbirth, or breast-feeding
(c) Benefits of ARV therapy if available
(d) Infant feeding options
(e) Availability of support services in the community (NACO, Handbook for Counsellors; CDC, 1995; UNAIDS, 1999)

Counselling of Couples: Counselling aims at promoting safe sexual practices and encouraging disclosure to the sexual partner, and thus reducing transmission of HIV. Counsellors should be aware that many individuals are reluctant to get tested with their partner. Both partners should give their consent to undergo HIV test. *Discordant couples* are those, who do not have the same HIV test result. The counsellor should help the couple cope with feelings of anger and resentment, encourage them to accept safe sex practices to prevent HIV transmission to the HIV-negative partner and discuss strategies to help the HIV-positive partner to live with the infection and also family planning (CDC, 1995).

Premarital Counselling: Voluntary premarital counselling and testing, with both individuals giving consent, can help the couples in deciding about having children and planning for the future (CDC, 1995).

Partner Violence: While dealing with cases of partner violence, the counsellors should be aware of how lack of empowerment of women and their low status in society vis-à-vis men can adversely affect their ability to protect themselves against HIV. They need to be trained to ask specific questions about partner violence (CDC, 1995).

Partners of HIV-Positive Persons: Encourage the sexual and/or drug-injecting partners of HIV-positive individuals to participate in prevention counselling. Discuss strategies for maintaining HIV-negative status by behavioural changes (CDC, 1995).

Indeterminate Test Results: If the result of the HIV test is *indeterminate* (inconclusive), ask the client to come back for a *repeat* test, about 3 months after his or her last exposure to HIV. The components of counselling are similar to that for post-test counselling for HIV-positive individuals.

Victims of Rape: Counselling helps to prepare victims of rape for a positive result of HIV test. The client should be informed about use of ARV drugs for PEP. The client may be referred to a counsellor who has experience of counselling rape victims on issues such as safe abortion and STI testing and treatment (CDC, 1995).

Occupational Exposure: Health care providers who have been exposed to HIV should be advised to
(a) Abstain from unprotected sex and to avoid pregnancy until a HIV seronegativity is confirmed
(b) Avail of ARV therapy for PEP
(c) Periodically undergo repeat HIV test as advised (NACO, Training Manual for Doctors)

Young People: Counselling services must cater to the specific needs of young people, and take into account the social context of their lives, since they comprise more than 50 per cent of newly infected persons worldwide. The counselling needs to be age-appropriate, using familiar language, and examples (CDC, 1995).

IDUs: Personalised interactive models of counselling that set goals for the client may be successfully use these models for bringing about behaviour changes and thereby reducing the risk of contracting HIV infection. IDUs ought to be referred to deaddiction and rehabilitation services (CDC, 1995).

Terminally Ill Patients and their Families: Many terminally ill patients may opt for *home care*. Counselling can help in providing emotional support to the terminally ill patient, his or her family members, relatives and friends and mentally prime them to deal with impending death. The counsellor should impart information on
(a) *Home care* of the patients, precautions while handling patient's body fluids and handling the patient's body after death
(b) Legal issues (Will, pensions, and power of attorney)
(c) Occupational issues (sick leave and loss of capacity to work)
Psychiatric interventions including specific counselling may be necessary.

20.8 – REFERRALS

20.8.1 – Types of Referrals

Source Referral: In this type of referral, HIV-positive clients are encouraged to counsel their partners about the possibility of their exposure to HIV. Health care providers are not directly involved, but they advise the affected person about the nature of information to be passed on to their partners, and the ways of doing it.

Provider Referral: The source client is encouraged to reveal the names of partners. The counsellor confidentially counsels the partner directly, without revealing the name of the source client to the partners.

Conditional Referral: The health care providers obtain the names of partners from the source client, who is then allowed a reasonable time period to counsel his or her partners. Failing this, the health care provider or counsellor directly counsels the partners, without naming the concerned client (UNAIDS/WHO, 2000).

20.8.2 – Procedure for Referrals

A client availing of counselling services may require referral to treatment, care, and support services. The counsellor should discuss client's stated needs and willingness to receive referral services, explain what the client should expect from the referral, ask the client if there are places he or she would prefer. If more than one facility provides the same service, a written referral note is given to the client, recording the services that the client has availed of. If any service has not been used, the reasons for not utilising services are listed. This will reveal how responsive specific services are to the needs of the clients (CDC, 1995).

20.9 – EVALUATION OF COUNSELLING

20.9.1 – Keeping Records of Counselling Sessions

Recording the client's consent and his or her HIV test results, securely storing confidential information in filing cabinets, keeping records of the counselling sessions, and jotting down the salient points or words are important aspects of record keeping. The salient points or words may be used to complete the records after the session. Writing during the counselling sessions may distract the client. Record keeping will permit continuity, even if there are different counsellors for subsequent sessions and prevent the client from having to repeat the same information from the previous session (CDC, 1995).

20.9.2 – Methods for Evaluation

Since direct observation of a counselling session is difficult due to the confidential nature of HIV counselling, following methods may be used for evaluation: taping the session with the client's consent, using "dummy" clients, role play, and using feedback from clients through "client satisfaction surveys" (CDC, 1995; UNAIDS, 2000).

20.9.3 – Areas for Evaluation

The quality of counselling is evaluated in the following areas:
(a) Interpersonal relationship
(b) Gathering information
(c) Giving correct and comprehensible information, and handling special situations (CDC, 1995; UNAIDS, 2000)

Where named-based or coded reporting is used, it is necessary to periodically assess the degree of incomplete reporting, incomplete diagnosis, and duplication (multiple reporting) of cases (UNAIDS/WHO, 2000).

20.10 – IMPROVING COUNSELLING SERVICES

20.10.1 – Strategies for Supporting Counsellors

1. Regular supervision of counselling sessions
2. Periodic training sessions and seminars for acquiring new skills
3. Conducting case conferences to address issues and difficulties that arise during counselling sessions
4. Providing opportunities for peer support/sharing of skills and experiences
5. Arranging for regular support from mentors (professionals carrying out similar work who provide support to staff) who can help in the professional growth of the counsellors
6. Preventing "burnout" of counsellors by limiting the number of clients a counsellor can see in a day or a week (CDC, 1995)

20.10.2 – Confidentiality and Security of Information

It is essential to develop infrastructure such as data storage and transmission systems that ensure the physical security of data and electronic security of computer files. Legislation to protect against breaches of confidentiality and prevent *non-public health use* of data will increase public confidence in surveillance (UNAIDS/WHO, 2000; CDC, 1995).

REFERENCES

Centers for Disease Control and Prevention (CDC), 1995, US Public Health Service Recommendations for HIV Counselling and voluntary testing for HIV of pregnant women. Morb Mort Wkly Rep 44: 1–15.
Gielen A.C., O'Campo P., *et al.*, 1997, A woman's disclosure of HIV status – experiences of mistreatment and violence in an urban slum setting. Women's Health 25(3): 19–31.
Gottlieb M.S., 2001, AIDS – past and future. N Engl J Med 344: 1788–1790.
National AIDS Control Organisation (NACO), 2001–2002. HIV/AIDS handbook for counsellors (2001–2002). New Delhi: Government of India.
National AIDS Control Organisation (NACO). Training manual for doctors. New Delhi: Government of India.
New Mexico AIDS Education and Training Center, 2006, Fact Sheet 151. Safer sex guidelines. University of New Mexico Health Sciences Center. www.aidsinfonet.org. Revised 18 July.
New Mexico AIDS Education and Training Center, 2006, Fact Sheet 207. Vaccinations and HIV. University of New Mexico Health Sciences Center. www.aidsinfonet.org. Revised 15 March.
New Mexico AIDS Education and Training Center, 2006, Fact Sheet 800. Nutrition. University of New Mexico Health Sciences Center. www.aidsinfonet.org. Revised 1 May.
New Mexico AIDS Education and Training Center, 2006, Fact Sheet 801. Vitamins and minerals. University of New Mexico Health Sciences Center. www.aidsinfonet.org. Revised 18 April.

New Mexico AIDS Education and Training Center, 2006, Fact Sheet 802. Exercise and HIV. University of New Mexico Health Sciences Center. www.aidsinfonet.org. Revised 14 February.

Semba R.D., *et al.*, 1994, Maternal vitamin A deficiency and mother to child transmission of HIV-1. Lancet 343: 1593–1597.

UNAIDS/WHO, 1999, HIV in pregnancy – a review. Occasional Paper No. 2. Joint United Nations Programme on HIV/AIDS 1999, pp 6–37.

UNAIDS/WHO, 2000, Opening up the HIV/AIDS epidemic. www.unaids.org. August.

UNAIDS, 1999, Counselling and voluntary HIV testing for pregnant women in high HIV prevalence countries – elements and issues. UNAIDS Best Practice Collection, 99.44E. www.undaids.org

UNAIDS, 2000, Tools for evaluating HIV voluntary counselling and testing. UNAIDS Best Practice Collection, 00.09E. www.unaids.org

United Nations, 1998, HIV/AIDS and human rights – international guidelines. New York: United Nations. HR/PUB/98/1:13.

WHO/UNAIDS, 1999. Questions and answers on reporting, partner notification, and disclosure of HIV and/or AIDS sero-status – Public Health and Human Rights implications. www.who.int

CHAPTER 21

VOLUNTARY COUNSELLING AND TESTING SERVICES

Abstract

VCT is a process by which an individual undergoes counselling, enabling him or her to make an informed choice about being tested for HIV. VCT strategy promotes knowledge and awareness about HIV infection and safer sexual practices, encourages community response in developing support systems for HIV-affected individuals, allows potential clients to decide whether to take the HIV test and promotes behaviour change to prevent the transmission of HIV. This can be an entry point for provision of various HIV-related services. In *functionally integrated* VCT models, most key services, including HIV testing are outsourced while *structurally integrated* VCT models provide key services, including in-house HIV testing. Programme managers need to know the potential challenges associated with integration of VCT in family planning settings, so that they can take these into account during the planning stages. For successful integration, both qualitative and quantitative methods should be used to assess the organisational objectives, internal capacity of the organisation, resources needed to integrate the services, community needs, and cultural, social, and economic barriers to accessibility of services.

It is essential to increase the access to VCT services through a variety of settings. Among other suggested models for VCT are provision of HIV counselling along with family planning counselling in antenatal and postnatal care settings, free-standing services, private sector models, home-testing, outreach programmes, and social marketing. Youth seem to prefer VCT along with other youth-friendly services such as skill building courses, and sports activities. Innovative approaches are necessary to reach groups such as pregnant women, out-of-school youth, and IDUs.

Key Words

Assessment of Community, Confidentiality, Counselling, Family planning, Models for VCT, Operational assessment, Organisational assessment, Shared confidentiality, Voluntary counselling and testing

21.1 – INTRODUCTION

VCT is a process by which an individual undergoes counselling, enabling him or her to make an informed choice about being tested for HIV (UNAIDS, 2000). Counselling is a confidential dialogue between the client and the health care provider, which enables the client to take realistic personal decisions. Any person (professionally trained counsellor, doctor, nurse, paramedical worker or volunteer) with necessary patience, dedication, commitment, aptitude, knowledge, and skills can undertake counselling. Pre- and post-test and follow-up counselling is offered to *any* client who is contemplating about taking the HIV test. The strategy of combining counselling services with voluntary testing gives prospective clients the *informed choice* to know their HIV status and to decide whether or not to take the HIV test.

21.1.1 – Issues in VCT

Ideally, each country should determine the procedures for informed consent for using VCT services. Involvement of parents or guardians while testing and reporting results of adolescents is a key issue. According to the *Kenyan National VCT Guidelines* issued in 2001, "mature minors" do not need parental consent. Mature minors include those individuals younger than 18 years who are "married, pregnant, parents, engaged in behaviour that puts them at risk, or are child sex workers". Though the HIV test results are to be disclosed only to the client, the guidelines say that counsellors should encourage those younger than 18 years to inform their parents about the results (NASCOP, 2001). Among 240 young people tested in Kenya and Uganda, fewer than one-fourth told their parents about their test results (Horizons Program, 2001). Another issue is that clients report to VCT centres after onset of non-herpes skin eruptions or herpes zoster. There were no significant gender differences in the distribution of these conditions in a study conducted in western India (Maredia *et al.*, 2004).

21.2 – UTILITY OF VCT SERVICES

Utility for the Community: VCT strategy promotes knowledge and awareness about HIV infection and safer sexual practices. It also encourages community response in developing support systems for HIV-affected individuals. It allows potential clients to decide whether to take the HIV test (UNAIDS, 2000; UNAIDS, 2001). It is an effective strategy for promoting behaviour change to prevent the transmission of HIV (UNAIDS, 2001). This can be an entry point for provision of various HIV-related services, such as prevention of MTCT (PMTCT), clinical management of HIV-related illnesses, and provision of support services (psychological, social, and legal).

Utility for HIV-Infected Clients: This strategy enables early initiation of desirable behaviour changes to prevent transmission of infection and reinfection. It empowers the individual to cope with the diagnosis, take *informed decisions* about informing sexual partners, sexual relationships, safe sex, pregnancy and breastfeeding, and plan his or her future. VCT strategy improves compliance to various interventions and provides information on the available care and support services.

Utility for HIV-Negative Clients: It enables early initiation of behaviour changes to prevent infection, helps in informing partners about the benefits of getting tested, and improves the adoption of family planning services.

21.3 – ESSENTIAL ELEMENTS OF VCT PROGRAMME

Client-centred counselling must be offered to *any* client who is contemplating about taking the HIV test. This comprises pre- and post-test counselling and follow-up sessions. After counselling, the client should be able to take an *informed decision*, whether to take the HIV test or not. Oral or written informed consent is to be obtained, without any duress. HIV testing is preferably done at the same site. Alternatively, the client may be referred elsewhere. The result of the HIV test is revealed only to the client. If the client reveals a desire for *shared confidentiality* (i.e. sharing the result with spouse, partner, a family member, or a close friend), the counsellor should offer guidance. Seropositive status is *not* to be used as a parameter for differentiation or discrimination between groups of clients/patients. The physical environment at the VCT centre ought to be conducive to confidential discussions between the client and the counsellor. Clients ought to have access to prevention, care and support services, with maintenance of the client's confidentiality and privacy when referral services are utilised. The VCT services should comply with the protocols and national laws related to the provision of HIV-related services (UNAIDS, 1999; Baggaley *et al.*, 2001).

21.4 – METHODS FOR ASSESSMENT

Factors to be Assessed: The following factors should be assessed by both *qualitative* and *quantitative* methods during the planning process itself.
(a) Organisational objectives
(b) Internal capacity of the organisation
(c) Resources needed to integrate the services
(d) Community needs
(e) Cultural, social, and economic barriers to accessibility of services
(f) Review of socio-demographic indicators and data on characteristics of the local population
(g) KAP of the potential clients regarding sexual and reproductive health (SRH)
(h) In-depth interviews, meetings or focus group discussions (FGD) with key informants, family planning staff, clients, community members, and other organisations providing HIV-related services (UNFPA, 2002)

Duration of Assessment: The time period for assessment would depend on size of the community, complexity of factors affecting the HIV prevalence rates, and number of other services available. Adequate time period for most organisations to compile data and assess the situation is 2–6 weeks (UNFPA, 2002). The IPPF/WHR self-assessment module can complete the assessment in approximately 1 week (IPPF/WHR, 2000).

21.5 – ASSESSMENT OF THE COMMUNITY

Community assessment is useful for building on existing services, rather than duplicating them; knowing the responses of local organisations to the idea of introducing VCT services; and compiling a list of referral centres for care and support services (UNFPA, 2002). Community assessment comprises the following:

Epidemiological Assessment: To decide the target groups for VCT services (high-risk groups and general population) and to calculate prevalence rates of HIV infection to prioritise allocation of resources for activities.

Knowledge and Attitude to HIV/STIs: To study the knowledge and attitudes of the community by listening surveys, observation of education sessions, and field-based research.

Estimating Level of Risk Among Potential Clients: These include gender-related power relations that affect negotiation of safer sexual relations, economic pressure resulting in commercial sexual activity, relations between young women and older men, and cultural practices contributing to high HIV prevalence rates (UNFPA, 2002; NACO, Training Manual for Doctors). If the risk factors are high in the *general* population, routine risk assessment is necessary. If the risks are high in certain *subgroups* in the population, specialised outreach services may be suitable.

Charting the Existing HIV/STI-Related Services: The following information should be collected by interviewing staff of local organisations and used for charting (or "mapping"): existing HIV/STI-related services, community needs, and gaps in services.

21.6   ORGANISATIONAL ASSESSMENT

Self-assessment module (IPPF/WHR, 2000) can be used to conduct an organisational assessment. The criteria for assessment include:
(a) Compatibility with the organisation's goals
(b) Criteria used by the organisation for deciding the services to be offered (expressed community needs, epidemiological data, opinions of staff, resource and cost–benefit analyses, and funding/income generation)
(c) Knowledge, attitudes, level of training, and skills of the staff for delivering VCT services
(d) Existing resources and infrastructure
(e) Compatibility of monitoring and evaluation plan with that for family planning services
(f) Review of relevant legislation, regulations, and national policies (UNFPA, 2002).

21.7 – OPERATIONAL ASSESSMENT

Clients, community members, and staff members from other organisations and family planning personnel are to be involved in operational assessment. Activities for integration are to be identified and operational issues discussed with these stakeholders. The issues for operational assessment are

21.7.1 – Operational Similarities and Differences

Greater degree of similarity between the existing services and VCT indicates that the required financial investment would be less, and that it would be easier to integrate. Provider skills required by both family planning and VCT services are clinical knowledge and skills related to reproductive health, communication skills related to influencing behaviour, and managerial skills for tasks such as condom distribution. If provider skills are similar, providers may be given limited training related to counselling tasks. Integration is facilitated if the setting already provides privacy and permits confidentiality. The existing educational materials can be adapted to suit VCT services. Similarly, the existing logistics system and MIS may be modified to suit the needs of integrated VCT services.

21.7.2 – Selecting Target Groups

In all areas, VCT services are offered to persons with signs and symptoms of HIV/AIDS; pre- and post-test counselling to those who are concerned about their HIV status. In *resource-poor* areas, VCT services may be targeted at selected clients. In areas with *high HIV prevalence*, prevention counselling should be given to all those who use family planning services, and then pre-test counselling may be offered to high risk groups or those interested in VCT. In *low-prevalence* areas, an initial risk assessment is carried out and VCT should be offered to those considered to be at "high risk".

21.7.3 – Identifying Additional Resource Needs

Budgeting (for capital and recurring costs) will depend on:
(a) Type of VCT model adopted (structural or functional integration)
(b) Prevalence of HIV infection in the area (large number of clients will lead to higher costs)
(c) Higher degree of similarity of existing services to VCT (fewer additional resources will be required)

Resources for Infrastructure: This includes space for HIV testing and storage space for equipment, kits, and other commodities; cost of renovation of existing site, with cost of disruption to existing sites during renovation; and counselling room that is well-ventilated, provides privacy, and has adequate space for two chairs, a table, and a filing cabinet for securing confidential client information.

Resources for Materials: For community mobilisation, additional resources may be required for preparing leaflets, posters, and audio-visual aids. If the VCT service adopts a structurally integrated model (that provides for on-site HIV testing), purchase of test kits would involve recurring costs. The requirement of HIV test kits would depend upon the number of clients.

Resources for Hiring Staff and Training: More staff members should be hired to avoid increasing the workload of the existing staff and to maintain the quality of existing services. Additional staff will be needed for HIV testing if the structural model (on-site HIV testing) is adopted and if the number of clients is high. Both new and existing staff will need basic training in counselling skills (for counsellors), skills for effective outreach education, skills in testing technology (for laboratory staff), and supervisory and managerial skills (for managers).

Resources for Referral Networks: Recurring costs in terms of travel costs and staff time for visits, joint training, networking, and attending meetings with other organisations providing HIV-related services.

Resources for Monitoring and Evaluation: Monitoring and evaluation requires additional staff time costs for developing tools for monitoring and evaluation, interviewing users and non-users of VCT services, attending regular meetings, and compiling and analysing data on monitoring and evaluation.

21.8 – MODELS FOR VCT SERVICE DELIVERY

In *functionally integrated* models, most key services, including HIV testing, are outsourced (provided through a referral service), and they require well-developed referral systems. *Structurally integrated* models provide key services, including in-house HIV testing (on the premises).

Choice of VCT Model:
Each model has its merits and demerits. There being no ideal model for provision of VCT services, the choice of model will depend on:
(a) Community needs and attitudes to HIV epidemic, political attitudes, and commitment
(b) Prevalence of HIV seropositivity and stage of the epidemic in the area
(c) Availability of financial and other resources
(d) Setting in which VCT services are offered (WHO/UNAIDS, 2001).

Classical Model for VCT:
This model offers the client pre- and post-test counselling along with HIV testing. Several variations within the *classical model*, which may be appropriate in different settings, include: group information sessions followed by brief individual pre-test counselling; pre- and post-test counselling for the couple, or family; and restricting post-test counselling, only to those who test HIV positive. Clients who test HIV negative are merely informed of their test result, *without* post-test counselling. This is done in some countries with low HIV prevalence.

21.8.1 – Integrated Model

The integration of SRH services was advocated by the International Conference on Population and Development (ICPD) held in Cairo (Egypt), in 1994. *Integration* envisages the provision of a constellation of services such as family planning,

MCH, and prevention and management of STIs including HIV (Fleishman *et al.*, 2002; Caldwell & Caldwell, 2002). Integration of HIV-related services within family planning settings is beneficial to both clients and family planning service providers. In this model, VCT services are incorporated into ongoing services provided in the family planning setting. However, it is not obligatory for the family planning service providers to provide all the VCT services. Therefore, organisations will have to decide as to which VCT-related services can be integrated in their setting. A variant of this model envisages addition of HIV counselling and testing services to existing family planning services, while VCT programmes start providing contraceptive counselling, particularly on correct and consistent use of condoms. This will address the dual risk of HIV infection and unintended pregnancy faced by women without integrating these two programmes.

21.8.1.1 – Justification for an integrated model

Both family planning and VCT have a common target group namely sexually active people, who are potentially at risk of contracting STI and HIV. Clients with STI are those who have engaged in unsafe sexual behaviour and therefore, they are vulnerable to both HIV infection and unwanted pregnancy. Family planning and VCT services require similar infrastructure because they have to maintain confidentiality and respect the client's privacy. Integrating VCT may require minimal changes to the existing infrastructure for family planning services, which reduces the cost of establishing VCT services.

Consequent to the integration of STI services in some family planning programmes, the service providers have already been trained in assessing clients' risks, managing STIs, and providing appropriate referrals. These skills are also required for providing VCT services. Both family planning and VCT services require the constant supply of similar services and commodities, such as male and female condoms. The same logistic system can be used to ensure the supply of other commodities such as HIV test kits. The existing family planning outreach workers such as community-based distribution agents and peer educators, can help in raising awareness about HIV and promote VCT services in the community. Integration of VCT in family planning setting may help in community approval of taking the HIV test as "normal" and ultimately, the stigma associated with HIV infection may diminish. Family planning clinics are accessible and already serve large sections of the population. Increased involvement of males in contraception and other family planning services is a possible outcome of provision of VCT services (UNFPA, 2002).

The programme for PMTCT alone cannot meet the goals for reducing HIV infection in infants. Use of condoms can prevent HIV infections in uninfected women and prevent unwanted pregnancies in women living with HIV (WHO/UNFPA, 2006). Family planning services in sub-Saharan Africa have been found to be more cost-effective in PMTCT of HIV than the provision of nevirapine (Sweat *et al.*, 2004). An expenditure of US$45,000 to increase contraceptive services would prevent 88 HIV-positive births whereas for the same cost, provision of nevirapine would prevent only 68 such births (Reynolds *et al.*, 2006).

21.8.1.2 – Potential challenges in integration

Programme managers should be vigilant about the potential challenges associated with integration of VCT in family planning settings, so that they can take these into account during the planning stage itself. Family planning clients may perceive a reduction in quality of family planning services, as a result of introduction of VCT services. For maintaining quality of family planning services after integrating VCT, it is essential to identify managerial capacity and requirement of financial and other resources. If the demand for VCT services is high, the existing family planning personnel who take on counselling duties may have an increase in workload. This problem is overcome by providing adequate training to the existing staff and by hiring more staff, if necessary. VCT counsellors should discuss sexual behaviour in detail and not just focus on contraceptive methods. With adequate training, counsellors can focus on STI/HIV and family planning issues. Non-health workers (people living with HIV/AIDS, social workers, and volunteers) have a role in VCT activities such as outreach education and counselling. In order to overcome the stigma attached to HIV, the family planning clients and the community should be informed about HIV infection and HIV testing (UNFPA, 2002).

21.8.2 – Other Models

Integrating family planning services with the ongoing services for PMTCT in 14 high-prevalence countries could double the number of HIV-positive births averted in addition to saving women's lives and averting child deaths (Stove *et al.*, 2003). Provision of HIV counselling along with family planning counselling in antenatal and postnatal care settings would help pregnant women in avoiding infection and help in identifying HIV-infected pregnant women (Best, 2004).

Among other suggested models are free-standing services, private sector models, and home-testing (FHI, 2002; WHO/UNAIDS, 2001). Outreach efforts and social marketing are also useful (Boswell & Baggaley, 2002). Surveys in many countries reveal that youth prefer VCT along with other youth-friendly services such as skill building courses and sports activities. Innovative approaches are necessary to reach groups such as pregnant women, out-of-school youth, and IDUs. More information is needed on coping mechanisms in persons who test HIV positive, with whom they share HIV test results, who provides emotional support, and long-term outcomes of VCT programmes (Boswell & Baggaley, 2002).

REFERENCES

Baggaley R., Kawaye I., and Miller D., 2001, Counselling, Testing and Psychological Support. In: HIV/AIDS prevention and care in resource-constrained settings – a handbook for the design and management of programmes (P.R. Lamptey and H. Gayle, eds.). Arlington, VA: Family Health International. www.fhi.org

Best K., 2004, Family planning and the prevention of mother-to-child transmission of HIV: a review of the literature. Arlington, VA: Family Health International. www.fhi.org

Boswell D. and Baggaley R., 2002, Voluntary counselling and testing: a reference guide – responding to the needs of young people, children, pregnant women and their partners. Arlington, VA: Family Health International.

Caldwell J.C. and Caldwell P., 2002, Is integration the answer for Africa? Int Fam Plann Persp 28(2): 108–110.

Family Health International (FHI), 2002, Models of HIV voluntary counselling and testing (VCT) service delivery. Arlington, VA: FHI. www.fhi.org

Fleishman F.K.G., Hardee K., and Arawal K., 2002, When does it make sense to consider integrating STI and HIV services with family planning services? Int Fam Plann Persp 28(2): 105–107.

Horizons Program, 2001, HIV voluntary counselling and testing among youth: results from an exploratory studying Nairobi, Kenya and Kampala and Masaka, Uganda. Washington, DC: Population Council.

IPPF/WHR, 2000, Self-assessment module on integrating STI/HIV/AIDS services into sexual and reproductive health programmes. June.

Maredia J., Bharmal R.N., Gurav R.B., and Kartikeyan S., 2004, Profile of clients in voluntary counselling and testing programme for HIV/AIDS. J Comm Health 6(2): 76–80.

NACO, Training manual for doctors. New Delhi: Government of India.

National AIDS and STD Control Programme (NASCOP), 2001, National guidelines for voluntary counselling and testing. Nairobi: Ministry of Health.

Reynolds H.W., Janowitz B., Homan R., and Johnson I, 2006, The value of contraception to perinatal HIV transmission. Sex Transm Inf 7 Feb [PubMed]. PMID Nr 16505747.

Stove J., et al., 2003, Adding family planning to PMTCT sites increases the benefits of PMTCT. Issue Brief: Population and Reproductive Health. Washington, DC: USAID, October.

Sweat M.D., et al., 2004, Cost-effectiveness of nevirapine to prevent mother-to-child HIV transmission in eight African countries. AIDS 18: 1661–1671.

UNAIDS, 1999, Counselling and voluntary HIV testing for pregnant women in high HIV prevalence countries – elements and issues. UNAIDS Best Practices Collection 99.44E. www.unaids.org

UNAIDS, 2001, The impact of voluntary counselling and testing – a global review of the benefits and challenges. UNAIDS Best Practices Collection. www.unaids.org

UNAIDS, 2000, Voluntary counselling and testing (VCT). Technical Update. UNAIDS Best Practices Collection. www.unadis.org

UNFPA, 2002, Integrating HIV voluntary counselling and testing (VCT) services within family planning settings to prevent HIV infection in women – guidelines for implementation. www.unfpa.org

WHO/UNAIDS, 2001, Technical consultation on voluntary HIV counselling and testing: models for implementation and strategies for scaling of VCT services. Harare, Zimbabwe. 3–6 July.

WHO/UNFPA, 2006, Glion consultation on strengthening the linkages between reproductive health and HIV/AIDS: family planning and HIV/AIDS in women and children. Geneva: WHO. WHO/HIV/2006.02. www.who.int

CHAPTER 22

PREVENTION OF MOTHER-TO-CHILD TRANSMISSION

Abstract

Intrauterine transmission can occur as early as 8 weeks of gestation. Many intervention strategies have been suggested during the antenatal, intranatal, and post-natal periods. After counselling, if the client decides to continue her pregnancy, she should be informed about post-partum contraception and infant feeding options. While advising infant feeding options, the health care provider should bear in mind the health and socio-economic status of the mother, the cost of breast milk substitutes, and the risks of not breastfeeding. After delivery, the baby should be screened for HIV status and this should be preceded, and followed, by counselling of the mother or both parents. In asymptomatic babies, all the vaccines are to be given as per the national schedule. If the baby is symptomatic, all vaccines *except* live vaccines should be given as per the national schedule.

Key Words

Antiretroviral therapy, Breastfeeding options, Elective caesarean section, HIV in women, Infant feeding options, Intrapartum transmission, Intrauterine transmission, Medical termination of pregnancy, Mode of delivery, Mother-to-child transmission, MTCT estimation, Post-natal transmission, Transplacental transmission, Vaginal cleansing, Vitamin A, Vertical transmission

22.1 – HIV INFECTION IN WOMEN

In 1985, India joined the ranks of the first few countries that initiated serosurveillance among high-risk groups. The first group of seropositive individuals (10 female sex workers) was detected in April 1986. Within a short period of 18 months, it became apparent that heterosexual promiscuity was the major mode of transmission in India and that the seropositivity was low (about 4 per 1,000). HIV seropositive pregnant women (first detected in 1986) and their infants were followed up and HIV-infected children were detected in 1987–1988 (Ramachandran, 1990). Most HIV-infected women live in developing countries and have limited access to health care facilities. Occurrence of HIV-related diseases and eventual death of the mother devastates the family. Most HIV-infected infants also die by the age of 5 years. The uninfected infants face the frightening prospect of becoming AIDS orphans, with all its associated adverse consequences.

22.1.1 – Estimating Seroprevalence

The WHO estimates the number of HIV-positive women from available national or regional estimates of HIV infection and the male to female ratio reported in AIDS cases or large community surveys (Chin, 1990). A two-step procedure is followed to estimate the number of HIV-positive pregnant women and infants:

1. Number of births among HIV-infected women is calculated from available data on seroprevalence age-specific fertility rates in each country.
2. Number of infected infants is calculated on the assumption that 25 per cent of infants born to HIV-infected women are infected at birth.

It is difficult to calculate mortality rates for women and children in developing countries since reliable cause and age-specific mortality rates are not available. The WHO has developed the following model for calculating the impact of the HIV epidemic on the under-5 mortality rate (U5MR) in a population with U5MR of 100 per 1,000.

(a) If 5 per cent of pregnant women are HIV-infected, the U5MR will increase by 9 per cent.
(b) If 10 per cent of pregnant women are HIV-infected, the U5MR will increase by 18 per cent.
(c) If 20 per cent of pregnant women are HIV-infected (as in sub-Saharan Africa), the U5MR will increase by 36 per cent (Chin, 1990). Thus, the HIV epidemic has wiped out the decline in maternal and child mortality achieved in the last few decades.

22.1.2 – Need for Routine Screening for HIV in Pregnancy

Many experts advocate routine screening of all pregnant women for HIV infection on the same lines as screening for syphilis. The rationale for screening for syphilis is to provide therapeutic intervention to prevent intrauterine infection. The justification for routine HIV screening is as follows:

• ARV drugs are available to prolong the asymptomatic period.
• If found seropositive, she could be advised to take steps for preventing transmission.
• It enables identification of children who may need follow-up and social support.
• Although no therapeutic intervention is available to prevent intrauterine HIV infection, the pregnant woman may be advised to undergo MTP, if found seropositive in early pregnancy.

22.1.3 – Problems with Routine Screening

Even in countries where MTP is legal, many HIV-infected women may refuse to undergo MTP. Studies have shown that most women will not change their reproductive choices even after they are informed about their HIV-positive status

(Ahluwalia *et al.*, 1998). In most *developing* countries, most HIV-infected women do not belong to any high-risk group. Screening of all pregnant women is impracticable since the vast majority of them do not attend antenatal clinics. HIV screening facilities are neither available nor affordable. Thus, most infected women would continue to remain undetected. In the *developed* countries, infected pregnant women usually belong to high-risk groups, know about HIV infection, and most avail of antenatal care. However, many may not consent to undergo HIV screening (Ramachandran, 1990).

22.2 – RISK FACTORS IN PREGNANCY

Almost 30–50 per cent of the neonates acquired infection during the antenatal period and about 50–70 per cent during the intrapartum period (Working Group on MTCT, 1995). The risk of mortality in HIV-positive babies is 50 per cent in first 2 years of life and 80 per cent in first 5 years (Working Group on MTCT, 1995). MTCT is also called "transplacental" or "vertical" transmission.

Biological Risk Factors: The rate of transmission from male to female is two to three times higher than that from female to male (European Study Group, 1996; Royce *et al.*, 1997). It has been suggested that the *Langerhans' cells* of the uterine cervix may provide a portal of entry for HIV and some HIV serotypes may have higher affinity for these cells, and therefore, may be more efficient in heterosexual transmission (Soto-Ramirez *et al.*, 1996). Vulval and vaginal inflammation or ulceration may facilitate entry of the virus. Inadequately treated or silent chlamydial and other STIs may also act as cofactors for HIV infection and transmission (Hoegsberg *et al.*, 1990; Mayaud, 1997; Sewankambo *et al.*, 1997; Laga *et al.*, 1993). History of *genital ulceration* has been established as a cofactor for HIV acquisition (Latif *et al.*, 1989; Johnson *et al.*, 1989; Plourde *et al.*; 1994). Other non-sexually transmitted lesions of the cervix such as schistosomiasis may also facilitate HIV infection (Feldmeier *et al.*, 1994). Women in primary infection stage and those in terminal stage (AIDS) have a higher risk of transmitting the virus. History of HIV-positive babies in previous deliveries is a risk factor. Repeated pregnancies, poor nutrition, and other infections result in worsening of the already lowered physiological immunity in pregnancy, leading to rapid progression of the disease (NACO, Training Manual for Doctors).

Social Risk Factors: In developing countries, where male resistance to use of condom is common, female barrier methods are expensive or unavailable (UNAIDS/WHO, 1999; Feldblum *et al.*, 1995; Drew *et al.*, 1990). The desire and societal pressure to reproduce often prevent women from taking necessary precautions to guard against infection. Even after diagnosis of HIV seropositive status, most women will not change their reproductive choices (UNAIDS/WHO, 1999).

22.2.1 – Effect of Pregnancy on HIV Infection

In normal pregnancy, there is a decrease in levels of Ig molecules, complement, and a more significant decrease in CMI. These normal changes have led to the concern that pregnancy would accelerate the progression of disease in HIV-infected women (Minkoff, 1995; Rich *et al.*, 1995; UNAIDS/WHO, 1999). However, studies comparing HIV-infected pregnant women with never-pregnant controls have revealed no difference in –
(a) Progression of natural history of HIV
(b) Death rate between the two groups
(c) Birth weight of babies (Markson *et al.*, 1996; UNAIDS/WHO, 1999).
Higher maternal mortality seen in some African countries appears to be due to more women with advanced disease becoming pregnant, resulting in higher rates of HIV-related complications. Pregnant HIV-infected women were more likely to develop bacterial pneumonia. Infections were more common during the post-partum period in HIV-positive women (UNAIDS/WHO, 1999).

22.3 – PROBABLE TIMINGS FOR TRANSMISSION

A substantial proportion of infection occurs late in pregnancy ("intrauterine transmission") or at the time of delivery ("intrapartum transmission"). Knowledge about the likely timing of MTCT is important for planning feasible interventions (UNAIDS/WHO, 1999). If the virus is detectable within 48 hours of birth, the infant is determined to be infected *in utero*. Intrapartum infection is assumed if the viral studies are negative during the first week of life, but positive between 7 and 90 days (Bryson *et al.*, 1992).

Intrauterine Transmission: Intrauterine transmission can occur as early as 8 weeks of gestation. HIV-1 and p24 viral antigens have been detected in foetal specimens and placental tissue. The rapid progression of infections in some infants suggests that the infection may have been acquired *in utero* (Viscarello *et al.*, 1992; Backe *et al.*, 1993; Langston *et al.*, 1995).

Intrapartum Transmission: Data from a register of twins reveal that the first-born twin had a twofold higher risk of contracting HIV-1, as compared with the second-born twin (Goedert *et al.*, 1991). It is believed that the vaginal delivery of the first twin reduces the exposure of the second twin to the virus in the cervical and vaginal secretions (UNAIDS/WHO, 1999). Elective caesarean section has been shown to reduce the risk of transmission (European Collaborative Study, 1994). *Prolonged* (more than 4 hours) rupture of membranes is shown to increase the risk of transmission (Women and Infants Transmission Study Group, 1996). It takes some days after infection for viral studies to become positive. Negative reports of viral studies in about half of the infected infants, at the time of birth indicate that the transmission occurred during labour or delivery (UNAIDS/WHO, 1999).

Post-Natal Transmission: Post-natal transmission through breastfeeding is *assumed* to explain most of the differences in transmission in rates between developed countries (no or short period of breastfeeding) and developing countries (prolonged breastfeeding) (UNAIDS/WHO, 1999). Some researchers have recovered HIV in both free and cellular portions of breast milk. Infection through breastfeeding has been associated with maternal immune suppression, maternal vitamin A deficiency, and a lack of IgM and IgA in breast milk (van de Perre *et al.*, 1992; Semba *et al.*, 1994; Nduati *et al.*, 1995).

22.4 – FACTORS AFFECTING MTCT

22.4.1 – Viral Factors

High levels of maternal viraemia (advanced disease, high levels of p24 antigens in blood) increase the risk of transmission (European Collaborative Study, 1992; Mayaux *et al.*, 1997; Thea *et al.*, 1997). The new technique of quantitative PCR for DNA and RNA has shown an increased association between maternal viral load and risk of transmission from mother to child. Local viral load in cervicovaginal secretions is also an important determinant of transmission during the intrapartum period (John & Kreiss, 1996; Loussert-Ajaka *et al.*, 1997). Little is known about the role of mucosal HIV-1 antibodies and viral shedding in the genital tract, in intrapartum transmission (John & Kreiss, 1996; John *et al.*, 1997). Maternal ARV therapy during pregnancy and postexposure prophylaxis in the child after birth is thought to reduce transmission by reducing the viral load (Newell *et al.*, 1997).

22.4.2 – Maternal Factors

MTCT is more likely with low CD4 counts or high CD4/CD8 ratios in maternal blood (Minkoff, 1995; Fowler & Rogers, 1996). Unprotected sex during pregnancy has been linked to an increased risk of MTCT. This is attributed to: (a) repeated exposure to different viral strains during pregnancy, probably from multiple sexual partners (Bulterys & Goedert, 1995); (b) repeated infection with the same strain of HIV, probably from the same sexual partner; (c) effect of cervical or vaginal abrasions or inflammation; (d) increase in chorioamnionitis (Naeye & Ross, 1983); and (e) increased viral shedding in cervicovaginal secretions due to STIs (Ghys *et al.*, 1997). Women with vitamin A levels below 1.4 micromoles per litre had a 4.4-fold increased risk of transmission (Semba *et al.*, 1994). It has been suggested that vitamin A may have immune stimulatory properties and a role in maintaining the integrity of vaginal mucosa or placenta (UNAIDS/WHO, 1999). The possible role of micronutrients like zinc and selenium has also been implied (UNAIDS/WHO, 1999).

22.4.3 – Placental Factors

Placental infections, chorioamnionitis, and non-infectious conditions such as abruptio placentae have also been implicated (St Louis *et al.*, 1993; Boyer *et al.*, 1994). Breaks in the placenta can occur at any stage of pregnancy and may be related to transmission (Burton *et al.*, 1996). The Hofbauer cells of the placenta and trophoblasts probably express CD4 and are thus susceptible to infection. Maternal use of tobacco and "hard" drugs may cause placental disruption and increase the risk of transmission. In malaria endemic areas, infection of placenta is common in pregnancy (UNAIDS/WHO, 1999).

22.4.4 – Obstetric Factors

Obstetric factors are important because the majority of MTCT occurs during the time of labour and delivery. The suggested mechanisms for intrapartum transmission of HIV-1 include: (a) direct contact of skin and mucous membrane of the infant with maternal cervicovaginal secretions during labour; (b) ingestion of virus from these secretions; (c) fourfold increase in HIV-1 levels in cervicovaginal secretions during pregnancy; and (d) ascending infection to the amniotic fluid (Reggy *et al.*, 1997; Henin *et al.*, 1993). The higher rate of infection in *first-born* twins may be due to their longer duration of exposure to infected secretions (Goedert *et al.*, 1991; UNAIDS/WHO, 1999). This effect is more pronounced in vaginally delivered twins, where a twofold increase in infection is seen in first-born twins over the second-born (Goedert *et al.*, 1991). The duration of rupture of membrane (more than 4 hours) is an important risk factor, while the duration of labour does not appear to be a risk factor (Women and Infants Transmission Study Group, 1996; UNAIDS/WHO, 1999). Obstetric factors such as the use of foetal scalp electrodes, episiotomy, and vaginal tears have been implicated in some studies, but not in others (UNAIDS/WHO, 1999). A randomised clinical trial in Europe has confirmed that delivery by elective caesarean section had a protective effect (European Mode of Delivery Collaboration, 1999).

22.4.5 – Fetal Factors

Concordance between infant and maternal human leucocyte antigen (HLA) has been associated with higher risk of transmission (MacDonald *et al.*, 1998). Other foetal factors include co-infection with other pathogens, foetal nutrition, and foetal immune status (Steihm, 1996).

22.4.6 – Infant Factors

Mucins, HIV antibodies, lactoferrin, and secretory leucocyte protease inhibitor (SLPI) are *protective factors* present in breast milk (Steihm, 1996; van de Perre *et al.*, 1993). The risk of transmission through breast milk depends on factors such as stage of maternal disease, maternal vitamin A levels, patterns of breastfeeding (exclusive or mixed), and presence of breast abscesses, mastitis, nipple cracks (van de Perre *et al.*, 1992). Decreased acidity, decreased mucus, lower IgA activity, and thin mucosa in the gastrointestinal tract of the newborn may increase susceptibility (Steihm, 1996). The immune system of the newborn may also be deficient in macrophage and T-cell immune response (Steihm, 1996).

22.5 – CHALLENGES IN DEVELOPING COUNTRIES

In developing countries, cost of counselling, HIV testing, ARV therapy, and breast milk substitutes for infant feeding is very high. The mother should take informed decision regarding the appropriate method for feeding her infant. Transmission of HIV through breast milk can be avoided by continuing ARV treatment in breastfed children. In some developing countries, low priority is still given to preventing MTCT and more attention and higher budgetary allocation is given to care of under-5 children; and prevention and treatment of acute respiratory infections and diarrhoeal diseases (Soucat & Knippenberg, 1999).

22.6 – POSSIBLE INTERVENTION STRATEGIES

Intervention strategies (proposed or under investigation) are as follows:
• Medical termination of pregnancy
• Behavioural interventions in pregnancy – lifestyle changes (avoiding tobacco and drug use), avoiding unprotected sexual intercourse and multiple sexual partners
• Therapeutic interventions – treatment of STIs, supplementation of vitamin A and other nutrients, immunotherapy (under trial), and ARV therapy
• Obstetric interventions – avoiding invasive tests, birth canal cleansing, delivery by elective caesarean section
• Modified infant feeding practices – heat treatment (Pasteurisation) of expressed breast milk, early cessation of breastfeeding (UNAIDS/WHO, 1999)

22.6.1 – Nutritional Interventions

Vitamin A deficiency has been associated with increased viral load in breast milk (Nduati *et al.*, 1995), and higher risk of MTCT (Semba *et al.*, 1994). Several randomised control trials of vitamin A and other micronutrients such as zinc and selenium are in progress in African countries. The advantages of supplementation would be its low cost, other potential health benefits for the mother, its practicability in most health care settings, and ability to implement without HIV testing (UNAIDS/WHO, 1999). In a randomised control trial in Tanzania, multivitamin (but not vitamin A alone) supplementation in HIV-positive pregnant women has been shown to reduce the risk of low birth weight, severe preterm birth (less than 34 weeks gestation), and small for gestational age at birth, and increase CD4, CD8, and CD3 counts. The effect on MTCT is yet to be determined in this study (Tanzania Vitamin and HIV Infection Trial Team, 1998).

22.6.2 – Mode of Delivery

Many studies have revealed that caesarean section was associated with a reduction in MTCT. A meta-analysis of prospective follow-up studies has shown that elective caesarean section reduced the risk of MTCT by more than 50 per cent, after adjusting for ARV therapy, birth weight, and stage of maternal infection (International Perinatal HIV Group, 1999). The use of elective caesarean section must take into account the risk of maternal morbidity and mortality, the availability of safe operating facilities, and the accessibility of maternal services for women in future pregnancies (UNAIDS/WHO, 1999).

22.6.3 – Vaginal Cleansing

It is hypothesised that the use of antiseptics (chlorhexidine or benzalkonium chloride) during labour and delivery could reduce the intrapartum transmission of HIV-1. The advantages are its low cost, other potential health benefits for the mother, its feasibility in most health care settings, and ability to implement without HIV testing (UNAIDS/WHO, 1999). Microbicides are anti-HIV substances that could reduce the risk of HIV infection and other STIs during vaginal or anal intercourse. Use of these products can be controlled by women without the need for cooperation from their male partners or without their partners' knowledge. Success of vaginal cleansing with microbicides depends on women remembering to use them correctly and consistently during each act of coitus. Microbicides can be incorporated in gels, foams, creams, thin films, or vaginal pessaries. The possible mechanisms of action include immobilising the virus, creating a barrier between the virus and epithelial cells, and preventing HIV from reproducing and establishing infection after it has entered the body. So far, no anti-HIV microbicide has been approved. Microbicides closest to approval are Carraguard, cellulose sulphate gel (also being studied as a contraceptive), PRO 2000 Gel, BufferGel, and Savvy (Fact Sheet 157, 2006).

22.6.4 – Modification of Infant Feeding Practices

Potential modifications of infant feeding include:
(a) Complete avoidance of breastfeeding
(b) Avoiding breastfeeding in the presence of breast abscesses or cracked nipples
(c) Early cessation of breastfeeding
(d) Pasteurisation of breast milk (Newell *et al.*, 1997; Kuhn & Stein, 1997)

The mothers should be given the information on the advantages and disadvantages of breastfeeding and replacement feeding. They should be encouraged to make a fully informed decision about infant feeding (UNAIDS/WHO, 1999).

22.6.5 – Discussing Breastfeeding Options

Experts have recommended that HIV-positive mothers in *developed* countries should not breastfeed in order to avoid the potential 5–20 per cent risk of transmission of HIV through breast milk. In the developed countries, breastfeeding is relatively uncommon and artificial feeding is safe and affordable (Ramachandran, 1990). However, in *developing* countries, breastfeeding is essential for the survival of the infant, irrespective of its HIV status because artificial feeding is neither safe nor affordable. Hence, in developing countries, health care personnel need to discuss infant feeding options; bearing in mind the health and socio-economic status of the mother, the cost of breast milk substitutes, and the risks of not breastfeeding. The final decision on breastfeeding is to be left to the mother.

Socio-Economic Factors: In Asian and African communities, stigma is associated with not breastfeeding. High cost of breast milk substitutes is important in developing countries (Soucat & Knippenberg, 1999).

Risk of Transmission of HIV: It is estimated that risk of infection of babies is higher if the duration of breastfeeding exceeds 6 months (Miotti *et al.*, 1999). Physiological barriers and protective factors like mucins, HIV antibodies, lactoferrin, and SLPI (van de Perre *et al.*, 1993; Steihm, 1996) are considered to be responsible for the low risk of transmission through breast milk. The risk of gastroenteritis, respiratory tract infections, and malnutrition is higher in poor countries (Soucat & Knippenberg, 1999). Continuation of ARV treatment in breastfed children has shown to be helpful in preventing the transmission of HIV through breast milk (Soucat & Knippenberg, 1999).

22.6.6 – Interventions Recommended by NACO

NACO has recommended the following interventions:
Antenatal Period:
(a) Explaining to the client, the risk of HIV transmission to baby
(b) Discussing option of MTP in the first trimester
(c) Safe practices in antenatal care if the client decides to continue the pregnancy

Intranatal Period: ARV therapy and safe practices in intranatal care.

Post-Natal Period:
(a) Educating the mother about post-partum contraception
(b) Discussing infant feeding options
(c) Screening the baby for HIV status and this should be *preceded*, and *followed* by counselling of the mother or both parents

(d) If the baby has *no symptoms* of HIV-related diseases, all the vaccines (live and killed) are to be given, as per the national schedule. If the baby is *symptomatic*, all the vaccines *except* live vaccines should be given as per the national schedule (NACO, Training Manual for Doctors).

22.6.7 – Antiretroviral Therapy

Long-Course Zidovudine Therapy: The use of long-course ZDV in pregnancy was recommended as the standard of care in Europe, United States, Brazil, and Thailand, and the introduction of this regimen has resulted in a dramatic reduction in rates of transmission. This regimen is based on Paediatric AIDS Clinical Trials Group (PACTG) trial, in which asymptomatic pregnant women in a non-breastfeeding population participated. The women are given ZDV *orally* (100 mg five times daily), after 14th week of pregnancy, and *intravenously* during labour. The infants receive *oral* doses of 2 mg per kg body weight, four times daily for 6 weeks (UNAIDS/WHO, 1999). Although viral resistance to ZDV monotherapy has been reported, it is not common, and there are concerns about using this regimen in any subsequent pregnancy (Srinivas *et al.*, 1996; Eastman *et al.*, 1998). Recent reports of ZDV toxicity in mice have renewed concern about the long-term effects of the drug (UNAIDS/WHO, 1999). This regimen is not feasible in *developing countries* because it is difficult to monitor blood parameters, drug reactions, intravenous infusions during delivery, and the treatment of the infant for 6 weeks. Women in developing countries have a higher prevalence of anaemia, which is aggravated by ARV therapy. Access to voluntary testing and counselling is poor in these countries (UNAIDS/WHO, 1999).

Short-Course Zidovudine Therapy: The regimen consists of ZDV during the antenatal period only. ZDV is given orally in a dose of 300 mg twice daily from 36th week of pregnancy to onset of labour and 300 mg every 3 hours from onset of labour until delivery. Randomised short-course ZDV trials have been conducted in Thailand and African countries, where all the participants were advised *not* to breastfeed and were provided with infant formula foods (UNAIDS/WHO, 1999).

Combination Therapy: Recent recommendations advise the use of a three-drug combination therapy, but with rapid advances in ARV therapy, such recommendations can change frequently. The long-term follow-up of the PETRA trial, which is coordinated by UNAIDS, is going on in *predominantly breastfeeding* populations in several African countries. Interim results at 6 weeks of age of the infant suggest that the lowest risk of transmission was seen with the following regimen – ZDV and lamivudine (3TC) from 36th week of pregnancy and during labour for the mother; and for 1 week post-partum for mother and child (UNAIDS/WHO, 1999).

Nevirapine: Nevirapine is a non-nucleotide reverse transcriptase inhibitor (NNRTI) with potent ARV activity and a high safety profile. The drug achieves high and long-lasting circulating levels and hence *one-dose treatment* is given during labour. However, there is rapid development of resistance to this drug.

REFERENCES

Ahluwalia I.B., De Villis R.F., and Thomas J.C., 1998, Reproductive decisions of women at risk for acquiring HIV infection. AIDS Educ Prev 10(1): 90–97.

Backe E., *et al.*, 1993, Foetal organs affected by HIV-1. AIDS 7: 896–897.

Boyer P.J., *et al.*, 1994, Factors predictive of maternal-fetal transmission of HIV-1. JAMA 217: 1925–1930.

Bryson Y.J., *et al.*, 1992, Proposed definition for in-utero versus intra-partum transmission of HIV-1. N Engl J Med 327: 1246–1247.

Bulterys M. and Goedert J.J., 1995, From biology to sexual behaviour – towards the prevention of mother-to-child transmission of HIV. AIDS 10:1287–1289.

Burton G.J., *et al.*, 1996, Physical breaks in the placental trophoblastic surface: significance in the vertical transmission of HIV. AIDS 10(11): 1294–1295.

Chin J, 1990, Current and future dimensions of the HIV/AIDS pandemic in women and children. Lancet ii: 221–222.

Drew W.L., *et al.*, 1990, Evaluation of the virus permeability of a new condom for women. STD 17: 110–112.

Eastman P.S., *et al.*, 1998, Maternal viral genotypic zidovudine resistance and infrequent failure of zidovudine therapy to prevent perinatal transmission of human immunodeficiency virus type-1 in Pediatric AIDS Clinical Trials Group Protocol 076. J Infect Dis 177(3): 557–564.

European Collaborative Study, 1992, Risk factors for mother-to-child transmission of HIV-1. Lancet 339: 1009–1012.

European Collaborative Study, 1994, Caesarean section and the risk of vertical transmission of HIV-1 infection. Lancet 343: 1464–1467.

European Study Group in Heterosexual transmission of HIV, 1996, Relationship to number of unprotected sexual contacts. J AIDS 11: 388–395.

Feldblum P.J., *et al.*, 1995, The effectiveness of barrier methods in preventing the spread of HIV. AIDS 9 (Suppl A): 585–593.

Feldmeier H., Krantz I., and Poggensee G., 1994, Female genital schistosomiasis as a risk factor for the transmission of HIV. Int J STD AIDS 5(5): 368–372.

Fowler M.G. and Rogers M.F., 1996, Overview of perinatal infection. J Nutr 126: 2602S–2607S.

Ghys P.D., *et al.*, 1997, The associations between cervicovaginal HIV shedding, sexually transmitted diseases and immunosuppression in female sex workers in Abidjan, Cote d'Ivoire. AIDS 11(12): F85–F93.

Goedert J.J., *et al.*, 1991, International registry of HIV-exposed twins. High risk of HIV-1 infection for first born twins. Lancet 338: 1471–1475.

Henin Y., *et al.*, 1993, Virus excretion in the cervicovaginal secretions of pregnant and non-pregnant HIV-infected women. J AIDS 6: 72–75.

Hoegsberg B., *et al.*, 1990, Sexually transmitted diseases and Human immuno-deficiency virus among women with pelvic inflammatory disease. Am J Obstet Gynecol 163: 1135–1139.

International Perinatal HIV Group, 1999, Mode of delivery and vertical transmission of HIV-1: a meta-analysis from fifteen prospective cohort studies. N Engl J Med 340: 977–987.

John G.C. and Kreiss J., 1996, Mother-to-child transmission of human immunodeficiency virus type-1. Epidemiol Rev 18(2): 149–157.

John G.C., *et al.*, 1997, Genital shedding of human immunodeficiency virus type-1 DNA during pregnancy: association with immunosuppression, abnormal cervical and vaginal discharge and severe vitamin A deficiency. J Infect Dis 175(10): 57–62.

Johnson M.A., *et al.*, 1989, Transmission of HIV to sexual partners of infected men and women. AIDS 3: 367–372.

Kuhn L. and Stein Z., 1997, Infant survival, HIV infection and feeding alternatives in less-developed countries. Am J Pub Health 87(6): 926–931.

Laga M., *et al.*, 1993, Non-ulcerative sexually transmitted diseases on HIV as risk factors for HIV-1 transmission in women – results from a cohort study. AIDS 7: 95–102.

Langston C., *et al.*, 1995, Excess intra-uterine foetal demise associated with maternal human immunodeficiency virus infection. J Infect Dis 172:1451–1460.

Latif A.S., *et al.*, 1989, Genital ulcers and transmission of HIV among couples in Zimbabwe. AIDS 3: 519–523.

Loussert-Ajaka I., *et al.*, 1997, HIV-1 detection in cervicovaginal secretions during pregnancy. AIDS 11(13): 1575–1581.

MacDonald K.S., *et al.*, 1998, Mother-child class I HLA concordance increases perinatal human immunodeficiency type-1 transmission. J Infect Dis 177(3): 551–556.

Markson L.E., *et al.*, 1996, Association of maternal HIV infection with low birth weight. J AIDS Hum Retrovirol 13(3): 227–234.

Mayaud P., 1997, Tackling bacterial vaginosis and HIV in developing countries. Lancet 350: 530–531.

Mayaux M.J., *et al.*, 1997, Maternal viral load during pregnancy and mother-to-child transmission of human immunodeficiency virus type-1: the French Perinatal Cohort Studies. J Infect Dis 175: 172–175.

Minkoff H., 1995, Pregnancy and HIV infection. In: HIV infection in women (H. Minkoff, J.A. De Hovitz, A. Duerr, eds.). New York: Raven Press, pp 173–188.

Miotti G., *et al.*, 1999, HIV transmission through breast-feeding – a study in Malawi. JAMA 282: 744–749.

Naeye R.L. and Ross S., 1983, Coitus and chorioamnionitis: a prospective study. Hum Devel 6: 91–94.

National AIDS Control Organisation (NACO). Training manual for doctors. New Delhi: Government of India.

Nduati R.W., *et al.*, 1995, Human immunodeficiency virus type-1-infected cells in breast milk: association with immunosuppression and vitamin A deficiency. J Infect Dis 172(6): 1461–1468.

New Mexico AIDS Education and Training Center, 2006, Fact Sheet 157. Microbicides. University of New Mexico Health Sciences Center. www.aidsinfonet.org. Revised 9 March.

Newell M.-L., Gray G., and Bryson Y.J., 1997, Prevention of mother-to-child transmission of HIV-1. AIDS 11 (Suppl A): S165–S172.

Plourde P.J., *et al.*, 1994, Human immunodeficiency virus type-1 seroconversion in women with genital ulcers. J Infect Dis 170: 313–317.

Ramachandran P., 1990, HIV infection in women. ICMR Bulletin 20(11&12): 111–119.

Reggy A., Simonds R.J., and Rogers M., 1997, Preventing perinatal HIV transmission. AIDS 11 Suppl A: S61–S67.

Rich K.C., *et al.*, 1995, CD4+ lymphocytes in perinatal human immuno-deficiency virus (HIV) infection – evidence for pregnancy-induced immune depression in uninfected and HIV-infected women. J Infect Dis 172: 1221–1227.

Royce R.A., *et al.*, 1997, Sexual transmission of HIV. N Engl J Med 15: 1072–1078.

Semba R.D., *et al.*, 1994, Maternal vitamin A deficiency and mother to child transmission of HIV-1. Lancet 343: 1593–1597.

Sewankambo N., *et al.*, 1997, HIV-1 infection associated with abnormal vaginal flora morphology and bacterial vaginosis. Lancet 350: 546–550.

Soto-Ramirez L.E., *et al.*, 1996, HIV-1 Langerhans' cell tropism associated with heterosexual transmission of HIV. Science 271: 1291–1293.

Soucat A. and Knippenberg R., 1999, Large scale implementation of prevention of mother-to-child transmission of HIV – issues for East Asia and Pacific. Technical Update No. 1. Bangkok: UNAIDS/UNICEF/WHO, p 17.

Srinivas R.V., *et al.*, 1996, Development of zidovudine-resistant HIV genotypes following postnatal prophylaxis in a perinatally infected infant. AIDS 10(7): 795–796.

St Louis M.E., *et al.*, 1993, Risk for perinatal HIV-1 transmission according to maternal immunologic, virologic and placental factors. JAMA 169: 2853–2859.

Steihm E.R., 1996, Newborn factors in maternal-infant transmission of pediatric HIV infection. J Nutr 126: 2632S–2636S.

Tanzania Vitamin and HIV Infection Trial Team, 1998, Randomized trial of effects of vitamin supplements on pregnancy outcomes and T-cell counts in HIV-1 infected women in Tanzania. Lancet 351: 1477–1478.

The European Mode of Delivery Collaboration, 1999, Elective caesarean section versus vaginal delivery in prevention of vertical HIV-1 transmission: a randomized clinical trial. Lancet 353: 1035–1039.

Thea D.M., *et al.*, 1997, The effect of maternal viral load on the risk of perinatal transmission of HIV-1. J Infect Dis 175: 707–711.

UNAIDS/WHO, 1999, HIV in Pregnancy – A Review. Occasional Paper No. 2. Joint United Nations Programme on HIV/AIDS 1999, pp 6–37.

Van de Perre P., *et al.*, 1992, Postnatal transmission of HIV-1 associated with breast abscess. Lancet 339: 1490–1491.

Van de Perre P., *et al.*, 1993, Infective and anti-infective properties of breast milk from HIV-1 infected women. Lancet 341(8850): 914–918.

Viscarello R.R., *et al.*, 1992, Fetal blood sampling in HIV-seropositive women before elective mid-trimester termination of pregnancy. Am J Obstet Gynecol 167: 1075–1079.

Women and Infants Transmission Study Group, 1996, Obstetrical factors and the transmission of human immunodeficiency virus type-1 from mother to child. N Engl J Med 334: 1617–1623.

Working Group on Mother-to-Child Transmission of HIV, 1995, Rates of mother-to-child transmission in Africa, America, and Europe – results from 13 perinatal studies. J AIDS Hum Retrovirol 8: 506–510.

CHAPTER 23

HUMAN RIGHTS, LEGAL, AND ETHICAL ISSUES

Abstract

In general, laws are not in tune with rapidly changing needs of the society. Legal ambiguities, contradictions, and prohibitions hinder health and educational programmes such as family life and sex education. Obsolete laws ought to be reformed and new laws enacted, to facilitate provision of care and support services for HIV-infected individuals. In most countries, medical treatment is legally viewed as an interference with the patient's body. Informed consent must be obtained from a mentally competent patient, over a stipulated age, before any treatment is administered. In view of the HIV epidemic, an increasing number of countries have enacted laws that prevent health care personnel from disclosing patient-related information acquired in the course of their duties. Modern medical management uses a *team approach* that requires rapid transfer of patient-related information between members of a team, which amplifies the risk of breach of confidentiality.

– Collaborative studies between developed and developing countries could give the impression of experimentation on the population of resource-poor countries by a developed country and could invite allegations of exploitation. Debates on this issue led to the revised Helsinki declaration of 2000, but the issue is still unresolved.

– In *voluntary euthanasia*, a doctor or other person acts directly to end a suffering person's life when the person specifically requests death. In *physician-assisted suicide* (PAS), a doctor provides the means and information necessary for a person to commit suicide. In some states of Australia, euthanasia is legal under specified circumstances and as per Australian law, it is not a crime to attempt or to commit suicide. Holland was the first country in the world, followed by Belgium, to legalise both PAS and physician-unassisted euthanasia. In January 2006, the US Supreme Court upheld State of Oregon's 8-year old *Death with Dignity Act*. In most countries of the world including India, euthanasia is *illegal*.

Key Words

Confidentiality, Declaration of Helsinki, Disclosure, Euthanasia, Informed consent, Partner counselling, Physician-assisted suicide, Privacy, Secrecy, Shared confidentiality

23.1 – CONCEPTS

1. Confidentiality: It is an ethical and/or legal duty of the health care professionals, not to disclose to anyone else, without prior informed consent of the client, any personal, physiological and psychological information that was

obtained by them in the context of their professional relationship with the client.

2. Shared Confidentiality: This involves encouraging people to get HIV tests done and to inform their sexual partners about the results, in order to prevent further infections.

3. Disclosure: It is the act of informing any third party about the HIV status of an infected person, with or without the consent of the infected individual.

4. Secrecy: This is a state of affairs, often resulting from fear, shame, and/or a sense of vulnerability. This depends on personal motives of the individual who holds the secret.

5. Right to Privacy: This right creates an obligation on governments to provide legal safeguards to protect an individual's right to privacy.

6. Informed Consent: This concept is based on the principle that competent individuals are entitled to make informed choices in connection with decisions that affect his or her body or health (UNAIDS/WHO, 2000).

23.2 – ETHICS IN PROVISION OF HEALTH CARE

All health care providers have a *moral duty* to care for any individual who seeks health care. However, no health care provider is *ethically* or *legally* obliged to put his or her life at risk while treating HIV cases where facilities for universal safety precautions do not exist (Muthuswamy, 2005). Hence, it should be made legally compulsory to provide the highest level of internationally recognised and recommended universal precautions at all health care facilities (Muthuswamy, 2005). A comprehensive programme is needed to

(a) Train health care personnel in preventing accidental occupational exposure to blood and body fluids, including needle stick injuries.

(b) Train health care personnel in maintaining confidentiality regarding HIV status of patients under their care. In case disclosure is necessary, informed consent is to be obtained. Only in exceptional circumstances, confidentiality may be breached.

(c) Provide PEP for all personnel at risk, if indicated.

23.2.1 – Ethical Issues Related to Health Care and Biomedical Research

1. Issues related to basic human rights
2. Provision of health care and standard of care
3. Prevention of harm of any kind
4. Protection – of privacy and confidentiality, of vulnerable groups, social stigma, and discrimination
5. Informed consent across cultures and consultation with communities
6. Mechanisms for ethical review
7. International collaboration to consider community benefits and needs of host country

8. Epidemiological studies, socio-behavioral studies and clinical trials of drugs, vaccines, microbicidal agents – standard of care, accessibility, affordability, post-trial benefits, equity, and sustainability of interventions.

23.3 – ETHICS IN PUBLIC EDUCATION

The response to the HIV epidemic ought to be mature and not impulsive, as in a panic situation. Allaying irrational public reaction while avoiding social or psychological harm to infected persons and those at high risk is the key to halting the spread of any epidemic. In view of public apprehensions without knowledge of facts, public education would be the best antidote, for fear is bred in the unknown (Vas & de Souza, 1991).

The HIV epidemic in India, which was restricted mainly to high-risk groups in the 1990s has started spreading to unsuspecting groups in rural populations, which includes women and children. India accounts for 10 and 65 per cent of the total burden of HIV-positive persons in the world and South and South-East Asia, respectively. Given such a scenario, it is still being debated whether educationists and counsellors should break taboos about discussing sexual behaviour. This raises issues pertaining to human rights, ethics, and the law (Muthuswamy, 2005).

IEC campaigns for the general population will promote tolerance, understanding, and reduce fear, stigma, and discrimination. Involvement of HIV-infected persons, politicians and noted personalities may help in changing public attitudes (UNAIDS/WHO, 2000).

Education campaigns need to include sensitisation of health care providers, employers, police, and members of the legal profession in attitudes of non-discrimination, acceptance, principles of confidentiality, and informed consent. This is a priority since HIV-infected individuals usually face discrimination at the hands of these very professionals (UNAIDS/WHO, 2000). Non-discriminatory practices are to be promoted in the workplace (UNAIDS/WHO, 2000).

23.3.1 – Conflicts of Interest

Conflicts of interest may influence HIV-related activities. Public health experts would like to know the prevalence of HIV infection. An individual client would be jittery when called upon to submit to HIV testing. Sex workers may feel that HIV testing may increase social stigma and affect their earnings. An educator may want to promote correct scientific information regarding sex, sexuality, and sexual behaviour but may be opposed by self-appointed "protectors" of morals and other vested interests having imaginary fears about increase in promiscuity (Vas & de Souza, 1991). The mass media can play a positive role by reporting on the HIV epidemic in a non-partisan, non-sensational, responsible, and non-discriminatory manner.

23.4 – ETHICAL ISSUES RELATED TO ACCESS

These issues are related to access to preventive measures like condoms, microbicidal agents, sterile injecting equipment, safe blood supply. Although access to ARV drugs has received attention, challenges pertain to sustainability of the programme, and ensuring that the intended beneficiaries get access to ARV drugs.

23.5 – LEGAL ASPECTS

A clear distinction should be made between ethics and the law. An action or deed that is legal need not be ethical. Human history is replete with instances where intolerant and prejudiced legislation has been framed and enforced (e.g. laws during the days of Apartheid in South Africa, and slave ownership laws in southern parts of the United States before abolition of slavery). Laws enacted for protecting the community need to preserve the dignity of individuals. Any legal sanction for targeting any group (known for high-risk and/or sexually "deviant" behaviour) for selective discrimination, isolation, or quarantine would be unjustifiable and unjust (Vas & de Souza, 1991).

In some countries, the urgency to contain the HIV epidemic has led to rather hasty decisions such as compulsory HIV testing of foreign students and immigrants with scant attention to ethical considerations. When protective measures tend to be overprotective, they may infringe human rights and dignity of the individual. Many of the created apprehensions are not supported by scientific facts and have tended to confuse relevant issues. It is unethical for lawmakers to consider abrogation of fundamental rights of HIV-infected individuals on the basis of concern for common good. This is especially when specific measures for containing other communicable diseases such as vaccination, isolation, and quarantine are not relevant to control of HIV infection (Vas & de Souza, 1991).

Unnecessary ambiguities, contradictions, and prohibitions in laws act as barriers to health and educational programmes because laws are not in tune with rapidly changing needs of society. Many former French colonies still follow the anti-birth control laws that were enacted by France in 1920. In these countries, it would be illegal to promote condoms as a protection against HIV. In Italy, an old law that prohibited any discussion on sexual matters was in existence till March 1975, when it was repealed (WHO, 1985). Hence, laws that are out of touch with ground realities and social needs ought to be repealed and new laws enacted, to facilitate provision of care and support services for HIV-infected individuals. Employment regulations need to be framed to prohibit HIV-related discrimination (UNAIDS/WHO, 2000). India is on the threshold of enacting a comprehensive law on HIV/AIDS, which will address issues related to treatment, insurance, social security, research ethics, blood transfusions, organ and tissue transplantation, and role of the media (Muthuswamy, 2005).

23.5.1 – Family Life Education and the Law

Lack of family life education and ignorance about sexual matters, even among educated individuals, is another factor that contributes to the spread of HIV infection. Family life education is known by various names, such as "sexuality education", or "family life skills" or "reproductive health education" or "responsible parenthood education" in some countries (WHO, 1985). Family life education helps in knowing one's body, structure and function of reproductive system, changes during adolescence, sex and sexuality, factors causing marital harmony and disharmony, sexual health problems, and STIs. Family life education is *compulsory* in schools in China, Scandinavian countries, Germany, Czech Republic, and the Philippines. It is an *elective subject* in some states of the United States. It is *legally prohibited* in Saudi Arabia, Iran, Argentina, and Chad. In the United Kingdom and Holland, there is no legislation but family life education is a component of school education (WHO, 1985).

23.5.2 – Age of Consent to Treatment

Laws in some countries stipulate that prior consent of parents/guardians must be obtained before adolescents/minors receive health care. An individual ought to have sufficient experience and intelligence to meet the basic requirements of consent. However, laws set an *arbitrary age* (usually 18 or 21 years) as age of attaining *adulthood* as if all individuals magically develop maturity on their 18th or 21st birthday. The minimum age of consent varies from country to country, depending on gender of the individual and purpose of the consent. The minimum age of consent varies for marriage, criminal responsibility, voting rights, military service, consent to sexual intercourse, and consent to health care (WHO, 1985).

In the United States, the Joint Commission of Institute of Judicial Administration and American Bar Association has recommended that parental consent should be *waived* when "the social utility of medical care outweighs the negative impact of not taking parental consent" (Joint Commission, 1979).

Some nations follow the *emancipated minor rule*, which upholds that adolescents who are nearing the age of majority and who are independent in lifestyle and finance should be able to give consent to health care. This rule recognises the fact that older adolescents have sufficient experience and intelligence to meet the basic requirements of consent. In some countries, minors do *not* require parental consent for treatment of STIs and drug abuse, contraceptive services, and health care in pregnancy (WHO, 1985). According to the *Kenyan National VCT Guidelines* issued in 2001, "mature minors" do not need parental consent. Mature minors include those individuals younger than 18 years who are "married, pregnant, parents, engaged in behaviour that puts them at risk, or are child sex workers". Although the HIV test results are to be disclosed only to the client, the Kenyan guidelines say that counsellors should encourage those younger than 18 years to inform their parents about the results (NASCOP, 2001).

23.6 – CONFIDENTIALITY

Keeping client-related information confidential is a legal and ethical obligation. Besides health care professionals, this duty also pertains to professions such as psychotherapy, psychiatry, legal practice, and social work. The individual's "right to privacy" enforces the ethical duty of confidentiality (UNAIDS/WHO, 2000). Professional codes of conduct need to be drafted and enforced to penalise breaches in confidentiality and informed consent. In view of the HIV epidemic, an increasing number of countries have enacted laws that prevent health care personnel from disclosing patient-related information acquired in the course of their duties (Magnusson, 1994). Health care providers have a legal obligation to take all reasonable steps to prevent disclosure of HIV-related information (Magnusson *et al.*, 1997).

It is absolutely essential to protect HIV-related information because an epidemic of discrimination and social stigma has accompanied the HIV epidemic. Since HIV status is a sensitive issue, health care providers should be careful in protecting HIV-related information. In the context of HIV infection, "confidentiality" also includes the expectation of an HIV-infected person that his or her status will not be disclosed without his or her consent (UNAIDS/WHO, 2000).

Unless an individual is hospitalised, request forms for HIV test should not contain information that can identify the individual. It is recommended that HIV test request forms use a *four-letter name code* constructed from the first two letters of surname, followed by the first two letters of first name. The same name code should be used on pathology forms. In countries where notification to public health authorities is compulsory, anonymous codified notification is used (Magnusson *et al.*, 1997).

Extending VCT services to urban slum dwellers, rural populations, and socially marginalised groups such as homosexuals, *hijras*, and prison inmates, will enable clients learn about their HIV status and receive support, if necessary (NACO, Training Manual for Doctors).

23.6.1 – Justification for Confidentiality

- Respecting confidentiality protects the privacy of clients/patients and breach of confidentiality can result in discrimination, harassment, and vilification (Magnusson *et al.*, 1997).
- A client, who is convinced of confidentiality, is more likely to confide personal information. Public health interventions are effective only if they are based on good provider–client relationship.

• On the other hand, if patients fear breach of confidentiality, they are unlikely to come forward for HIV testing, counselling, and HIV-related services. Studies reveal that assurance of privacy encourages voluntary HIV testing (Fehrs *et al.*, 1988; Hirano *et al.*, 1994).

23.6.2 – Breach of Confidentiality

The principle of confidentiality is *not* absolute and may be breached if
1. There is a need for sharing information or discussion among health care providers for care, treatment, and counselling of the client/patient. The details should remain confidential and restricted to that group only (UNAIDS/ WHO, 2000; Chapman, 1997).
2. Disclosure to public health authorities is permitted if a patient places "health of the public at risk" through irresponsible behaviour (Fehrs *et al.*, 1988). Such legal provisions may vary from state to state within the same country, but they do not clarify whether a patient's spouse or sexual partner falls within the definition of "the public" (Chapman, 1997).
3. A Court of Law orders such information to be disclosed (Chapman, 1997).
4. As per directives of the Supreme Court of India, all HIV-infected individuals need to be encouraged to disclose their HIV status to their spouse/sexual partners. This is to be done only after counselling of spouse/sexual partners. *Partner counselling* has been included as a component of India's National HIV Policy (NACO, Training Manual for Doctors).
5. Disclosure to spouse/sexual partner is permitted in many countries if failure to disclose HIV status may harm an innocent third party (Chapman, 1997). In such situations, ethical principle of "confidentiality" may be in conflict with the ethical principle to "do no harm". It is necessary to weigh the potential harm and benefits to the parties involved, before opting for the future plan of action (UNAIDS/WHO, 2000).

23.6.3 – Consequences of Breach of Confidentiality

Except in the circumstances mentioned above, disclosure of a client's HIV status *without* his or her consent is hazardous and is fraught with social and legal consequences, which may ultimately jeopardise health-related activities.

Social Consequences: These include social ostracism and rejection. Female clients have very valid reasons such as physical violence, harassment, or abandonment, for fearing to disclose their HIV status to their husbands/partners (United Nations, 1998; Gielen *et al.*, 1997)

Legal Consequences: In countries where privacy is protected by law, a person can sue a health care provider who has disclosed his or her HIV status. The legal aspects are complicated since laws pertaining to HIV-related information and HIV medicine may differ from state to state within the same country (Fehrs *et al.*, 1988). In some countries, migrant workers may face termination of employment and deportation if their HIV-positive status is revealed.

Effect on other Health-Related Activities: Besides HIV-related activities, registration for antenatal care and blood donation would also be hindered because individuals would be afraid of getting tested if they fear that the results will be disclosed against their will (UNAIDS/WHO, 2000).

23.6.4 – Problems

Modern medical management uses a *team approach* that requires rapid transfer of patient-related information between members of a team (doctors, nurses, psychologists, and physiotherapists). Patient records are centralised and widely available. This amplifies the risk of breach of confidentiality. There are numerous glaring examples of disclosure of sensitive patient-related information to persons not involved in patient care (Magnusson, 1994). It is the counsellor's duty to conceal the identity of the client, and to ensure support, to prevent family disruption, violence, and desertion by spouse (WHO/UNAIDS, 1999). Some organisations have created separate systems for HIV-related records with protocols for access, while others keep separate HIV-related records that are never filed in the centralised hospital record. Ultimately, it is self-discipline on part of health care providers that will protect the privacy of patients (Magnusson *et al.*, 1997).

23.7 – INFORMED CONSENT

Health care providers must obtain informed consent *before* subjecting an individual to HIV testing, or disclosing HIV status to others (see exceptions), or initiating health care interventions. In most countries, the law views medical treatment as an interference with the patient's body. Informed consent must be obtained from a mentally competent patient, over a certain age (varies from country to country), before any treatment is administered. A provider, who fails to do so, may be liable for charges of assault (van Reyk, 1997). Obtaining informed consent is not obligatory only under exceptional circumstances
(a) When the law provides for mandatory testing, as in some countries (UNAIDS/WHO, 2000)
(b) When testing donors of blood, organs, tissues, semen for artificial insemination, and ova for *in vitro* fertilisation (Joint Commission, 1979; NACO, Training Manual for Doctors)
(c) During anonymous unlinked testing of blood samples for epidemiological purposes (van Reyk, 1997).

23.7.1 – Beneficial disclosure

This is the disclosure about a client's HIV status, *with* the informed consent of the client (UNAIDS/WHO, 2000). It is voluntary in nature, and respects the autonomy and dignity of the affected individual, and maintains confidentiality. The merits of beneficial disclosure are discussed below:

(a) Helps affected individual, spouse/sexual partner, and family
(b) Promotes greater openness in the community about HIV
(c) Meets ethical obligations to prevent onward transmission of HIV infection
(d) An increase in number of people, who are willing to disclose their HIV status, will lead to an increase in the "critical mass" of people who are openly involved in combating the epidemic and will diminish discrimination, stigma, and secrecy
(e) Protects those not yet infected and provide more care and support for those who are already infected (UNAIDS/WHO, 2000).

23.8 – HIV TESTING AND REPORTING

Counselling is essential before and after HIV testing. It would be unethical to test an individual's HIV status and leave him or her in the lurch by not providing any advice (Vas & de Souza, 1991). If reporting of HIV infection to public health authorities is essential for epidemiological purposes, confidentiality should be scrupulously maintained. Otherwise, individuals may refuse HIV testing for fear of disclosure. Some countries have recently considered whether they should implement a policy of "named HIV case reporting". Under such a policy, health care providers would be required to report by name, all HIV cases diagnosed by them, to a health authority (UNAIDS/WHO, 2000). Other forms of case reporting include:

(a) Sending reports without patient-identification information
(b) Using a code, also called "unique identifier", for each patient (UNAIDS/WHO, 2000)

Neither the policy of reporting of HIV-infected persons, nor that of reporting AIDS cases, has been useful in low-income countries, largely because less than 10 per cent of HIV-positive persons have been tested for HIV. Thus, most people are unaware of their HIV status. Many people are reluctant to get tested for HIV due to social stigma, discrimination, and lack of access to services including ARV therapy, if they test positive. In countries where "named case reporting" is adopted, health care providers are required to report cases by name. This may lead to breach of confidentiality and privacy. There is also a problem of underdiagnosis and under-reporting. Where "unnamed case reporting" (also called "anonymous case reporting") is employed, multiple reporting of the same cases is a problem. This is because HIV-infected individuals tend to disbelieve their HIV test report and may get the HIV test repeated at different laboratories hoping for a negative test result. Resource-poor countries lack sufficient capacity or infrastructure for maintaining confidentiality. Case reporting requires expensive infrastructure and trained manpower. By itself, case reporting does not necessarily improve access to supportive services, nor does it lead to any additional benefit for either the client or the community. It may divert resources from more cost-effective strategies such as public information campaigns, VCT, and education of youth (UNAIDS/WHO, 2000).

23.9 – ISSUES PERTAINING TO TREATMENT

The physician's primary responsibility is to treat the patient without discrimination if the condition is within the physician's current domain of competence. All human experimentation (related to clinical trials of drugs or vaccines) should strictly follow guidelines laid down in the Declaration of Helsinki (WMA, 1964). The motto to be followed is *primum non nocere* (Latin: "first, do no harm"). Respect for confidentiality, privacy, human rights, and dignity is essential (Vas & de Souza, 1991). Ethically and legally, each patient has a right to be treated with all available resources till his or her death. But, the best available palliative care may not provide pain relief for some patients. Some may decide to forego palliative care in hospitals and prefer care by family members (van Reyk, 1997).

23.10 – ISSUES PERTAINING TO DYING

Health care providers may come across patients in terminal stages of AIDS who may harbour thoughts about their impending death. Some may accept reality and want to talk about it, while others may deny that they are dying. Health care providers ought to strive to create an environment in which issues pertaining to death can be freely discussed (van Reyk, 1997). Desire to talk about death may be caused by depression (which should be diagnosed and treated by a specialist), concern about becoming a burden to family members, feeling of guilt, and dissatisfaction with the care received (van Reyk, 1997).

Euthanasia (Greek: *eu* = good; *thanatos* = death) refers to voluntary euthanasia in which, a doctor or other person acts directly to end a suffering person's life when the person specifically requests death (van Reyk, 1997). In most countries of the world including India, euthanasia is *illegal*. In PAS, a doctor provides the means and information necessary for a person to commit suicide (van Reyk, 1997). The attitude of psychiatrists to PAS and euthanasia also seems baffling. Although a significant number of psychiatrists endorsed their support for PAS, only a small minority of them were willing to get involved in assessing patients requesting for PAS (Ranjan *et al.*, 2005). The legal status in several countries is given below:

Australia: In some states of Australia, laws stipulate conditions under which a health provider can agree to a patient's request to stop or withhold "futile treatments" (van Reyk, 1997). In Northern Territory, euthanasia is legal under specified circumstances (van Reyk, 1997). As per Australian law, it is not a crime to attempt or to commit suicide. But inciting or counselling someone to commit suicide is a crime. Australian law does not specify as to what constitutes "inciting" or "counselling" (van Reyk, 1997).

Switzerland: Assisted suicide has been legal since the 1940s. In December 2005, a hospital allowed assisted suicide rather than at the home of the terminally ill patient (DNA, 2006).

Netherlands: The Dutch were the first in the world to legalise both PAS and physician-unassisted euthanasia. In the Netherlands, both are approved for mentally ill patients and Dutch law does not make psychiatric assessment mandatory (DNA, 2006). As per Dutch law, the concerned patient is sent for psychiatric assessment only if the treating physician feels that the patient may be psychiatrically ill (Kissane & Kelly, 2000). Terminally ill patients who may request for PAS may also be suffering from depressed mental status, where one's judgment can be severely impaired. Many terminally ill patients have withdrawn their request for PAS after successful treatment of their mental depression (Kelly & McLoughlin, 2002). Dutch guidelines for termination of life of mentally ill patients require an opinion from an independent psychiatrist about the *incurable* nature of the illness from a prognostic point of view. With the current state of medical knowledge, no one can truly claim the curability of severe mental illnesses. Currently available medications can bring about drastic improvements to the point of sustained remission. But, curability remains a distant dream (Ranjan *et al.*, 2005).

Belgium: Belgians followed the Dutch and passed assisted dying laws (DNA, 2006).

United States: In January 2006, the US Supreme Court upheld State of Oregon's 8-year-old Death with Dignity Act by a 6–3 vote. Under Oregon law, a doctor can prescribe a lethal dose of medication to a terminally ill patient of sound mind, who makes the request in writing in the presence of witnesses. The written request is to be repeated at least 2 weeks later. The patient has to meet a few other requirements. The patient must swallow the prescribed lethal medication and it cannot be administered by anyone. The "right to die" issue became a topic of household debate when repeated Court rulings permitted the husband of Terri Schiavo (a terminally ill Florida woman) to order the removal of her feeding tube that had kept her alive for 15 years. Most Americans disapproved of the measures taken by the US Federal Government to defy the repeated Court rulings (DNA, 2006).

23.11 – HIV AND MARRIAGE

Maintaining confidentiality is essential to any public health strategy. However, the law has recognised *exception* to the rule of confidentiality when public interest outweighs private or personal interest. The Supreme Court of India (8 SCC 296 of 1998) upheld the action of a physician who informed the status of a HIV seropositive prospective groom to a would-be bride, when the groom moved the Court for breach of confidentiality (Kumar, 1998; Muthuswamy, 2005). The

Supreme Court of India also ruled that an HIV-positive man who marries and transmits the infection to the spouse would be criminally liable under Sections 269 and 270 of the Indian Penal Code – these sections pertain to negligent or malignant acts, likely to spread a disease dangerous to life. However, the Court did not lay down any protocol by which, such a disclosure was to be made. Human rights activists in India opine that the Court's ruling has overlooked situations where HIV-infected or HIV-discordant couples wish to marry each other with the knowledge of each other's HIV status (Shreedhar, 2002). In certain situations, such as those related to employment or insurance, privacy should be maintained. However, in the absence of anti-discriminatory legislation in most countries, discriminatory practices continue (Muthuswamy, 2005).

23.12 – ETHICAL ISSUES IN HIV/AIDS RESEARCH

23.12.1 – Background Information

In June 1964, the World Medical Association adopted the "Declaration on human experimentation" at Helsinki (Finland). This came to be known as the "Declaration of Helsinki" (WMA, 1964). The expansion of HIV/AIDS research into other developing countries has increased the awareness regarding adapting ethical standards created in the developed countries to resource-poor settings of the developing countries. The issues pertaining to standard of care and use of placebo in control arm of the trial have dominated debates on clinical trials in developing countries.

In the developed countries, any clinical trial of a new ARV drug requires the standard ARV drug in the control arm. However, when a drug trial is conducted in the resource-poor countries, a *placebo* is used instead of the standard ARV drug in the control arm. Apparently, placebo is used because *no treatment* is the norm in the resource-poor countries either because standard drugs are not available or because they are too expensive. Further, clinical trials in the resource-poor countries were designed to use lower dosage or reduced duration of treatment to suit affordability or feasibility factor. In 1997, Marcia Angell's article in *Lancet* questioned the perinatal transmission studies (to prevent MTCT of HIV infection) being conducted in Africa with US collaboration (Angell, 1997). This article generated heated debates that ultimately resulted in revision of the Declaration of Helsinki in 2000.

In October 2000, the 32nd Assembly of the World Medical Association at Edinburgh (Scotland) came up with a revised declaration titled "Ethical Principles for Medical Research involving Human Subjects" (WMA, 2000). The revised declaration stipulated that all trials should aim for achieving the *highest attainable standards* with no placebo in the control arm, if a standard treatment was available. This declaration became instantly controversial because of sensitivities involved in HIV/AIDS research and because ground realities in developing countries cannot allow adherence to these guidelines. These debates led to development of guidelines by National Bioethics Advisory Council (USA), Nuffield Council

(UK), and the European Union. India, the Philippines, and South Africa have devised their own guidelines to suit local requirements (Muthuswamy, 2005). These multiple guidelines indicate the dilemma faced by researchers in the developed world while collaborating with developing countries on various research programmes. The world has still not reached a consensus on this issue.

The HIV Preventive Trial Network (HPTN) Ethics guidance document prepared by HPTN Ethics Working Group of Family Health International, North Carolina, USA emphasises

(a) Protection of vulnerable groups from exploitation
(b) Non-discriminatory access to benefits of research
(c) Minimisation of research-related harms (medical, psychological, social, and economic)
(d) Improvement of local access to care and capacity building for that care so that accessibility can be sustained once the research ends (FHI, 2003)

In India, almost all ARV drugs are available in the market, but they are not provided by the public health system. In 2004, NACO announced that ARV drugs will be provided *free of cost* to all women who participated in prevention of MTCT studies. Use of standard ARV treatment in the control arm in a trial situation creates a dilemma. Even if the ARV drugs are provided by sponsors during the trial period, it is not known whether it will be sustained after the trial period is over (Muthuswamy, 2005).

23.12.2 – Ethical Issues in Experimental Studies

Collaborative studies could give the impression of experimentation on the population of resource-poor countries by a developed country and could invite allegations of exploitation. Debates on this issue led to the revised Helsinki declaration of 2000, but the issue is still unresolved (Muthuswamy, 2005). International collaborations need to consider

1. Needs of the host country and strategies for capacity building in host countries so that they function as equal partners with the sponsors.
2. Involving the host countries in design, implementation and monitoring of trials.
3. Distributing benefits including disseminating research findings.
4. Protecting dignity, safety, and welfare of participants in trials.
5. Respecting laws, guidelines, rules and regulations of both the host and sponsoring countries; this includes laws related to intellectual property rights, exchange of biological materials, data transfer, and protection of confidentiality (ICMR, 2000; Muthuswamy, 2005).

23.12.3 – Ethical Issues in Epidemiological Studies

- Extent of community consultation – when, where and whom to consult
- Informed consent – individual consent is to be insisted upon, in addition to consent of community leaders, tribal chiefs, or "gate-keepers"
- Privacy and confidentiality of all collected data

- Disclosure to third parties, including partner notification
- Prevention of stigma and discrimination at all levels
- Screening high-risk groups in addition to the general population
- Screening persons with diminished autonomy – individual consent is to be taken, where possible, in addition to consent of the legal guardians (Muthuswamy, 2005).

23.12.4 – Ethical Issues Related to Behavioural Studies

Although no specific cure or preventive vaccine is available as yet, it is generally agreed that behavioural changes may alter the rate of transmission of HIV/AIDS. These behavioural changes include risk reduction measures such as condom use and single sexual partner at *individual level*, and promotion of social norms at *community level*. The *ethical requirements* for biomedical research also apply to behavioural studies. Mobilisation of community and mass media may help behavioural researchers in tackling *psychosocial* issues such as stigma and discrimination that participants may face. Participants in some behavioural studies may face *legal risks* in some countries. For example, MSM are liable to face criminal charges under Section 377 of the Indian Penal Code.

23.12.5 – Ethical Sssues Related to Clinical Trials on Antiretroviral Drugs

These include approval from Institutional Ethics Committee (or Institutional Review Board), clearance from regulatory bodies, informed consent and partner notification, protection of vulnerable populations, tackling discordant couples, and sustained availability of successful products.

23.12.6 – Ethical Issues Related to Clinical Trials on HIV Vaccine

In May 2000, UNAIDS released a guidance document which outlined the strategies to be adopted for capacity building, community involvement, protection of vulnerable population, inclusion of women and children with adequate safeguards, informed consent, justice and equity in selection of subjects for the clinical trials, use of placebo or any other vaccine in the control arm, continued counselling for risk reduction, and monitoring mechanisms, care and support for HIV-infection or any associated complication that may occur during the course of the trial (UNAIDS, 2000).

Continued Counselling for Risk Reduction: There is a possibility that participants in clinical trials of HIV vaccines might abandon safer sexual practices and indulge in risky behaviour due to a *false sense of security*. Subsequently, if the vaccine is proved to be ineffective, this may lead to increased transmission of HIV. Hence, risk reduction measures include: (a) continued counselling programme on safer sexual practices; (b) condom promotion; (c) control of RTIs and STIs;

and (d) provision of microbicidal agents for women who are unwilling or unable to use condoms (Kent *et al.*, 1997; Muthuswamy, 2005; UNAIDS, 2000).

Provision of Antiretroviral Drugs: The consensus is that ARV drugs should be provided to subjects who become HIV seropositive during the vaccine trial. The vaccine trial protocol should clearly mention that ARV drugs would be provided for those subjects who may become HIV-positive during the course of the vaccine trial. This is in addition to a constellation of services that go beyond what is locally available at the vaccine trial sites (Muthuswamy, 2005).

23.12.7 – Ethical Issues Related to Clinical Trials on Vaginal Microbicidal agents

Vaginal microbicial agents offer an alternative method for individuals and couples who are unable or unwilling to use condoms. This method empowers women in protecting themselves. Behavioural scientists facilitate in designing behavioural interventions and for collecting data on sexual behaviour.

Ethical Requirements: Consent of the sexual partner, which may be difficult to obtain in all cases; provision of ARV drugs to those participants who may become HIV-positive or pregnant during the course of the trial; and provision to be made for post-trial availability and accessibility of successful microbicidal agents.

Requirements of Trial Design: Use of condoms in both the arms of the trial with active microbicidal agents and placebo; involvement of women from all strata of society; and trials ought to include women from high-risk groups as well as monogamous women with single sexual partners who perceive themselves as low-risk.

23.12.8 – Ethical Issues Related to Studies on Perinatal Transmission of HIV

Most children living with HIV were born to HIV-infected mothers and many of them will die before they reach their teens (UNAIDS, 2004). In resource-poor countries, a large number of HIV-positive children are being born each year to HIV-infected mothers due to high prevalence of HIV infection and high birth rates, lack of counselling and ARV, and lack of safe alternatives for breastfeeding. There is an urgent need for alternative regimens that can be used in resource-poor countries. Studies conducted in these countries have been criticised (Angell, 1997). This criticism has sparked a debate on standard of care in different settings. Ethical issues in poor countries pertain to

- Late reporting of pregnant women to health facilities for antenatal care while some women directly come for delivery
- Occurrence of many deliveries without the help of trained health care providers
- Non-availability of perinatal HIV counselling and testing programmes in many resource-poor countries

- Inability to afford even the short-course ARV therapy
- Inability to prevent breastfeeding by HIV-infected mothers due to psychosocial reasons and non-affordability of breast milk substitutes

Criticisms Regarding Perinatal Studies: Though highly debatable, critics have opined that policies need to be developed to save both the mother and the child, so that a treated mother cares for her healthy (uninfected) child. It is argued that

(a) The current obsession to prevent MTCT of HIV does not consider the effect on the HIV-infected mother or the fate of untreated mothers.
(b) The current strategy is resulting in saving children from HIV infection but the child has to lead the life of an orphan.
(c) More grandmothers are now caring for orphans than they did a decade ago (Muthuswamy, 2005).

REFERENCES

Angell M., 1997, The ethics of clinical research in the Third World. N Engl J Med 337: 847–889.
Chapman S., 1997, Legal issues in HIV medicine. In: Managing HIV (G.J. Stewart, ed.). North Sydney: Australasian Medical Publishing.
Daily News Analysis (DNA), 2006, Mumbai. 19 January, p 16.
Family Health International (FHI), 2003, Ethics Working Group of HIV Preventive Trial Network (HPTN) Ethical Guidance for HIV Preventive Trials. North Carolina, USA: FHI. www.fhi.org
Fehrs L.J., Fleming D., Foster L.R., *et al.*, 1988, Trial of anonymous versus confidential human immunodeficiency virus testing. Lancet 2: 379–381.
Gielen A.C., O'Campo P., *et al.*, 1997, A woman's disclosure of HIV status – experiences of mistreatment and violence in an urban slum setting. Women's Health 25(3): 19–31.
Hirano D., Gellert G.A., Fleming K., *et al.*, 1994, Anonymous HIV testing: the impact of availability on demand in Arizona. Am J Pub Health 84: 2008–2010.
Indian Council of Medical Research (ICMR), 2000, Ethical guidelines for biomedical research involving human subjects. New Delhi: ICMR.
Joint Commission of Institute of Judicial Administration and American Bar Association, 1979, Standards for minor's consent to medical services. Cambridge, MA: Ballinger.
Kelly B.D., and McLoughlin D.M., 2002, Euthanasia, assisted suicide and psychiatry: a Pandora's Box. Br J Psychiatry 181: 278–279.
Kent S.J., Clancy R.L., and Ada G.L., 1997, Prospects for a preventive HIV vaccine. In: Managing HIV (G.J. Stewart, ed.). North Sydney: Australasian Medical Publishing, pp 179–182.
Kissane D.W., and Kelly B.D., 2000, Demoralisation, depression, and desire for death – problems with the Dutch guidelines for euthanasia of the mentally ill. Aust NZ J Psychiatry 34: 325–333.
Kumar S., 1998, Medical confidentiality broken to stop marriage of man infected with HIV. Lancet 352: 1764.
Magnusson R.S., Scott M., and Irvine S.S., 1997, HIV and confidentiality. In: Managing HIV (G.J. Stewart, ed.). North Sydney: Australasian Medical Publishing.
Magnusson R.S., 1994, Privacy, confidentiality and HIV/AIDS health care. Aust J Pub Health 18: 51–58.
Muthuswamy V., 2005, Ethical issues in HIV/AIDS research. Indian J Med Res 121(4): 601–610.
National AIDS Control Organisation (NACO). Training manual for doctors. New Delhi: Government of India.
National AIDS and STD Control Programme (NASCOP), 2001, National Guidelines for Voluntary Counselling and Testing. Nairobi: Ministry of Health., Government of Kenya.
Ranjan S., Kumar M., and Saraswat N., 2005, Euthanasia and psychiatry. JAPI 53: 997–998.

Shreedhar J., 2002, India's HIV. www.fhi.org/NR/Shared/enFHI/

UNAIDS/WHO, 2000, Opening up the HIV/AIDS epidemic. August.

UNAIDS, 2000, Ethical considerations in HIV preventive vaccine research. UNAIDS Guidance Document. Geneva: UNAIDS. www.unaids.org

UNAIDS, 2004, Children on the brink – 2004. Global orphan numbers would be falling without AIDS. Geneva: UNAIDS. www.aegis.com/news/unaids/2004/UNO40709.html

United Nations, 1998, HIV/AIDS and Human Rights – International Guidelines. New York: United Nations. HR/PUB/98/1:13.

Van Reyk P., 1997, HIV and choosing to die. In: Managing HIV (G.J. Stewart ed.). North Sydney: Australasian Medical Publishing Co. Ltd.

Vas C.J., and de Souza E.J., 1991, Ethical concerns in AIDS. Mumbai: FIAMC Biomedical Ethics Centre.

WHO, 1985, Reproductive Health and the Law. WHO Chronicle. www.who.int/en/

WHO/UNAIDS, 1999, Questions and Answers on Reporting, Partner Notification, and Disclosure of HIV and/or AIDS sero-status – Public Health and Human Rights implications. www.who.int/en/

World Medical Association (WMA), 1964, Declaration on human experimentation. Helsinki (Finland): WMA, June 1964.

World Medical Association (WMA), 2000, Ethical principles for medical research involving human subjects – Revised Declaration. Edinburgh (Scotland): WMA, October.

CHAPTER 24

ROLE OF HEALTH CARE PROVIDERS

Abstract

Although the management of HIV-infected individuals is somewhat similar to that of patients with other chronic ailments, HIV infection is incurable and carries the additional burden of social stigma. Therefore, health care providers should consider the psychological and emotional aspects. With a non-judgmental attitude, health care providers can create a safe environment for discussing client's lifestyle, sexual and drug-using practices, fears, and anxieties. Health providers need to update themselves about the availability and location of various resources like medical facilities, legal aid, and social support services in order to help clients who may need them. Family members and friends of HIV-infected persons need access to accurate information, social support and referral services. The providers should accept the emotions and reactions of clients when they realise that they are HIV-infected. In countries where it is legally obligatory for doctors to report HIV infections and/or AIDS cases, the reporting should be done with due respect to privacy concerns. Patient support organisations provide significant support which can complement medical care. Health care personnel can assist patients by providing advice on what questions to ask before they choose a particular organisation.

Key Words

Access to information, Attitudes, Confidentiality, Continuing HIV education, Counselling, Family, Family of choice, Mutual aid groups, Online organisations, Patient support organisations, Peer support organisations

24.1 – EXPECTATIONS FROM HEALTH CARE PROVIDERS

In many aspects, management of HIV-infected individuals resembles management of patients with other chronic ailments. Patients and their family members should be educated about the need for behavioural changes and the need for long-term care. However, since HIV infection is incurable and carries the additional burden of social stigma, health care providers should also consider the psychological and emotional aspects. HIV-infected individuals and their families have the following expectations from their health care providers.

Counselling: Counselling is a difficult task that requires patience, dedication, and commitment. Health care providers who have the necessary aptitude, values, attitude, knowledge, and skills can undertake counselling.

Confidentiality: Health care facilities must have mechanisms for protecting privacy and all categories of health care personnel should be sensitised on issues pertaining to confidentiality and privacy. The client and their family members should be assured that the disclosed personal, intimate feelings, or events, would be kept confidential. This helps in building trust and rapport between the client and the health care provider.

Access to Accurate Information: The manner in which HIV- positive test result is conveyed can affect the HIV-infected person's pattern of coping with stress (Pergami *et al.*, 1994). During post-test counselling, the counsellor/provider should willingly listen to the client's queries and provide explanations that are easily understood. Preventive measures and the disease process ought to be clarified in a language devoid of medical jargon (Scott & Irvine, 1997). Since the news of HIV seropositivity produces emotional shock, many clients may not be able to recall what was discussed during post-test counselling. Hence, written information should be provided and follow-up counselling sessions should be arranged (Scott & Irvine, 1997).

Attitudes: Negative attitudes or uneasiness of health care providers may subsequently discourage HIV-infected persons from seeking health care. Health care providers should not judge or condemn the client's sexual preferences, behaviour, and lifestyle. Moralising or preaching may add to the feeling of guilt or "self-blame" in HIV-affected individuals. With a non-judgemental attitude, health care providers can create a safe environment for discussing client's lifestyle, sexual and drug-using practices, fears, and anxieties (Mansfield & Singh, 1993).

Tolerance to Alternative Therapies: Basic knowledge of common alternative therapies is essential for doctors because ignoring or rejecting alternative therapies will not prevent patients from trying them. Intolerant attitude to alternative treatments may stop patients from informing their providers. Open discussion about realistic treatment and choices should be promoted. Combination of modern medicine with alternative therapies should not be discouraged unless they are expensive or interfere with prescribed treatment (Scott & Irvine, 1997).

Referrals: Health care providers, who are uncomfortable while discussing patients' anxiety about death, should refer patients to other professionals who can openly deal with these concerns (Scott & Irvine, 1997).

Family and Friends: Involvement of family members and friends of HIV-infected persons is essential in advanced stages of the disease or if cerebral involvement develops. They need access to accurate information, social support and referral services. *Caution* – some family members may seek details, which the HIV-infected person may not like to be revealed. In such situations, doctors should strive to strike a balance, which is in the best interests of the patient. *Family of choice* – in case of persons belonging to socially marginalised groups, e.g. drug users and homosexuals, the "family" of choice may be different from their family of origin. Doctors should recognise significance of these relationships (Scott & Irvine, 1997).

Acceptance: The providers should accept the emotions and reactions of clients (including hostility) when they realise that they are HIV-infected. This acceptance of clients should not be affected by the counsellor's subjective feelings about their high-risk behaviour, lifestyle, sexual preference, or social background.

Information on Available Resources: Health care providers should be well informed about the availability and location of various resources like medical facilities, peer support groups, legal aid, and social support services in order to help clients who may need them.

24.2 – CONTINUING HIV EDUCATION

24.2.1 – Objectives of HIV Education for Doctors

1. For doctors who may not be involved in caring for HIV-infected persons, HIV education should aim at inclusion of HIV in differential diagnosis and practising universal precautions.
2. HIV education for doctors involved in treating HIV-infected patients should aim at enhancing their skills for better patient care and taking safe precautions to prevent HIV transmission.
3. HIV education ought to build on existing skills and should be evidence-based since doctors need strong reasons to change established practices (Stewart *et al.*, 1997).
4. Patient-mediated strategies are increasingly used in continuing HIV education programmes in order to change performance of physicians (Stewart *et al.*, 1997).

24.2.2 – Need for Continuing Education

Doctors need to update themselves by acquiring information from a variety of sources (e.g. medical journals, the Internet, lectures, workshops, and conferences) because HIV medicine is a rapidly developing multidisciplinary branch of medicine. Due to the rapid expansion of knowledge, development of new drugs, and occurrence of new opportunistic infections, there is an urgent need for continuing HIV education. The emergence of HIV epidemic has led to new ways of working, such as multidisciplinary team approach, and use of universal biosafety precautions. Traditionally, doctors have been teachers and educators within the health team. Some HIV-infected patients are well informed and acquire HIV-related information through self-help groups, the Internet and other sources. In the United States, HIV medicine is outside mainstream medical services and educational programmes are directed towards sharpening skills of a small group of primary care physicians, rather than increasing competence of all health care providers on a broad basis. A study shows that HIV education for primary care physicians has resulted in improved outcomes for HIV-infected persons (Volberding, 1996).

24.2.3 – Misconceptions Among Doctors

Misconceptions among doctors act as barriers to continuing HIV education. These include perception that HIV medicine is too complex and is best left to specialists, fear of spread of HIV to self, staff and other patients, and disapproval of high-risk behaviour that spreads HIV (Stewart et al., 1997). Except for care of terminally ill patients, HIV medicine is not complex and does not require expertise of HIV specialists. Outcomes are also known to improve when patient care is shared between primary care physician and specialist in HIV medicine, provided there is good communication between the patient, primary care physician, and the specialist (Stewart et al., 1997).

24.3 – ROLE IN PUBLIC HEALTH

In some countries, it is obligatory for doctors to report HIV infections and/or AIDS cases. Reporting should be done with due respect to privacy concerns. The names and addresses of the infected persons should not be revealed (Kaldor & Crofts, 1997). In the past, decisions on public health matters were left to politicians and bureaucrats, while doctors have been reluctant to advise on public health matters. In response to the HIV epidemic, doctors have taken a more active role in advising governments on public health issues and measures for epidemic control (Penny et al., 1997).

24.4 – PEER SUPPORT ORGANISATIONS

Peer support organisations are also known as mutual aid groups, patient support organisations, consumer organisations, or self-help groups. These organisations provide peer support and can optimise health outcomes (Trojan, 1989). Peer support organisations are of two types: face-to-face and online groups.

24.4.1 – Face to Face Organisations

Face-to-face organisations often operate around particular illnesses. Many patient support groups form at the local level and gradually expand to operate in larger geographical areas. Some organisations have federated at the national level while some may be associated with charitable foundations (Newell, 2004).

24.4.2 – Online Organisations

Many organisations have a presence on the Internet. Sponsorship of websites may be offered by corporate sponsors such as pharmaceutical companies (Newell, 2004). Online groups combine the advantages of self-help and the accessibility of computer networks (Finn, 1999). Online self-help groups can be of help to disabled individuals (Newell, 2004), those with rarer conditions and for those who need information on sensitive personal issues that they may not like to discuss with a health care worker (Hopkins & Fogg, 2002).There is also significant activity by health consumers on the Internet (Milio, 1996). Unfortunately, online peer support groups are not useful in resource-poor nations where poverty and illiteracy are among the several factors that preclude access to the Internet. Prospective members of online support groups ought to note that consumer protection and privacy laws may not exist in all countries (Newell, 2004).

24.4.3 – Choosing a Peer Support Organisation

Patients and their families need to independently explore the suitability of a particular organisation for them. Suitability of an organisation can play a vital role in the therapeutic relationship (Newell, 2004). When choosing a particular patient support organisation, patients and their families need to enquire about

- Origins of the organisation – whether formed by individuals opposed to a particular treatment or by researchers trying to recruit a cohort of patients for study?
- Rules of the organisation – whether the Constitution or Memorandum of Association allows participation by members and ensures their rights including ability to participate in governing the organisation?
- Source of funds – whether the source of funds has been revealed in its publicly available literature such as an annual report?
- Presence or absence of contractual agreements with pharmaceutical companies or other health care providers – these arrangements may influence the information provided and may include the likelihood of recruitment to a particular trial to the exclusion of others.
- Fees and charges – potential members need to know whether what they get in return for any membership fees represents value for money for them. Free or subsidised membership may indicate corporate sponsorship of the organisation.
- Type of organisation (whether treatment- or disease-specific) – organisations that are centred on a particular treatment may limit information options about other treatments and support.
- Helping mechanisms offered – social and emotional support, problem solving, and open environment for exchange of ideas.
- Experiences and perceptions of others, including that of health care personnel, but excluding that of persons associated with the organisation.
- Suitability and quality of information (accurate and reliable, or questionable) provided about the disease or disability (Newell, 2004).

24.4.4 – Role of Health care Personnel

Patient support organisations provide significant support that can complement medical care. Health care personnel can assist patients by providing advice on what questions to ask before they choose a particular organisation. Eventually, it is for the patients to decide which organisation will suit them best. Health care personnel need to ask patients how effective and helpful such organisations are because many organisations change over time. The gathered information can be used to inform other patients seeking information about or referral to patient support organisations.

REFERENCES

Finn J., 1999, An exploration of helping processes in an on-line self-help group focussing on issues of disability. Health Soc Work 24: 220–231.

Hopkins H. and Fogg S., 2002, Assessing health information on the Internet. Aust Health Consum 2: 21–23.

Kaldor J.M. and Crofts N., 1997, Epidemiological surveillance for HIV/AIDS. In: *Managing HIV* (G.J. Stewart, ed.). North Sydney: Australasian Medical Publishing, pp 169–172.

Mansfield S. and Singh S., 1993, Who should fill the care gap in HIV disease? Lancet 342: 726–728.

Milio N., 1996, Engines of empowerment: using information technology to create healthy communities and challenge public policy. Chicago: Health Administration Press.

Newell C., 2004, Finding a patient support group. Aust Prescr 27(1): 19–21.

Penny R., Grimes D., and Baume P., 1997, HIV and the doctor's role in public health. In: Managing HIV (G.J. Stewart, ed.). North Sydney: Australasian Medical Publishing, p 168.

Pergami A., Catalan J., and Hulme N., *et al.*, 1994, How should a positive HIV test result be given? the patient's view. AIDS Care 6: 21–27.

Scott M. and Irvine S.S., 1997, What do people with HIV, their carers and families want of their medical carers? In: Managing HIV (G.J. Stewart, ed.). North Sydney: Australasian Medical Publishing, pp 156–157.

Stewart G.J., Bollen M.D., and van der Weyden M.B., 1997, Continuing HIV education for health care workers. In: Managing HIV (G.J. Stewart, ed.). North Sydney: Australasian Medical Publishing, pp 166–167.

Trojan A., 1989, Benefits of self-help groups: a survey of 232 members form 65 disease-related groups. Soc Sci Med 29: 225–232.

Volberding P., 1996, Improving the outcomes of care for patients with human immunodeficiency virus infection. N Engl J Med 334: 729–731.

CHAPTER 25

PREVENTIVE HIV VACCINE

Abstract

Development of a preventive vaccine offers the best prospect for containing the HIV epidemic. A realistic goal of a preventive HIV vaccine would be to reduce viral load. Various approaches to developing an effective vaccine include delivering live vector vaccines to mucosal surfaces, incorporating multiple HIV proteins into recombinant live vector or DNA vaccine preparations, injecting purified DNA encoding for HIV proteins into skin or muscle, and incorporating cytokines in vaccine preparations. Liposomes offer multiple advantages as carriers of vaccine antigens.

Scientific obstacles to vaccine development include lack of a suitable animal model, antigenic diversity and hypervariability of HIV, its transmission by mucosal route and by infected host cells, resistance of wild-type virus to seroneutralisation, integration of HIV genome into the host cell chromosomes, latency of HIV in resting memory T-cells, rapid emergence of viral escape mutants in the host, and downregulation of MHC class I antigens.

Programme-related obstacles in developing an AIDS vaccine include inadequate political leadership, insufficient allocation of funds for AIDS vaccine research, insufficient coordination, lengthy approval process and delays in starting trials in developing countries, and insufficient standardisation of assays and reagents. Recruitment of volunteers is difficult in populations exposed to stigma, discrimination, rumours, misunderstandings, and media opinion. Due to decreasing HIV incidence rates, vaccine efficacy trials will have to be conducted at multiple centres or countries in order to achieve the necessary sample size.

In human trials of potential AIDS vaccines, counselling on safer sexual practices is essential. Phase I trials of TBC-M4 Modified Vaccine Ankara (MVA) HIV-1 multigenic subtype C vaccine and tgAACo9 began in December 2005 in India. The Aventis-Pasteur live recombinant vaccine is in phase III trial in Thailand. At present, there is no effective vaccine that can prevent HIV infection in uninfected individuals or halt the progress of the disease in HIV-infected persons.

Key Words

Ethical issues, Indian vaccine trials, Obstacles, Stealth liposomes, Vaccine trials, Vaccine vector

25.1 – INTRODUCTION

The most important discoveries of the next 50 years are likely to be ones of which, we cannot now even conceive

John Maddox, former editor of *Nature*

The development of a safe and effective preventive vaccine offers the best hope for containing the HIV epidemic since attempts to modify high-risk behaviour have met with variable results. ARV therapy neither destroys the virus nor cures HIV infection. The objective of ARV treatment is to retard the progress of illness in many patients with advanced disease, and to prevent onset of symptomatic HIV disease in asymptomatic or relatively healthy HIV-positive individuals (Harries *et al.*, 2004; Wig, 2002). A prospective vaccine should limit the initial infecting dose at mucosal surface and produce a rapid systemic cytotoxic T lymphocyte (CTL) response. CTL response can clear a lower viral burden. In such a case, dissemination of HIV to blood stream or organs would not occur (Kent *et al.*, 1997).

HIV infection spreads from a mucosal entry site through regional lymph nodes, to multiple organs. Theoretically, preformed neutralising antibodies, if present in high titres at the site of entry of HIV, may prevent infection of the host. But, the antigenic envelope region of the virus is highly variable. Viruses with variation in the envelope region may escape action of neutralising antibodies (Kent *et al.*, 1997). The replication of HIV to uncontrollably high levels occurs *before* the formation of anti-HIV immune response. Once HIV replicates to high levels within regional lymph nodes and disseminates throughout the body, the task of eliminating HIV becomes difficult for the following reasons:
1. Soon after infection, HIV replicates in sites, such as the brain, which are not easily accessible to the immune system
2. With ongoing replication, diverse mutants are formed, which escape recognition by CTL (Philips *et al.*, 1991)
3. HIV-infected cells provide a reservoir of HIV, which is difficult to eliminate
4. Immunodeficiency progressively reduces host's ability to respond to HIV
Therefore, HIV should be cleared from the body before it replicates to high levels. A realistic goal of a preventive HIV vaccine would be to reduce viral load (Kent *et al.*, 1997).

25.1.1 – Phases in Vaccine or Drug Trials

In general, new drugs or candidate vaccines are studied in the following phases:
1. Phase zero – preclinical research on pharmacology and toxicology
2. Phase I – randomised controlled clinical trials for safety, pharmacokinetics and pharmacological effects in healthy human volunteers
3. Phase II – randomised controlled double-blind clinical trials for safety and effectiveness in patients

4. Phase III – efficacy and safety studies in patients with other associated genetic disorders or ailments and in special groups (women of reproductive age, infants, children)
5. Phase IV – extended field trials when the drug or vaccine is in widespread use to study long-term side effects and side effects in special groups

25.2 – STUDY OF IMMUNE RESPONSES

25.2.1 – Immune Responses in those with "Natural Immunity" to HIV

Studies have been conducted in persons who were not infected though they were exposed to HIV perinatally (Rowland-Jones *et al.*, 1993), sexually (Rowland-Jones *et al.*, 1995), or occupationally (Clerici *et al.*, 1994). Exposed infants and adults, who remain uninfected, have high levels of HIV-specific cytotoxic T-cells (Rowland-Jones *et al.*, 1994; Rowland-Jones *et al.*, 1995). There is also a report of a HIV seropositive baby (born to an HIV-infected mother) who subsequently eliminated HIV (Bryson *et al.*, 1995). The above-mentioned findings indicate that it is possible to clear HIV infection and HIV-specific CD8 CTLs (rather than antibodies) seem to be the major effector mechanism for clearing HIV. While early approaches to develop HIV vaccine concentrated on generating neutralising antibodies, recent approaches have paid attention to generation of HIV-specific CTL responses (Kent *et al.*, 1997).

25.2.2 – Immune Responses in those with "Resistance" to HIV

A mutation in a gene that produces CKR5 (or CCR5 or R5) co-receptor is common among 1 per cent of persons of European descent. CKR5 co-receptor is required for infection by macrophage-tropic strains of HIV. Homozygotes (those possessing one pair of the mutant gene, inherited from both parents) do not have CKR5 co-receptors on their cells and are virtually immune or highly resistant to HIV infection in spite of multiple exposures to the virus (Liu *et al.*, 1996). Thus, macrophage-tropic strains of HIV seem to have a role in initiating infection and these strains could be used as antigens in vaccine preparation (Zhu *et al.*, 1993).

25.2.3 – Study of Local Mucosal Defences

The Peyer's patches in the gut generate most mucosal activated T- and B-lymphocytes. These lymphocytes subsequently migrate to other mucosal sites in the gut and genital mucosa. This intermucosal movement of activated lymphocytes constitutes the "common mucosal immune system". As compared with systemically delivered vaccines, vaccines delivered to any mucosal surface produce more efficient immune response due the common mucosal immune system. Mucosal T-lymphocytes inhibit viral replication (Kent *et al.*, 1997). IgA secreted by relocated B-lymphocytes can prevent absorption of virus into epithelial cells, interfere with assembly of new viruses, and drive IgA–virus complexes back into the lumen (Kent *et al.*, 1997).

Since there are apprehensions regarding the efficacy of current vaccines based on purified viral proteins and safety of live attenuated vaccines, new approaches are being tried in animal models; however, there is no suitable animal model for HIV-1 infection. These new approaches include:

- Delivering live vector vaccines to mucosal surfaces (e.g. recombinant adenovirus that affects respiratory mucosa; poliovirus, which affects the gut mucosa)
- Incorporating multiple HIV proteins into recombinant live vector or DNA vaccine preparations to broaden anti-HIV immune response
- Injecting purified DNA (that encodes for HIV proteins) into skin or muscle, which produces both antibody and CTL responses
- Incorporating cytokines in vaccine preparations to increase and influence the vaccine-induced immune responses (Kent *et al.*, 1997)

25.2.4 – Role of "Stealth" Liposomes

Liposomes offer multiple advantages as carriers of vaccine antigens as they are biodegradable, non-toxic, synthetic, and elicit both humoral and cell-mediated immunity (CMI). A wide variety of molecules (e.g. drugs, peptides, hormones, enzymes, and genetic materials) could be encapsulated within the aqueous spaces or intercalated into the bilayer membranes or carried on the surface of the liposomes. Manipulations of structural variables (lipid to antigen mass ratio, bilayer fluidity, vesicle size, surface charge, and mode of antigen association with the vesicles) usually induce variation in the level of immune response to a factor of 3. In order to ensure prolonged exposure to immune cells, the liposomes should survive for prolonged periods *in vivo*. Prolonged survival is achieved by shielding the surface of liposome with polyethylene glycol (PEG) and other hydrophilic polymers, or chemically modifying the hydrophobic part of the phospholipids. This shield fools the phagocytes into ignoring the liposomes. PEG-grafted liposomes are also called "stealth" liposomes. PEG is a synthetic non-toxic polymer. Its molecules get heavily hydrated due to its chemical affinity for water. To phagocytes, this molecular "cloak" of water of hydration makes the PEG-grafted ("stealth") liposomes appear like watery blobs and tend to ignore them. Consequently, the circulatory life of PEG-grafted liposomes increases and enhances the availability to the immune cells. Stealth liposomes are potentially useful in developing more effective vaccines. PEG-grafted stealth liposomes carrying antigenic epitopes of gp41 (a transmembrane protein of HIV-1) showed about twofold higher immune response and prolonged persistence of antibodies as compared to that of liposomes without PEG moieties (Singh & Bisen, 2006).

25.3 – SCIENTIFIC OBSTACLES

The development of vaccines against HIV poses a formidable challenge. There is inadequate understanding of immunological parameters, which protect against HIV infection or disease, and relationship of genotypic variations to the expression of antigens. Some strains of HIV-1 may evade vaccine-induced cellular immunity by genetic diversity, and nullifying anti-HIV cellular immune responses. Occurrence of genotypic mutation and recombination during replication of HIV (Kent *et al.*, 1997). Due to antigenic diversity and hypervariability of HIV, a vaccine protecting against one serotype may not be effective against another subtype or inter-subtype recombinant. Other obstacles to vaccine development include transmission by mucosal route, transmission of HIV by infected cells, resistance of wild-type virus to seroneutralisation, integration of HIV genome into the host cell chromosomes, latency of HIV in resting memory T-cells, rapid emergence of viral escape mutants in the host, and down-regulation of MHC class I antigens (Excler, 2005).

Mucosal Transmission: Since most cases of HIV are due to transmission by the sexual (mucosal) route, efforts are being made to develop a vaccine that would confer protective immunity at the mucosal level. It is necessary to validate the measurement of immune responses at the mucosal level. The CMI-oriented vaccine strategies may not be adapted to the situation since HIV shedding in semen seems to be poorly correlated with systemic CD8 CTL response in humans (Kozlowski & Neutra, 2003).

Animal Models: Lack of appropriate animal models for HIV-1 infection. HIV-1 does not reliably cause AIDS in animal models that can be infected (Kent *et al.*, 1997). In non-human primates, candidate vaccines have elicited immune responses ranging from highly effective to poorly effective in their ability to mitigate infection after challenge with HIV (Desrosiers, 2004). The animal models will be validated when the results in these models are compared with results of phase IIb trials (or "proof of concept" trials) and phase III trials (or "efficacy" trials) in humans (Excler, 2005).

Vaccine Vectors: There is a need to develop vaccine vectors. These vectors "deliver" the genes encoding the target viral immunogens to the antigen-presenting cells of the host's immune system. However, tests conducted on monkeys vaccinated with SIV vaccines have demonstrated SIV-specific CTLs. SIV vaccines seem to delay the onset of AIDS in SIV-infected monkeys (Hirsh *et al.*, 1994). Thus, if a preventive vaccine cannot produce an immune response that can clear HIV infection, it may still achieve the secondary goal of prolonging the disease-free period after HIV infection (Kent *et al.*, 1997).

25.4 – PROGRAMME-RELATED OBSTACLES

The programme-related obstacles in developing an AIDS vaccine include inadequate political leadership, insufficient allocation of funds for AIDS vaccine research, insufficient coordination, lengthy approval process and delays in starting trials in developing countries, and insufficient standardisation of assays and reagents.

Problems in Recruiting Volunteers: Recruitment of *lower-risk* volunteers for phase I trials and of *at-risk* volunteers for phase II and phase III efficacy trials is rendered difficult in less educated populations that are exposed to stigma, discrimination, rumours, misunderstandings, and media opinion (Excler, 2005).

Sample Size: Due to aggressive intervention programmes over the past decade, the incidence rates for HIV infection have shown a decreasing trend, except in some high-risk groups. As a consequence of decreasing incidence rates, vaccine efficacy trials will need to be multicentric or multi-country, in order to achieve the sample size needed at analysis for a given mode of transmission (Excler, 2005).

25.5 – VARIOUS APPROACHES

Classical vaccine strategies based on live attenuated or whole-inactivated HIV has severe limitations (Chertova *et al.*, 2002; Whitney & Ruprecht, 2004). Therefore most efforts to develop AIDS vaccine have focussed on newer vaccine approaches.

25.5.1 – Subunit Vaccine Based on Recombinant Envelope

Soluble glycoprotein of recombinant HIV and SIV have been explored as candidate subunit vaccines. The integrity of envelope glycoprotein seems to be necessary for inducing neutralising antibodies. The current envelope-based candidate vaccines elicit high gp120-binding antibody titres. Antibodies induced by gp120 can neutralise the homologus CXCR4 virus strain but are usually incapable of neutralizing the primary CCR5 isolates (Excler, 2005).

25.5.2 – Subunit Vaccine Based on Tat Protein

Subunit vaccines based on *Tat* protein have demonstrated partial protective efficacy in animal models and are being tested in humans (Fanales-Belasio *et al.*, 2002).

25.5.3 – Lipopeptides

Lipopeptides have the capability to induce a MHC class I-restricted CD8 response. Long lipopeptides with a fatty acid tail can induce broad cellular immune responses in humans and animals. Lipopeptides with a monolipid tail and a tetanus toxoid peptide appeared to be superior to double lipid tail peptide (Gahery-Segard *et al.*, 2003).

25.5.4 – Live Recombinant Vectors

A live attenuated viral or bacterial strain is used as a *vector* to carry HIV genes encoding the antigens of interest. Successive injections with the same vector will induce immunity to the vector (Excler, 2005).

Vaccinia Virus: Though *vaccinia virus recombinants* were the first vectors to be tried in animals and humans, their possible lack of safety in immune compromised persons has led to the use of *attenuated vaccinia virus* strains, such as NYVAC HIV-1 subtype C (an attenuated slow-replicative genetically engineered *vaccinia virus*), and Modified Vaccinia Ankara (MVA) – a highly attenuated, host range-restricted strain of *vaccinia virus* (Excler, 2005). In non-human primates, MVA recombinants were found to induce potent CTL responses which face pathogenic viral challenge (Im & Hanke, 2004).

Other Pox Viruses: Although canary pox (ALVAC) or fowl pox vectors are very safe in humans (being non-replicative in mammalian cells), they are less immunogenic, as compared to *vaccinia virus*.

Adenovirus: Recombinants of human adenovirus types 4, 5, and 7 (designated Ad4, Ad5 and Ad7) can be administered orally and intranasally and can induce both systemic and humoral immunity. Adeno-associated viruses (AAVs) are also used.

Alpha Viruses: Defective alpha virus or *replicons* of Venezuelan equine encephalitis (VEE) virus, Sindbis virus, and Semliki forest virus (SFV) offer the advantages of important amplification of viral message after infection and are able to target dendritic cells (Davis *et al.*, 2002).

Other Viruses: *Flavivirus:* yellow fever virus; *rhabdovirus:* vesicular stomatitis virus and rabies virus; *myxovirus:* influenza virus; *paramyxovirus:* Sendaï virus, measles virus; *picorna virus:* polio virus

Bacteria: Bacteria that have been explored as vectors for HIV vaccine include – *Bacillus Calmette Guérin* (BCG), *Salmonella*, *Shigella*, *Lactobacillus*, *Streptococcus*, and *Listeria monocytogenes* (Fouts *et al.*, 2003).
Limitations of vectors – In individuals previously exposed to the vector and who developed a residual immunity to the vector, most live recombinant viral and bacterial vectors show decreased immunogenicity, as compared to previously unexposed individuals. This limitation is observed with adenovirus type 5 (Ad 5), BCG, poliovirus, MVA and NYVAC (in populations vaccinated against smallpox). However this limitation is not seen with canary pox or fowl pox vectors. In order to overcome the problem of pre-existing immunity to a vaccine vector, efforts now focus on using adenoviruses that rarely affect humans, e.g. Ad11 and Ad35.

25.5.5 – Naked DNA

Intradermal or intramuscular injection of a purified plasmid DNA that carries a gene encoding an antigen usually results in an immune response of Th-1 type (Excler, 2005).

25.6 – ETHICAL ISSUES

In May 2000, UNAIDS released a guidance document which outlined the strategies to be adopted for capacity building, community involvement, informed consent, protection of vulnerable population, justice and equity in selection of subjects for the clinical trials, use of placebo or any other vaccine in the control arm, continued counselling for risk reduction, inclusion of women and children with adequate safeguards, monitoring mechanism, and care and support for HIV-infection or any associated complication that may occur during the course of the trial (UNAIDS, 2000).

25.6.1 – Continued Counselling for Risk Reduction

There is a possibility that participants in clinical trials of HIV vaccines might abandon safer sexual practices and indulge in risky behaviour due to a *false sense of security*. Subsequently, if the vaccine is proved to be ineffective, this may lead to increased transmission of HIV. Hence, it is strongly recommended that counselling on safer sexual practices should be carried out along with vaccine trials (Kent *et al.*, 1997; Muthuswamy, 2005). The UNAIDS Document (UNAIDS, 2000) also recommends counselling programme, condom promotion, provision of microbicidal agents for women who are unwilling or cannot use condoms, and control of RTIs and STIs.

25.6.2 – Provision of Antiretroviral Drugs

The vaccine trial protocol should clearly mention that ARV drugs would be provided for those subjects who may become HIV-positive during the course of the vaccine trial. This is in addition to a constellation of services that go beyond what is locally available at the vaccine trial sites (Muthuswamy, 2005).

25.6.3 – Paediatric Trials

Though ARV drugs have shown considerable efficacy in preventing MTCT of HIV, it is not known whether ARV drugs would prevent the transmission of HIV-1 through breast feeding beyond the neonatal period (Gaillard *et al.*, 2004). In addition, infants who escape MTCT are again at risk for HIV infection when they become sexually active as adolescents. Vaccination of infants, begun at birth, is an attractive immunisation strategy. However, development of protective active immunity may take time (Excler, 2005).

25.6.4 – Adolescents

Participation of adolescents in vaccine efficacy trials raises ethical and legal concerns. It is well known that some high-risk groups such as sex workers and MSM, also include adolescents. The growing consensus is that once efficacy is demonstrated in adults, bridging studies should be conducted in adolescents (Thorne & Newell, 2004).

25.7 – INDIA'S ROLE IN HIV VACCINE TRIALS

In 2002, it was decided to undertake parallel exploration of several scientific approaches and to embark on parallel implementation of clinical trials so ensure that an effective vaccine is available at the earliest (Excler, 2005).

25.7.1 – Centres of Excellence

Centres of excellence have been set up for clinical and laboratory evaluation of candidate HIV vaccines. This strategy will also help in capacity building for research and development in India. The centres of excellence are National AIDS Research Institute (NARI), Pune where a phase I clinical trial with AAV-based vaccine began in 2005; and Tuberculosis Research Centre, Chennai where a phase I clinical trial with MVA vaccine began in the same year.

25.7.2 – Various Regulatory Mechanisms

Mechanisms have been put in place to ensure transparency and accountability and to address ethical and safety concerns. These include National AIDS Vaccine Advisory Board; Informed Consent Group (to develop a template for informed consent documents to be used in phase I trials); Gender Advisory Board and Training (to incorporate gender concerns in AIDS vaccine trials); NGO Working Group (to ensure that communities are better informed and to obtain community support and representation); and National Consultation on HIV care and treatment (to define policy and technical guidelines for care and treatment of trial participants, including those who may become HIV-positive during the course of the trials).

25.8 – CONCLUSION

The worldwide pursuit for an effective vaccine against AIDS symbolises an unprecedented scientific and human challenge. A preventive vaccine would be the ultimate preventive tool that will complement the existing strategies for prevention. Parallel exploration of several scientific approaches and clinical trials is probably the only way to reach this goal. This unprecedented long-term attempt requires a strong political commitment, flexibility of processes, medical and scientific dedication, and collaboration on a mission mode along with community participation (Excler, 2005).

In 1987, the first candidate HIV vaccine entered clinical trials. Since then, more than 40 candidate vaccines have been evaluated in safety and immunogenicity trials. Between 1987 and 2003, more than 10,000 HIV-negative volunteers have participated in these trials. All vaccines tested so far in humans have proven to be safe. One candidate vaccine has progressed to efficacy (phase III) trials (Excler, 2005). Databases have been established by the International AIDS Vaccine Initiative (www.iavi.org) and the United States National Institutes of Health Vaccine Research Center (www.vrc.nih.gov). Several vaccine concepts, immunisation schedules, routes of administration, and adjuvants have been tested (Girard et al., 2004).

In 2005, 13 new trials of preventive candidate AIDS vaccines began in nine countries. India, China, and Russia started their first AIDS vaccine trials in the same year. A randomised, placebo-controlled, dose-escalating, double-blinded trial to evaluate the safety and immunogenicity of TBC-M4 MVA HIV-1 multigenic subtype C vaccine began in December 2005 in India. The antigens used are env, gag, tat-rev, and nef-RT of HIV-1 subtype C. A similar study on tgAACo9 (containing gag, protease and reverse transcriptase of HIV-1 subtype C) has also begun in India. These studies are in phase I. The Aventis-Pasteur live recombinant prime (ALVAC vCP1521) with boost (vax Gen gp120 B/E) is in phase III trial in Thailand (IAVI Report, 2006).

REFERENCES

Bryson Y.J., Pang S., Wei L.S., et al., 1995, Clearance of HIV infection in a perinatally infected infant. N Engl J Med 332: 833–838.

Chertova E., Bess J.W. Jr., Crise B.J., et al., 2002, Envelope glycoprotein incorporation, not shedding of surface envelope glycoprotein (gp120/SU), is the primary determinant of SU content of purified human immunodeficiency virus type-1 and simian immunodeficiency virus. J Virol 76: 5315–5325.

Clerici M., Levin J.M., Kessler H.A., et al., 1994, HIV-specific T-helper activity in sero-negative health care workers exposed to contaminated blood. JAMA 271: 42–46.

Davis N.L., West A., Reap E., et al., 2002, Alpha virus replicon particles as candidate HIV vaccines. IUBMB Life 53: 209–211.

Desrosiers R.C., 2004, Prospects for an AIDS vaccine. Nature Med 10: 221–223.

Excler J.L., 2005, AIDS Vaccine development: perspectives, challenges and hopes. Indian J Med Res 121(4): 568–581.

Fanales-Belasio E., Cafaro A., Cara A., et al., 2002, HIV-1 Tat-based vaccines: from basic science to clinical trials. DNA Cell Biol 21: 599–610.

Fouts T.R., De Vico A.L., Onyabe D.Y., et al., 2003, Progress toward the development of a bacterial vaccine vector that induces high-titer long-lived broadly neutralizing antibodies against HIV-1. FEMS Immunol Med Microbiol 37: 129–134.

Gahery-Segard H., Pialoux G., Figueiredo S., et al., 2003, Long-term specific immune responses induced in humans by a human immunodeficiency virus type-1 lipopeptide vaccine: Characterization of CD8+ T-cell epitopes recognized. J Virol 77: 11220–11231.

Gaillard P., Fowler M.G., Dabis F., et al., 2004, Use of anti-retroviral drugs to prevent HIV-1 transmission through breast-feeding: from animal studies to randomized clinical trials. J AIDS 35: 178–187.

Girard M.F., Mastro T.D., and Koff W., 2004, Human Immunodeficiency Virus. In: Vaccines (S.A. Plotkin and W.A. Orenstein eds.). Philadelphia: Saunders, pp 1219–1258.

Harries A., Maher D., and Graham S., 2004, TB/HIV – a clinical manual 2nd edn. Geneva: WHO. pp 137–154.

Hirsh V.M., Goldstein S., Hynes N.A., et al., 1994, Prolonged clinical latency and survival of macaques given whole inactivated simian immunodeficiency virus vaccine. J Infect Dis 170: 51–59.

IAVI Report, 2006, Ongoing trials of preventive AIDS vaccines. VAX 4(1): 1–2. www.iavireport.org

Im E.J. and Hanke T., 2004, MVA as a vector for vaccines for HIV-1. Expert Rev Vaccines 3 Suppl 1: S89–S97.

Kent S.J., Clancy R.L., and Ada G.L., 1997, Prospects for a preventive HIV vaccine. In: Managing HIV (G.J. Stewart ed.). North Sydney: Australasian Medical Publishing, pp 179–182.

Kozlowski P.A. and Neutra M.R., 2003, The role of mucosal immunity in prevention of HIV transmission. Curr Mol Med 3: 217–228.

Liu R., Paxton W.A., Choe S., et al., 1996, Homozygous defect in HIV-1 co-receptor accounts for resistance in some multiple-exposed individuals to HIV-1 infection. Cell 86: 367–377.

Muthuswamy V., 2005, Ethical issues in HIV/AIDS research. Indian J Med Res 121(4): 601–610.

Philips R.E., Rowland-Jones S.L., Nixon D.F., et al., 1991, HIV genetic variation that can escape CTL recognition. Nature 354: 453–457.

Rowland-Jones S.L., Nixon D.F., Ariyoshi K., et al., 1993, HIV-specific cytotoxic T-cell activity in an HIV-exposed but uninfected infant. Lancet 341: 860–861.

Rowland-Jones S.L., Sutton J., Ariyoshi K., et al., 1995, HIV-specific cytotoxic T-cell activity in HIV-exposed but uninfected Gambian women. Nature Med 1: 59–64.

Singh S.K. and Bisen P.S., 2006, Adjuvanticity of stealth liposomes on the immunogenicity of synthetic gp41 epitope of HIV-1. Vaccine 24: 4161–4166.

Thorne C. and Newell M.L., 2004, Prevention of mother-to-child transmission of HIV infection. Curr Opin Infect Dis 17: 247–252.

UNAIDS, 2000, Ethical considerations in HIV preventive vaccine research. UNAIDS Guidance Document. Geneva: UNAIDS.

Whitney J.B. and Ruprecht R.M., 2004, Live attenuated HIV vaccines – pitfalls and prospects. Curr Opin Infect Dis 17: 17–26.

Wig N., 2002, Anti-retroviral therapy – are we aware of adverse effects? JAPI 50: 1163–1171.

Zhu T., Mo H., Wang N., et al., 1993, Genotypic and phenotypic characterization of HIV-1 patients with primary infection. Science 261: 1179–1181.

CHAPTER 26

RESPONSE TO THE HIV EPIDEMIC IN INDIA

Abstract

The first phase of NACP (NACP-I) was launched in September 1992 in all states and union territories as a 100 per cent centrally sponsored project. NACP-I was completed in March 1999. Innovative grassroots programmes have been launched at the community level, but due to India's size and diversity, these community-based efforts need to be increased to face the challenge posed by the HIV epidemic. Networks of HIV-infected persons were established in the 1990s. These groups are striving to break the stigma and discrimination within Indian society. Some private sector companies have initiated HIV-related workplace programmes and Indian pharmaceutical companies have reduced the price of ARV drugs, adopted dual pricing strategies, and launched drug donation schemes.

Key Words

Activism, Advocacy, Community response, Corporate response, Government's response

26.1 – GOVERNMENT'S RESPONSE

NACP was launched in 1987. When the gravity of the epidemic became evident, NACO was established in 1992, to manage and coordinate the programme (MOHFW, 1999; NACO, Training Manual for Doctors). A Strategic Plan was prepared for the 5-year period 1992–1997, which received support from World Bank, WHO, and other international agencies. NACP phase I (NACP-I) was launched in September 1992 in all states and union territories as a 100 per cent centrally sponsored project, which was completed in March 1999. The objectives of NACP-I were to prevent HIV transmission, to decrease mortality and morbidity associated with HIV infection, and to monitor the socio-economic impact of HIV epidemic (MOHFW, 1999). The initial years of the programme focussed on reinforcing programme management capacities, targeted IEC activities, and surveillance in the epicentres of the HIV epidemic. The actual preventive activities that gained momentum only in 1992 were education and awareness programmes, ensuring blood safety, hospital infection control, condom promotion, and strengthening of clinical services for controlling STIs and HIV. NACP-I has utilised the entire credit of US$84 million provided by the World Bank and its supervisory mission monitored the project (MOHFW, 1999).

The contribution of the Government of India for the second phase of NACP (NACP-II) is US$38.8 million, while the International Development Association (IDA) has provided an interest-free credit of US$191 million to cover the total cost, estimated at US$229.8 million (about Rs. 11,550 million). In addition, US Agency for International Development (USAID) has provided Rs. 1,660 million for the 7-year (1999–2006) AVERT project in Maharashtra. DFID has also offered financial assistance of Rs. 1,540 million for a 5-year (1999–2004) project in the states of Andhra Pradesh, Gujarat, Kerala, and Orissa (MOHFW, 1999).

26.2 – NATIONAL LEVEL PROGRAMME MANAGEMENT

For managing NACP at the national level, the following organisations were established:

National AIDS Committee: This was constituted in 1986 under the Chairmanship of Union Minister for Health and Family Welfare in order to bring together various ministries, NGOs, and private institutions for coordinated implementation of the programme. This committee is the apex body thdt provides policy directions, monitors programme performance, and forms multisectorial collaborations (MOHFW, 1999).

National AIDS Control Board: Constituted under the Chairmanship of Secretary (Health), Ministry of Health and Family Welfare, Government of India, the Board has been entrusted with functions, which were being performed by Technical Advisory Committee under the chairmanship of the Director General of Health Services. It is not necessary to approach the Ministry of Finance for funding planned activities since it is represented on the Board. The Board meets at least four times a year or when required by Chairman of the Board or Project Director of NACO. The functions of the Board include: (a) reviewing policies, approving annual operational plan budget, and reallocating funds between programme components; (b) exercising financial and administrative powers beyond the power of Additional Secretary and Project Director (NACO); (c) expediting sanctions, approving procurement and awarding contracts to private agencies; and (d) forming managerial teams and appointing senior programme staff (MOHFW, 1999).

National AIDS Control Organisation (NACO): This was established by the Government of India in 1992 to function as an executive body in the Ministry of Health and Family Welfare, New Delhi. An Additional Secretary as its Project Director heads this executive body. Its Secretariat consists of an Additional Project Director, subject specialists, technical, and administrative staff. The WHO has provided a technical assistance of US$1.5 million to NACO (MOHFW, 1999). NACO aims to:
(a) Create an expanded response to the HIV epidemic and monitor the epidemic in all States by effective surveillance
(b) Provide technical support to State AIDS Control Societies

(c) Ensure public awareness about the HIV epidemic and its prevention and target interventions at high-risk groups

(d) Reduce vulnerability of people to HIV by ensuring blood safety and promoting condoms for safer sex

(e) Promote community and family-based care for HIV-infected persons in a conducive environment free of stigma and discrimination and to develop support services for HIV-infected individuals

(f) Alleviate adverse socio-economic impact due to the HIV epidemic

Twelve technical groups were constituted for priority areas of the programme such as epidemiology and HIV testing, targeted interventions, legal and ethical issues, HIV prevention in the workplace, and blood safety and transfusion services. The technical groups act as a technical resource and are mainly responsible for providing operational know-how and to recommend action plans for improving the quality of programme design (MOHFW, 1999).

26.3 – STATE-LEVEL PROGRAMME MANAGEMENT

AIDS control societies were first created in Tamil Nadu and Pondicherry on experimental basis. Due to successful functioning of these societies, the Union Government advised other states/union territories to constitute a registered society, exclusively devoted to implementation of NACP, under the chairmanship of the Secretary (Health). These broad-based societies have members representing various ministries NGOs.

26.4 – FIRST PHASE OF NACP

26.4.1 – Components of NACP-I

Programme Management: Establishing managerial organisations to strengthen programme management at national and State levels, as mentioned above.

IEC and Social Mobilisation: The objectives of IEC programmes are to improve knowledge of HIV and STIs by mobilising all sections of society, promote desirable practices and behaviour (avoiding multiple sexual partners, use of condom, sterilisation of needles and syringes, and voluntary blood donation), and create a conducive environment to care for and support HIV-infected persons. NACP envisages a multisector approach, involving diverse ministries. NGOs have been provided financial support to take up awareness and targeted intervention programmes for vulnerable groups such as commercial sex workers, truckers, intravenous drug users, street children, and migrant labourers. A National Counselling Training Programme has been launched to train grassroots-level counsellors. *National AIDS helpline* has been set up with a toll-free telephone number "1097" so that the caller can avail of counselling services in an atmosphere that maintains privacy and confidentiality of the caller. School AIDS Education Programme has been launched on a pilot basis in 15 states to provide lifestyle education and HIV-related information to high school students (MOHFW, 1999; NACO Training Manual for Doctors).

Blood Safety: As per the directives of the Supreme Court of India, national and state-level blood transfusion councils have been constituted as registered societies to supervise all aspects of blood safety programme. Zonal blood testing centres provide linkage with other blood banks and test blood samples from blood banks in the zone and send the HIV test reports on the same day. Blood component separation facilities have been set up to reduce the wastage of whole blood transfusion. Licensing of all blood banks has been made mandatory. Unlicensed blood banks are not permitted to provide blood transfusion services. Currently, as per *national blood safety policy*, it is mandatory to test every unit of blood for infections such as HIV-1 and HIV-2, hepatitis B/C, malaria, and syphilis. Drugs and Cosmetics Rules have been made more stringent regarding procedures for collecting, processing, storing, and distributing blood and blood products. Blood donation by professional blood donors has been completely banned in India since 1 January 1998. National HIV testing policy containing key issues such as HIV testing procedures, mandatory screening of all blood units, confidentiality has been framed (MOHFW, 1999; NACO Training Manual for Doctors).

Condom Promotion: Unprotected and multipartner sexual activity is a major cause of HIV transmission in India. Disadvantages of free distribution of condoms are difficulty in ensuring supplies and doubts about their actual use. NGOs and voluntary organisations are involved in condom promotion. The target group for condom promotion for STI prevention differs from that for family planning. Schedule R of the Drugs and Cosmetic Rules has been modified in 1995 and NACO has introduced quality control parameters as specified by the WHO. Condoms are available at a subsidised rate from many types of outlets, including the recently introduced condom-vending machines. Social marketing increases acceptability and provides easy accessibility. Condom promotion programme also involves informing potential clients that condoms prevent deposition of semen and urethral discharge, exposure to penile lesions, and unwanted pregnancies; and demonstrating the correct method of use. Potential clients are also informed about the common causes of failure of condoms: damage to condoms due to exposure to heat, moisture, sunlight, and prolonged storage, failure to expel air while wearing condoms (causes tearing), and incorrect use, particularly under influence of alcohol and/or drugs (NACO Training Manual for Doctors).

Surveillance: Voluntary testing centres have been established in all government medical colleges. Regular nationwide sentinel surveillance monitor the trends in HIV infection in different risk groups in the population. MIS has been established for monitoring trends in HIV infection and progress of NACP (MOHFW, 1999).

Control of Sexually Transmitted Infections: Early diagnosis and treatment of STIs is one of the main strategies in prevention and control of HIV infection because the virus is transmitted more easily in the presence of other STIs and similar high-risk behaviours lead to transmission of both STI and HIV (NACO, Training Manual for Doctors).

Targeted Interventions: Projects for targeted interventions include those for sex workers (ASHA project of Mumbai Municipal Corporation and APAC Project in Tamil Nadu), projects for IDUs in Manipur, Nagaland, and Assam, and for truck drivers in Rajasthan (MOHFW, 1999; NACO, Training Manual for Doctors). NGOs have been encouraged to undertake targeted intervention programmes for high-risk groups such as sex workers, migrant labourers, transport workers, and out-of-school youth. Bilateral agencies such as the USAID have supported targeted interventions (MOHFW, 1999).

26.4.2 – Outcome of NACP-I

Adult HIV prevalence rate was maintained at below 1 per cent. Capacity building in managerial and technical aspects has been successful nationwide in government organisations and NGOs. There has been an expansion in capacity to treat STIs and opportunistic infections. A network of blood surveillance centres has been established along with sentinel sites to monitor the epidemic. The estimated blood safety is nearly 100 per cent, as compared to only 30 per cent in 1992. Condom use has increased from less than 10 per cent to a range of 50–90 per cent in targeted groups of sex workers and the volume of condoms distributed through social marketing has increased by 50 per cent. Knowledge of prevention of HIV has reached a range of 54–78 per cent in urban areas. Laws concerning humane treatment of HIV-infected persons have been enacted. Groups and networks of HIV-infected persons have emerged as forces for advocacy and activism (World Bank, 2002; Motihar & Sharma, 2003).

26.4.3 – Limitations of NACP-I

As a consequence of centralised planning and implementation, regional differences were not considered. In the early years of NACP-I, State AIDS Cells were handicapped by administrative bottlenecks and procedural delays, which hampered the pace of programme implementation. Hence these Cells were subsequently converted into broad-based State AIDS Control Societies. Initially, sentinel surveillance was not conducted in all the states/union territories resulting in inadequate information regarding the progress of the HIV epidemic. In most states, targeted interventions were not implemented for groups practising high-risk behaviour (MOHFW, 1999). Due to the high school-dropout rates in India, out-of-school youth need targeted intervention programmes tailored to their needs (Motihar & Sharma, 2003). Procurement of stores and equip-ment was slow due to delays in formulation of specifications for equipment and - inadequate understanding of the purchase procedures of the World Bank (MOHFW, 1999).

26.4.4 – Lessons Learnt from NACP-I

Control of HIV epidemic was conceived as a public health programme. But, spread of the epidemic is driven primarily by individual behaviour, and has human rights, legal and socio-economic implications. Hence, the epidemic is to be treated as a national calamity that impinges on multiple sectors and its control would call for coordination between different arms of the government, voluntary organisations, and the community. Activities need to be decentralised at planning and implementation levels. Surveillance data is important for tracking prevalence of HIV infection in general population and in specific high-risk groups. Best practices for targeted interventions and care of HIV-infected persons have been documented for action. A comprehensive system for monitoring, evaluation, and quantitative measurement of performance would be integrated in NACP-II. Technical resource groups have been established to strengthen technical support and capacity for research. High priority has been given to mobilising the community, private sector, and other stakeholders from health and other sectors. NACO has assigned responsibility of timely purchase of stores and equipment to a professional procurement agency (MOHFW, 1999).

26.5 – SECOND PHASE OF NACP

NACP-II aims to reduce the spread of HIV infection in India and to strengthen the country's capacity to respond to HIV epidemic on a long-term basis.

Targeted Interventions: Participatory methods are to be used to identify high-risk groups such as sex workers, IDUs, migrant workers, MSM, and patients attending STI clinics. Services provided for these groups will include counselling, client-based IEC, treatment, and condoms.

Community-Based Interventions: Locally appropriate IEC and awareness campaigns, VCT, strengthening infrastructure for ensuring blood safety, and preventive interventions in hospitals and occupational environment.

Low-Cost Community-Based Care for HIV-Infected Persons: Improving the quality and cost-effectiveness of treatment of opportunistic infections such as tuberculosis and establishing new sources of support in partnership with NGOs.

Strengthening Capacity of Institutions: NACP-II would assist in improving the technical, managerial, and financial sustainability of programmes at various levels by enhancing managerial capacity for planning and implementation; training in areas of management and provision of interventions; enhancing quality, quantity, and timeliness of surveillance for HIV, STIs, and behavioural patterns; augmenting capacity for ongoing monitoring and evaluation; and aiding high-quality, peer-reviewed operational research agencies.

Intersectoral Collaboration: Strengthening HIV-related programmes in the public, private, and voluntary sectors and supporting social mobilisation and sharing of information among Central Government ministries, private, and voluntary sectors.

26.5.1 – Targets for NACP-II

Blood Safety: Reducing blood-borne transmission of HIV to less than 1 per cent of the total transmission, introducing mandatory screening test for HCV, establishing 10 new modern blood banks in uncovered areas, upgrading 20 major blood banks, establishing 80 new district-level blood banks in uncovered districts, setting up 40 additional blood component separation units, and increasing share of voluntary blood donation to at least 60 per cent of total blood donation, and to increase the total blood collection in India to 5–5.5 million units by the end of the project.

Training: To train at least 600 NGOs in conducting targeted intervention programmes among high-risk groups and to achieve condom use of not less than 90 per cent, and to promote control of STIs through NGOs.

Awareness: To attain awareness level of not less than 90 per cent among youth and persons in reproductive age group with priority given to folk arts and street theatre in rural areas, to cover all schools students studying in classes IX and X through school education programmes, to cover all universities through *University talk AIDS* programme, and to reduce prevalence of STIs and RTIs in the general population.

Voluntary Testing: Every district in the country is to have at least one voluntary testing facility.

Self-Help Groups: Provision of financial support to organisations of HIV-infected persons.

26.5.2 – Lacunae in NACP-II

Almost 60 per cent of NACP-II funds have been allocated for prevention and control. Paradoxically, preventive activities for general population have been allocated more funds as compared to that for high-risk groups. Only 14 per cent of the funds are earmarked for care and support of HIV-infected persons (Motihar & Sharma, 2003). There is no national information grid that collects HIV test reports from private laboratories. Hence, prevalence estimates are exclusively based on sentinel surveillance (Solomon & Ganesh, 2002).

26.6 – NATIONAL HIV POLICY

The thrust areas include reduction of MTCT and future load of HIV-infected children in the community, counselling and HIV testing of pregnant women, and confidentially perform tests to establish HIV status, and provision of ARV therapy to already identified infected women. The prevention of transmission of HIV/AIDS is the main stay of the national response to the epidemic. However, care and support for those infected and affected by HIV/AIDS are inseparable elements of an effective national response and must be integrated in the comprehensive approach to deal with the epidemic (NACO, Training Manual for Doctors).

26.6.1 – Salient Features of HIV Policy

• HIV status of a person should not come in the way of his or her right to education and employment.
• Mobilisation of resources from government and private sources to build a continuum of care and support services, comprising clinical management, nursing care, counselling, and psychosocial support, through home-based care.
• CME to sensitise health care providers to prevent discrimination of HIV-infected persons, and to train health personnel in diagnosis, treatment and follow-up of HIV-related illnesses.
• Since there is no public health rationale for mandatory HIV testing, the policy promotes voluntary testing with pre- and post-test counselling.
• As per decision of the Supreme Court of India, *partner notification* has been included as a policy. All HIV-infected individuals should be encouraged to disclose their HIV status to their spouse/sexual partners. However, the attending physician should disclose the HIV status to spouse/sexual partners, only after counselling (NACO, Training Manual for Doctors).

26.7 – COMMUNITY RESPONSE

India's tradition of socio-political activism has yielded some innovative grass-roots programmes. The Sonagachi project in Kolkata (one of the world's largest programmes for sex workers) has empowered sex workers, who refuse unprotected sex with clients. Given India's size and diversity, these community-based efforts should be scaled up to meet the challenge posed by the HIV epidemic. Networks of HIV-infected persons that emerged in the 1990s are striving to break the stigma and discrimination within Indian society (Motihar & Sharma, 2003). Affected and Infected Women Association, Churachandpur (AIWAC) operates in Churachandpur district of Manipur. CPK+ is a network for people living with HIV in the state of Kerala. Manipur Network of Positive People (MNP+) provides care and support to HIV-infected persons in the state of Manipur. Telugu Network of People living with HIV/AIDS (TNP+), registered in Vijayawada (Andhra Pradesh) is a support group for HIV-infected persons and their families (UNDP, 2006). Other patient support groups include Indian Network of People Living with HIV (INP+), Association of People Living with AIDS (APWA), and Positive Women's Network (PWN) (Shreedhar, 2002).

26.8 – PRIVATE SECTOR'S RESPONSE

A mid-2005 study of the World Economic Forum revealed that only 11 per cent of Indian corporates had a written policy to combat discrimination based on HIV status, 14 per cent had an active policy to protect workers, with 29 per cent providing access to voluntary testing facilities (UNDP, 2006). According to a Reuters Report, many business organisations in India unsure about how much money they should spend on HIV/AIDS prevention and education programmes or how investments in HIV prevention programmes will save them money in the future. Some companies may have effective HIV prevention programmes for their core staff that do not trickle down to the informal workforce (usually migrant workers hired on a contractual basis). As per estimates of the International Labour Organization (ILO), the cost of treating one HIV-positive employee in India with ARV drugs is about US$431 per year, which does not include the costs of absenteeism and treating opportunistic infections. The ILO estimates that later intervention is 3.5–7.5 times as costly as prevention (Kaiser Network, 2006).

The Confederation of Indian Industries (CII) and leading corporates such as the Tata Group, Mahindra & Mahindra, and Bajaj have initiated HIV-related workplace programmes. In Septermber 2005, CoRE-BCSD India, a forum of 52 companies aiming to achieve sustainable development, identified three possible areas where Indian corporates could initiate or scale up HIV-related activities. These areas included

(a) HIV prevention programmes at the workplace
(b) Upgrading clinical facilities and training clinical personnel
(c) Supporting or scaling up HIV prevention efforts among high-risk or vulnerable communities such as transport and migrant workers, with whom companies interact regularly

In March 2006, CII initiated a project involving business process outsourcing companies (UNDP, 2006).

Indian pharmaceutical companies have started producing cheaper ARV drugs (Motihar & Sharma, 2003). The Indian pharmaceutical industry has responded by reducing prices of ARV drugs, adopting dual pricing strategy, and launching drug donation schemes. Since tuberculosis is a major opportunistic infection in HIV-infected persons, the Bill and Melinda Gates Foundation pledged US$ 89 million for research on tuberculosis vaccine. In 2003, the pharmaceutical company AstraZeneca has established a Tuberculosis Research Centre in Bangalore (Bebaruah, 2004).

26.9 – CONCLUSION

HIV prevention and education efforts in India are complicated by social stigma and presence of numerous languages and dialects. India has multiple diverse HIV epidemics. HIV prevalence of over 1 per cent has been reported among pregnant women in Andhra Pradesh, Karnataka, Maharashtra, and Tamil Nadu. In the north-eastern states of Manipur, Nagaland, and Mizoram, all of which lie adjacent to the drug trafficking "Golden Triangle", the HIV epidemic is driven by injecting drug use. Though there is no national policy for harm reduction, some states such as Manipur, have adopted their own harm reduction policies (Fredriksson-Bass & Kanabus, 2006).

India is one of the biggest producers of cheap generic ARV drugs that are sold to many countries all over the world. Ironically, millions of Indians in need of ARV treatment are not receiving it because free-of-charge treatment is available only in selected government institutions in certain cities. It is alleged that as the epidemic continues to spread in India, persons in positions of power are confused about the "right strategy" to be adopted and that some still advocate promotion of "abstinence and faith". The Indian Government has been criticised for clinging to the idea that the epidemic is limited to high-risk groups and that targeting them is the best strategy to contain the epidemic. Harassment of AIDS outreach workers and peer educators by law enforcement agencies was reported in 2002. Some government officials have supported mandatory pre-marital testing for HIV and have proposed related legislation (Fredriksson-Bass & Kanabus, 2006).

REFERENCES

Bebaruah S., 2004, The new weapons. India Today. 31 May, pp 60–62.

Fredriksson-Bass J. and Kanabus A., 2006, HIV/AIDS in India. www.avert.org. Last updated 19 July.

Kaiser Network, 2006, Reuters Kaiser Daily HIV/AIDS Report. www.kaisernetwork.org. 17 July.

Ministry of Health and Family Welfare (MOHFW), 1999, Annual Report (1999–2000). New Delhi: Government of India.

Motihar R. and Sharma M.V., 2003, Strengthening India's response to HIV/AIDS. www.kit.nl/ils/exchange/

National AIDS Control Organisation (NACO). Training manual for doctors. New Delhi: Government of India.

Shreedhar J., 2002, India's HIV. Arlington, VA: Family Health International. www.fhi.org/NR/Shared/enFHI/

Solomon S. and Ganesh A.K., 2002, HIV in India. In: Topics in HIV medicine. International AIDS Society – USA 10(3): 19–24.

World Bank, 2002, Project update. www.worldbank.org/aids

CHAPTER 27

RESPONSE TO THE HIV/AIDS EPIDEMIC IN THAILAND

Abstract

In Thailand, a politically supported, well-funded, comprehensive HIV prevention programme has substantially reduced incidence of new HIV infections, raised condom usage, and decreased prevalence of STIs. Widespread transmission of HIV began in the late 1980s among high-risk groups but the initial response to the epidemic was limited. In 1991, HIV prevention and control became a national priority. HIV prevention education was made compulsory in schools, legal reforms carried out and the "100 per cent condom programme" was launched to distribute free condoms and enforce compulsory and consistent use of condoms in all brothel and non-brothel settings. Achievements need to be actively sustained by adopting strategies to match the shifts in the epidemic and by focussing on preventing transmission among IDUs and their sex partners.

Key Words

Bangkok, Chiang Mai, Hundred per cent condom programme, Thailand

27.1 – INTRODUCTION

Thailand is among the few developing countries where public policy has been effective in preventing the spread of HIV/AIDS. The country's politically supported, comprehensive HIV prevention programme has substantially reduced incidence of new HIV infections, raised condom usage, reduced visits to commercial sex workers, and decreased prevalence of STIs (Kanabus & Fredriksson, 2006). In late 2005, the estimated number of HIV-infected adults and children was 560,000 and 16,000, respectively. The country's population in July 2005 was 64.23 million. The prevalence of HIV infection in adults was 1.4 per cent. The number of AIDS-related deaths in 2005 was 21,000 (UNAIDS/WHO, 2006).

The first case of AIDS was reported in Thailand in 1984 (Phanuphak *et al.*, 1985). Widespread transmission of HIV began in the late 1980s among high-risk groups. HIV infection among IDUs rose from almost zero to 40 per cent during 1988–89. In 1989, it was reported that 44 per cent of sex workers in Chiang Mai in North Thailand were HIV-infected (Weniger *et al.*, 1991). The infection soon spread from male clients of sex workers to their wives and partners, and their children (Kanabus & Fredriksson, 2006).

27.2 – RESPONSE TO THE EPIDEMIC

The initial response was limited in the 1980s because the prevailing opinion was that the disease was imported from abroad and would be confined to individuals in high-risk groups such as gay men and IDUs. Legislators proposed that all foreigners should be tested for HIV infection before being admitted into the country. The Thai Government spent only US$180,000 on HIV prevention in 1988 (World Bank, 1997).

In 1991, HIV prevention and control became a national priority. The National Plan for Prevention and Alleviation of HIV/AIDS in Thailand was brought under the Office of the Prime Minister and the budget was increased almost 20 times to US$44 million in 1993 (Owens, 1991). A public information campaign was launched and HIV prevention education was made compulsory in schools. The "100 per cent condom programme" was launched to distribute free condoms and enforce compulsory and consistent use of condoms in all brothel and non-brothel settings where commercial sex was available. Legal reforms were carried out and proposals for mandatory reporting of names and addresses of AIDS patients were overlooked (World Bank, 2000). By 1996, the funding for the programme had increased to US$80 million annually (Kanabus & Fredriksson, 2006).

The second National Plan for Prevention and Alleviation of HIV/AIDS in Thailand, which covered the period from 1997 to 2001, mobilised the communities and people living with HIV/AIDS. In the 1990s, a randomised controlled trial carried out in Bangkok to study the effect of short-course ZDV in PMTCT of HIV showed that the transmission was reduced by 50 per cent (Shaffer *et al.*, 1999). By 1999, ZDV was being used for preventing perinatal transmission of HIV in most hospitals in Thailand (Kanshana & Simonds, 2002). Due to the Asian Financial Crisis in the late 1990s, funding for ARV therapy, treatment of opportunistic infections and condom distribution was reduced. In 2000, the funding for the National Programme was US$65 million annually (Kanabus & Fredriksson, 2006).

27.3 – CURRENT SITUATION

The third National Plan for Prevention and Alleviation of HIV/AIDS in Thailand was launched in 2001 and covers the period from 2002 to 2006. This plan envisaged the reduction of adult HIV prevalence to less than 1 per cent by the end of the plan period and the provision of access to care and support for at least 80 per cent of the people living with HIV/AIDS. Work on prevention and alleviation of HIV/AIDS was to be planned and implemented by local administrations and community organisations throughout the country. By 2003, the national adult HIV prevalence fell steeply to about 1.5 per cent (UNAIDS/WHO, 2006). Triple combination ARV therapy was introduced in 2000. Due to availability of cheaper generic drugs and external funding, the programme expanded more than eightfold between 2001 and 2003, with only a 40 per cent increase in budget (Ford, 2004). Between 72,000 and 91,000 persons in need of ARV treatment were receiving it by the end of 2005 (WHO, 2006).

27.4 – CONCLUSION

Thailand's epidemic is currently more diverse than it was a decade ago and the routes of transmission have been changing. This indicates that Thailand's achievements have to be actively sustained by effective programmes that adopt strategies to match the shifts in the epidemic. As many as half of new HIV infections every year are occurring within marriage or steady relationships where condom use tends to be very low (UNAIDS, 2002). The "100 per cent condom" programme aimed at 100 per cent condom use for all commercial and casual sex. But, the programme has not had much impact on transmission of HIV from infected male clients of female sex workers and from infected male IDUs to their wives and regular sex partners (WHO, 2002). There are concerns regarding the decreasing rates of condom use in non-brothel settings by indirect sex workers and rising HIV prevalence in some parts of Thailand, especially in Bangkok (World Bank, 2000). Over 10 per cent of brothel-based sex workers and 45 per cent of IDUs attending treatment clinics were found to be HIV-infected in 2003 (UNAIDS/WHO, 2006).

There has been a huge increase in the number of non-brothel sex service establishments such as massage parlours and clubs. This will require revamping safe sex campaigns in circumstances where patterns of commercial sex have changed. The neglected dimensions of Thailand's epidemic are premarital sex among sexually active young people, MSM, and IDUs. There has been an increase in HIV prevalence among IDUs in every region of the country. An estimated one-fifth of new HIV infections in the previous decade have been attributed to injecting drug use. Yet only a small proportion of Thailand's HIV prevention efforts are focussed on this high-risk group (Kanabus & Fredriksson, 2006).

Till date, Thailand's HIV prevention programmes have not focussed on preventing transmission among IDUs and their sex partners. In 2002, UNAIDS recommended that giving clean needles to IDUs should be considered to combat the spread of the epidemic (Bhatiasevi, 2002). HIV prevalence among previously imprisoned IDUs and that among their never-imprisoned counterparts was 49 and 20 per cent, respectively. Thus, incarceration appears to be a significant risk factor for HIV infection among drug injectors in Thailand, as in Indonesia. Many of the drug users were probably infected in prison (UNAIDS/WHO, 2006).

It is estimated that use of condoms by sexually active adolescents is less than 50 per cent. Thus, adolescents form a high-risk group. The Health Ministry in Thailand is planning to add more condom vending machines in public places. However there are also reports about authorities refusing to install condom vending machines fearing they promote promiscuity (World Bank, 2000).

REFERENCES

Bhatiasevi A., 2002, Thailand – Call to provide clean needles to drug users. Bangkok Post November 27. Cited by Kanabus A., and Fredriksson J., 2006.

Ford N., 2004, The role of civil society in protecting public health over commercial interests: lessons from Thailand. Lancet 363. Cited in: Kanabus A., and Fredriksson J., 2006.

Kanabus A. and Fredriksson J., 2006, HIV and AIDS in Thailand. www.avert.org/aidsthai.htm. Last updated 13 July 2006.

Kanshana S. and Simonds R.J., 2002, National Programme for preventing mother-to-child HIV transmission in Thailand: successful implementation and lessons learned. AIDS 16: 953–959.

Owens C., 1991, Alarming spread of AIDS virus in Thailand may threaten country's recent economic gains. Asian Wall Street Journal 13(51): 4.

Phanuphak P., *et al.*, 1985, A report of three cases of AIDS in Thailand. Asian Pac J Allergy Immunol 3: 195–199.

Shaffer N., *et al.*, 1999, Short-course zidovudine for preventing perinatal HIV-1 transmission in Bangkok, Thailand: a randomised controlled trial. Lancet 353: 773–780.

UNAIDS, 2002, AIDS epidemic update. Geneva: UNAIDS, p 10.

UNAIDS/WHO, 2006, 2006 Report on the global AIDS epidemic. Geneva: UNAIDS/WHO.

Weniger B.G., *et al.*, 1991, The epidemiology of HIV and AIDS in Thailand. AIDS 5 (Suppl 2): S71–S85.

World Bank, 1997, Confronting AIDS: public priorities in a global epidemic. Oxford: Oxford University Press, pp 275–276.

World Bank, 2000. Thailand's response to AIDS: Building on success, confronting the future. Thailand Social Monitor 5: 1–11. Cited by Kanabus A. and Frediksson J., 2006.

WHO, 2002, HIV/AIDS in Asia and the Pacific region. Geneva: WHO, pp 21–23.

WHO, 2006, Progress on global access to HIV anti-retroviral therapy – Report on '3 by 5' and beyond. Geneva: WHO, 28 March.

CHAPTER 28

RESPONSE TO THE HIV/AIDS EPIDEMIC IN CHINA

Abstract

The response to the HIV epidemic in China has been relatively slow for want of political commitment, openness to confronting the epidemic, adequate resources, and a favourable legal and institutional environment. During the first phase of the epidemic (1985–1988), cases of AIDS were primarily among foreigners or Chinese people who had travelled overseas. The second phase (1989–1993), began with the detection of HIV infection among IDUs; AIDS and drug addiction were perceived as consequences of contact with the West and AIDS was known as *aizibing*, or the "loving capitalism disease". During the third phase (1994–2001) a major scandal resulted in blood or plasma donation-related HIV infection. The fourth phase started in 2001, when a Plan of Action to contain, prevent, and control HIV/AIDS was published. Subsequently, the "Four Frees and One Care Policy" and "China CARES", a community-based HIV treatment, care and prevention programme were announced. Recently introduced interventions include a policy of 100 per cent condom use (in some regions), detection and treatment of STIs, peer education, and VCT.

Key Words

Aizibing, China, China Comprehensive AIDS Response, Four Frees and One Care Policy

28.1 – INTRODUCTION

The first case of AIDS was reported in Beijing, China in 1985 (Kanabus & Noble, 2006). In January 2006, the Chinese Government, along with the WHO and UNAIDS, jointly estimated that 650,000 people were HIV-infected in China, including 75,000 AIDS patients (Ministry of Health *et al.*, 2006). However, the overall prevalence of HIV infection in China is considered "low" because China's population is estimated at about 1.3 billion. The 2006 estimate is lower than the 2003 estimate of 840,000 HIV-infected persons due to improved methods of estimation. Estimates of prevalence of HIV infection are difficult due to under reporting, shortage of testing equipment and trained staff. Estimations forecast a generalised epidemic in China with between 10 and 20 million HIV-positive Chinese (Kanabus & Noble, 2006).

28.2 – PHASES OF CHINA'S EPIDEMIC AND RESPONSE

The epidemic in China has been described in four distinct phases (Zhang & Ma, 2002).

Phase I: The first phase, from 1985 to 1988, was characterised by few cases of AIDS in coastal cities primarily among foreigners or Chinese people who had travelled overseas. In 1986, the Health Ministry announced plans to test all foreign students for HIV infection if they had been in the country for more than 1 year and students entering China would require a certificate from their country of origin affirming that they were not HIV-infected. The authorities believed that homosexuality and "abnormal" sexuality were a "limited" problem (Kanabus & Noble, 2006).

Phase II: The second phase, from 1989 to 1993, began with the detection of HIV infection in 146 drug users in southwest Yunnan province. At the end of 1989, 153 Chinese and 41 foreigners were reported as HIV-infected. AIDS and drug addiction were perceived as consequences of contact with the West and AIDS was known as *aizibing*, or the "loving capitalism disease" (Kanabus & Noble, 2006).

Phase III: The third phase began in late 1994, when HIV infection spread beyond Yunnan province. By the year 1998, HIV infection was reported from all 31 provinces, autonomous regions and municipalities. Heterosexual transmission accounted for 7 per cent of infections while 60–70 per cent of reported HIV infections were among IDUs (Zhang & Ma, 2002). In November 1998, the State Council published the Medium and Long Term Plan for AIDS Prevention and Control with the specific objectives of stopping HIV transmission through blood supply by 2002, controlling spread of HIV among IDUs, restricting number of China's HIV infections to fewer than 1.5 million by 2010, providing information about HIV/AIDS and preventing STIs to over 70 per cent of the population, including 45 per cent of rural inhabitants and 8 per cent of persons with high-risk behaviour by the year 2002 (Kanabus & Noble, 2006).

Phase IV: The fourth phase started in 2001, when events suggested that the silence surrounding HIV/AIDS in China was beginning to end. In June 2001, the "China Plan of Action to contain, prevent and control HIV/AIDS (2001–2005)" declared that blood for clinical use would have to undergo complete HIV testing. In August 2001, there were between 600,000 and 800,000 HIV-infected persons, with about 6 per cent of these infections caused by contaminated blood. In December 2003, the "Four Frees and One Care Policy" was announced. The new policy had the following objectives:
1. Free ARV treatment for all rural residents or poor urban residents who need treatment.
2. Free VCT.
3. Free ARV treatment for HIV-infected pregnant women to prevent MTCT and HIV testing of newborn babies.

4. Free schooling for children orphaned by AIDS.
5. Care and economic assistance to households of people living with HIV/AIDS.

In 2003, the Chinese Government launched "China CARES" (China Comprehensive AIDS Response), a community-based HIV treatment, care and prevention programme. By October 2003, a pilot programme was launched in seven provinces to provide free domestically produced ARV drugs. However, more than one-fifth of the 5,000 patients receiving ARV treatment dropped out of the programme, primarily because these drugs caused strong side effects. About 1,100 health care providers were trained in HIV care during 2004 (WHO, 2005a). ARV treatment has been introduced in 28 provinces and autonomous regions and by June 2005, about 20,000 people were receiving these drugs (UNAIDS/WHO, 2005). In 2004, local governments were instructed to undertake mass education campaigns for the general population and to counter stigma and discrimination. The Chinese Government also increased its support for harm reduction among IDUs and condom use among high-risk groups (Kanabus & Noble, 2006).

28.3 – GEOGRAPHICAL VARIATIONS

Though HIV infection has been reported from all 31 provinces, autonomous regions, and municipalities, there are wide variations in HIV prevalence across the country. In some areas, more than 1 per cent of pregnant women are HIV-positive, implying that these areas have entered the stage of generalised epidemic (Kanabus & Noble, 2006). The majority of HIV infections have been detected in urban areas of Guangdong, Yunnan, and Henan provinces and Guangxi autonomous region. Currently, Qinghai province and Tibet autonomous region are the least affected (UNAIDS/WHO, 2005). The epidemic in high-prevalence areas is described below.

Yunnan Province: HIV infection rates are very high (50–80 per cent) among IDUs in certain cities of Yunnan province. Sexual transmission accounted for 15 per cent of infections in 2000. Most of the infections are reported in young people aged between 15 and 30 years and are equally distributed among minority populations and the majority Han population. The male to female ratio has shifted from 4:1 in 1997 to 3:1 in 1998 (UNAIDS, 2002).

Henan Province: In 2000, a major scandal among paid plasma donors received wide coverage in the international media. Though the actual number of persons infected with HIV through contaminated blood transfusions is unknown, it is estimated that there could be more than one million victims in Henan province alone (Watts, 2003a). However, other estimates of HIV infection range from below 150,000 to above one million (UNAIDS, 2002).

Guangdong Province: In late 2002, injecting drug use and sexual transmission accounted for 82 per cent and 2 per cent of HIV infections, respectively.

Guangxi Zhuang Autonomous Region: In some parts of this region, HIV infection among IDUs is between 20 and 70 per cent. Condom use is low among commercial sex workers and HIV infection has been detected among non-injecting patients with STIs. There is a risk of an impending heterosexual HIV epidemic in this region (UNAIDS, 2002).

Xinjiang Uygur Autonomous Region: While the majority of HIV infections in this region are related to injecting drug use, local female sex workers were also found to be infected in 1998. The reported number of HIV infected persons more than doubled from 2,125 in 1998 to 4,416 in 2000. In 2000, the male to female ratio was 6:1 (UNAIDS, 2002).

Sichuan Province: This province lies on a major drug trafficking route. Since 1996, HIV infection has been reported among IDUs. By 2000, injecting drug use, transfusion-related infections, and sexual transmission accounted for 68 per cent, 23 per cent, and 6 per cent, respectively, of all new HIV infections in this province. In 2000, the male to female ratio was 4:1 and 93 per cent of new HIV infections were among persons under 30 years (UNAIDS, 2002).

28.4 – HIGH-RISK GROUPS

HIV/AIDS epidemic in China has been associated with high-risk groups. But, as HIV infection spreads among the general population, heterosexual transmission would become the predominant route (Kanabus & Noble, 2006).

Injecting Drug Users: It is unofficially estimated that there are between six and seven million drug users in China, of which about three to three and a half million inject drugs. But only 860,000 persons were registered drug users in the country in 2000. Chinese law prescribes harsh punishments for manufacturing, trafficking, and supplying illicit drugs and drug users are sent to compulsory rehabilitation centres. During 2004, there was a change in attitude towards preventing HIV transmission among IDUs and a pilot programme for providing methadone treatment was launched. A study found that the rates of injecting drug use and drug use-related crime had decreased in the areas where the pilot programme was operational. In the same year, clean needle exchange programmes were established at about 50 sites in several provinces (Kanabus & Noble, 2006). The Chinese Government has announced plans to establish more than 1,400 needle exchange sites and over 1,500 drug treatment clinics in seven provinces in southern and western China, where an estimated two million IDUs live (UNAIDS/WHO, 2005).

Paid Blood/Plasma Donation: In 1988, importing of blood and blood products was prohibited by the Chinese Government to prevent blood-borne transmission of HIV. This ban provided opportunities for local commercial blood collecting companies. These companies operated illegally and collected blood/plasma from blood collection centres in remote and poor areas to avoid interference from the authorities (UNAIDS, 2002). Poor people sold blood to commercial blood processing companies to increase their income. Frequently, the blood was simultaneously collected from several persons and mixed together in a container and plasma was removed. The remaining blood, mixed with the blood of others, was transfused back to the donors. Such procedures, along with the reuse of needles and non-sterile equipment, favoured the transmission of HIV infection. By September 2003, blood or plasma donation-related HIV infection had been reported in all provinces, autonomous regions, and municipalities, except Tibet autonomous region. Since the mid-1990s, illegal blood collection companies have been closed. But it is reported that even some of the official blood collection centres fail to meet the necessary standards and some people are still paid to donate blood (Kanabus & Noble, 2006).

Commercial Sex Workers: Most commercial sex workers in China (estimated at between four and six million in 2000) are women from rural areas who have migrated to big cities seeking better incomes. The Chinese Government has established "re-education centres" in every province where "re-education" emphasises mainly on the "social evils" of prostitution. Limited information is provided on sexual health and safer-sex practices. Commercial sex is illegal in China. The infrequent use of condoms is made more difficult because a woman carrying condoms may be arrested by the local police because this is viewed as "proof" of prostitution. This is despite a 1998 regulation repealing the previous rules (Settle, 2003).

In 1999, China's first-ever condom advertisement was shown on Chinese television and was promptly banned because condom advertisements were illegal. In 2002, to make condom promotion legitimate, the Ministry of Health redefined the condom as a "medical device" rather than a "sexual commodity". Inspired by Thailand's success in reducing its number of new HIV infections, some regions in China have recently introduced a policy of 100 per cent condom use. Other recently introduced interventions include detection and treatment of STIs, peer education, and VCT for HIV (Kanabus & Noble, 2006).

Men having Sex with Men: In 2001, homosexuality ceased to be illegal and it was deleted from the official list of "mental disorders". An estimated 2–6 million MSM (Zhang & Ma, 2002). There are concerns that the high rates of unprotected sex (49 per cent) among MSM could result in a significant rise in prevalence of HIV infection (Choi *et al.*, 2003). Homosexual men in China face severe social pressure to hide their sexual orientation and to get married (UNAIDS, 2002). Stigmatisation of homosexual behaviour impedes open discussion of risky behaviour, precludes adoption of safer-sex practices and fuels an HIV epidemic (Pilcher, 2003). The prevalence of HIV infection among MSM is unknown. Lack of approved organisations or networks to support gay men adversely affects HIV prevention programmes. Access to the Internet may help sharing of information on HIV/AIDS with gay organisations in other countries (UNAIDS, 2002). However, there are allegations that the Chinese Government censors educational websites (Kanabus & Noble, 2006).

Migrant Workers: The total number of temporary and permanent migrant workers in China is estimated at around 120 million. Many migrants are young unmarried males with higher disposable incomes. Due to peer pressure and the relative anonymity in larger cities, young male migrants are likely to visit commercial sex workers. Since most STIs are notifiable, migrant workers are frequently unwilling to seek health care from public health care facilities fearing discrimination and loss of their jobs. HIV epidemic among migrant workers can be fuelled by factors such as lack of family and community support, limited access to HIV prevention services, and lack of information on HIV/AIDS. Returning or visiting migrant workers, many of whom do not know their HIV status, may infect their wives or other sex partners in their own community (Kanabus & Noble, 2006).

28.5 – CONCLUSION

The relatively slow response to the HIV/AIDS epidemic in China is attributed to several factors: (a) lack of political commitment; (b) lack of openness in confronting the epidemic; (c) lack of adequate resources; (d) severe stigma and discrimination against people infected by HIV/AIDS; and (e) unfavourable legal and institutional environment. China is facing a challenge of educating people on the correct and consistent use of condoms and of providing adequate supply of quality condoms (Settle, 2003). In early 2003, the sales of condoms in China were reported to be two billion per year, though the male population is 650 million (Kanabus & Noble, 2006). Since 1998, various plans announced the commencement of HIV prevention education in schools. However, the extent of implementation of this plan is not known. Poor public awareness about the HIV epidemic and the stigma and discrimination experienced by people living with HIV are among the constraints that hinder a more effective response to the epidemic in China (UNAIDS/WHO, 2005).

The continuing stigma and discrimination has been attributed to failure of the Government to educate the public in the early years of the epidemic and the continued inadequacy of correct information on HIV/AIDS. Consequently, HIV infection remains hidden and implementation of effective HIV prevention programmes is hindered. Till recently, HIV-infected persons in China did not have any legal protection against discrimination. In June 2005, it was announced that a law banning discrimination against people with HIV/AIDS would be effective by the end of 2005 (Kanabus & Noble, 2006).

MTCT of HIV remains low in China. However, once increasing numbers of women in reproductive age group are infected by their partners, perinatal transmission would increase. It is unofficially estimated that about one million children in Henan province alone would be orphaned as a result of the blood collection scandal. However, the Chinese Government estimates that 260,000 children may be orphaned by 2010. Many school-going children have allegedly been taken out of school because they must work and care for their sick parents (Kanabus & Noble, 2006).

Tuberculosis is the leading infectious cause of death in China. Every year, about 1.3 million persons are newly infected and 150,000 die due to tuberculosis. It is necessary to implement effective tuberculosis control programmes since the combination of HIV and tuberculosis epidemics would have disastrous effects (WHO, 2005b). There are concerns that the HIV epidemic could diminish economic progress in China (Watts, 2003b).

REFERENCES

Choi K-H, *et al.*, 2003, Emerging HIV-1 epidemic in China in men who have sex with men. Lancet 361: 2125–2126.

Kanabus A., and Noble R., 2006, HIV and AIDS in China. www.avert.org/aidschina.htm. Last updated 2 August 2006.

Ministry of Health, UNAIDS/WHO, 2006, 2005 update on the HIV/AIDS epidemic and response in China, 24 January.

Pilcher H.R., 2003, Stigmatization fueling Chinese HIV. Nature June 20.

Settle E., 2003, AIDS in China: An annotated chronology 1985–2003. www.casy.org/chron/ AIDSchron _111603.pdf

UNAIDS, 2002, HIV/AIDS: China's titanic peril. Geneva: UN Theme Group on HIV/AIDS in China, pp 1–87.

UNAIDS/WHO, 2005, Aids epidemic update. Geneva: UNAIDS/WHO, pp 1–5, 31–44.

Watts J., 2003a, Hidden from the world, a village dies of AIDS, while China refuses to face a growing crisis. Cited in: Kanabus A. and Noble R., 2006.

Watts J., 2003b, HIV could blunt progress in China, Clinton warns. Lancet 362: 1983.

WHO, 2005a, Summary Country Profile for HIV/AIDS treatment scale-up. Geneva: World Health Organization. www.who.int

WHO, 2005b, Global tuberculosis control: Surveillance, Planning, Financing. Geneva: World Health Organization.www.who.int

Zhang K-L., and Ma S-J., 2002, Epidemiology of HIV in China, BMJ 324: 803–804.

CHAPTER 29

EPILOGUE

Abstract

The HIV/AIDS epidemic has polarised society and its institutions since it affected marginalised groups and was linked to sexual behaviour and drug use. Initially, the HIV epidemic was typically blamed on foreigners and foreign behaviour. The global response to the epidemic was slow and inadequate for want of socio-political will to deal with the taboos that cause stigma, discrimination, and denial, and to tackle the social precursors of the epidemic such as gender inequalities, poverty, and human rights abuses.

The introduction of highly active anti-retroviral therapy (HAART) represented a turning point. However, HAART created numerous treatment challenges including frequent adverse effects. AIDS activists played a crucial role in driving the agenda for drug research and early access to new medications for critically ill patients and agitated for low-cost ARV drugs. The epidemic has strongly influenced decisions regarding blood-banking procedures. In the developed countries, HIV/AIDS has become a chronic manageable condition and the focus is on managing the toxic effects of ARV therapy and drug resistance. However, in the world's poorest countries, access to effective ARV therapy is minimal and the epidemic continues to erode the foundations of society with social and economic consequences that would affect future generations. Preventing new infections, developing an effective vaccine, bringing about legal and social reforms, ensuring access to ARV treatment, and strengthening health-care systems of poor countries are among the several challenges for the future.

Key Words

AIDS activism, ACTG, Blood banking, Doha Agreement, Donor support, Food and Drug Administration, Global response, Global Programme on AIDS, HAART, '3 by 5' Initiative, International funding, PEPFAR, Red ribbon, UNAIDS, World Trade Organisation.

29.1 – EARLY YEARS OF THE EPIDEMIC

On 5 June 1981, the (CDC reported five cases of *Pneumocystis* pneumonia among previously healthy homosexual men (CDC, 1981a). At that time, few would have realised that it was the forerunner of a pandemic that would cause more deaths than the estimated 25 million caused by Black Death in the 14th century. On 3 July 1981, 26 additional cases of the new disease were reported in New York City and California (CDC, 1981b). On 10 December 1981, the New England

Journal of Medicine published three consecutive articles on a disease with acquired cellular immune deficiency (Gottlieb *et al.*, Siegal *et al.*, and Masur *et al.*, 1981). On 18 June 1982, the first report indicating sexual transmission of the disease was published. Opportunistic infections and Kaposi's sarcoma were reported in 34 Haitians in the United States and *Pneumocystis* pneumonia in 3 haemophiliacs in July 1982 (CDC, 1982a; Sepkowitz, 2001). The first case of the new disease was reported from Africa in 1982 and in the following year, a heterosexual epidemic was reported in Central Africa (Kaiser Network, 2006).

In September 1982, the CDC announced the name 'acquired immunodeficiency syndrome' and the official acronym 'AIDS'. Four risk factors were cited: male homosexuality, intravenous drug use, Haitian origin, and haemophilia A. In 1983, the CDC added a fifth risk group: female sexual partners of men with AIDS. The risk groups soon included infants, prisoners, and Africans (Sepkowitz, 2001). On 13 January 1984, AIDS was declared a notifiable disease in the United States for the first time (CDC, 1984). At least one AIDS case was reported from each region of the world by 1985. In 1986, the first HIV cases were reported in Russia and India (Kaiser Network, 2006). The first comprehensive needle exchange programme was established for IDUs in Tacoma, Washington, USA in 1988 (Kaiser Network, 2006).

29.2 – THEORIES ABOUT AETIOLOGY

Numerous theories were put forward by the scientific community regarding the cause of the new disease. These included cytomegalovirus (a potentially immunosuppressive virus); use of amyl nitrite and isobutyl nitrite as sexual stimulants (both known immunosuppressive agents); repeated exposure to another's sperm triggering an immune response resulting in a condition resembling chronic graft versus host disease and ultimately, opportunistic infections; and general overloading of the immune system and wearing out of the immune system (Sepkowitz, 2001).

People from outside the scientific community suggested that the disease was a punishment for homosexual men and IDUs (Shilts, 1987; Sepkowitz, 2001). Some were unwilling to accept that the disease could be transmitted through heterosexual contact. For example, the spread of the disease in Haiti was postulated to be a result of voodoo practices rather than heterosexual intercourse (Leonidas & Hyppolite, 1983). In 1983, Luc Montagnier and his associates from Pasteur Institute isolated a retrovirus (later to be called "human immunodeficiency virus" or HIV) from patients with AIDS and in the following year, Robert Gallo and his colleagues from the National Cancer Institute (USA) reported that the retrovirus caused AIDS (Sepkowitz, 2001).

Doubts about the viral cause persisted. In 1987, the American philosopher Louis Pascal came up with a highly debatable theory. All the early AIDS cases originated in the Central African countries – Congo, Rwanda, and Burundi. This area was subjected to trials of a live attenuated polio vaccine on 300,000 men, women, and children. Pascal propounded that the vaccine containing the attenuated virus, which was grown in monkey kidney cells, may have carried the virus. This theory was based on Dr. Albert Sabin's report that an unknown virus contaminated such a batch (www.lifepositive.com/aids.html). As late as 1998, there were persons who doubted that HIV causes AIDS. It was postulated that a disease caused by use of recreational drugs was being blamed on a passenger virus (Duesberg & Rasnick, 1998). Improvements in ARV therapy have para-doxically intensified the debate (Stewart *et al.*, 2000; Sepkowitz, 2001). A study in Texas, USA, has reported that about 30 per cent of persons of Latin American or African descent believed that HIV was a government conspiracy to kill minorities (Fact Sheet 158, 2006).

29.3 – SOCIO-POLITICAL CONSEQUENCES

From its very inception, the HIV/AIDS epidemic polarised society and its insti-tutions since the epidemic affected groups that were already marginalised and was linked to sexual behaviour. Till universal precautions in handling blood and body fluids (CDC, 1982b) were introduced, clinicians feared for their own safety in caring for patients with AIDS. Discussions on public health strategies to con-tain the epidemic created anxieties in the high-risk groups and caused difficul-ties in striking a balance between the rights of infected persons and the rights of other members of society to be protected from the fatal disease (Gottlieb, 2001). In 1990, NGOs boycotted the International AIDS Conference in San Francisco to protest against US immigration policy that excluded HIV-infected persons. In 1993, US President Bill Clinton signed HIV immigration exclusion policy into law (Kaiser Network, 2006).

As the epidemic continued to devastate one country after another, the lessons learnt about HIV prevention and control in one country were not put to use in others. The HIV epidemic was typically blamed on foreigners and foreign behav-iour, just as the French once called syphilis 'the Italian disease' and Italians considered it 'the French disease'. By the time the scale of the epidemic was fully appreciated, the cost in lives and money had increased exponentially (Sepkowitz, 2006).

29.4 – AIDS ACTIVISM

Many agents were studied in the early years of the epidemic for their effective-ness as ARV drugs. The slow nature of the formal clinical trials provoked patients to criticise the medical-industrial complex as uncaring and uncoopera-tive (Arno & Feiden, 1992; Sepkowitz, 2001). Many agents such as Compound Q (Chinese cucumber plant root) and peptide T entered formal clinical trials under pressure from AIDS activists seeking effective therapy. When studied, the agents proved to be ineffective. Charlatans and quacks exploited the growing despair and frustration among AIDS patients by promoting or selling fraudu-lent and unproven remedies, a phenomenon called "HIV fraud" (Fact Sheet 206, 2006). Due to the absence of effective ARV drugs, the early 1980s were demoralising and fatalism settled in. In March 1985, the US FDA licensed an HIV test that detected anti-HIV antibodies (Sepkowitz, 2001).

In 1986, the National Institutes of Health organised the ACTG, now the largest clinical trials group in the United States. The US FDA granted approval for the first ARV drug, ZDV in March 1987. After the initial enthusiasm, many AIDS patients turned against ZDV and alleged that cheap and simple treatments had been overlooked in favour of a mediocre, expensive and toxic agent (Arno & Feiden, 1992; Sepkowitz, 2001). The concept of placebo-controlled trials, overall pace, and sincerity of scientific investigation were among the points of contention between the community of AIDS patients and the medical establishment. In the years after the introduction of ZDV, progress in drug research was very slow, further deteriorating the relationship between physicians and the community of AIDS patients. In the 1990s, the community of AIDS patients and the medical establishment had begun a productive collaboration that remains the hallmark of AIDS care today (Sepkowitz, 2001).

29.5 – THE RED RIBBON

In early 1991, Visual AIDS Artists Caucus in New York conceived the idea for a global symbol in the fight against the HIV/AIDS epidemic. At that time, yellow ribbons were popular in the United States as a symbol of solidarity with soldiers deployed in the First Gulf War (www.hiv.bg). The red ribbon was introduced as the international symbol of AIDS awareness in 1991 (Kaiser Network, 2006). The Red Ribbon International was founded in London on Easter Monday in 1992. The Red Ribbon Foundation was founded in 1993 in memory of Paul Jabara, a singer, song writer, actor, producer, and music theatre composer, who died of AIDS. UNAIDS has incorporated the Red Ribbon into its own logo. Red colour is the symbol of passion, care, and concern towards those affected and for those who care for and support those affected. Red like blood represents the pain caused by deaths due to AIDS. It also represents the anger and helplessness of confronting a disease for which there is no cure. It is also a sign of warning not to ignore one of the biggest problems of our time. The Red Ribbon is intended to be a symbol of hope that the search for a vaccine and cure is successful and it offers symbolic support to those living with HIV. At the International AIDS Conference in Toronto in August 2006, Red Ribbon Awards were presented to five organisations in Thailand, Ukraine, Zimbabwe, Bangladesh, and Zambia to appreciate outstanding community leadership and action that has helped curtail the spread and impact of HIV/AIDS (www.hiv.bg).

29.6 – ADVENT OF HAART

The introduction of HAART represented a turning point (Gottlieb, 2001). The high cost of HAART was offset by savings in other areas, particularly hospital and home-care charges (Sepkowitz, 2001) and quality of life years regained

(Palmer, 2003). In 1989, the successful prophylaxis of *Pneumocystis* pneumonia with cotrimoxazole was reported (Shafer *et al.*, 1989). By early 1997, the annual AIDS-related mortality rate declined by 75 per cent since 1995 in the developed world and the incidence of several opportunistic infections had also declined (Pallela *et al.*, 1998; Gottlieb, 2001). ARV therapy also reduced MTCT of HIV (Connor *et al.*, 1994) in countries where it is available. This achievement demonstrates how an investment in basic research can change the natural history of a fatal disease. ARV therapy continues to evolve rapidly in response to the evolution of HIV in a kind of "biologic chess game" (Gottlieb, 2001).

Initially it was hoped that HAART taken continuously for a number of years might lead to the eventual eradication of HIV from the body. HAART was more complex than mono or dual therapy and posed numerous treatment challenges including pill burden, frequent dosing intervals, food restrictions, and numerous adverse effects (Palmer, 2003; Nwokike, 2005). Adherence to prescribed schedules was difficult and sustained adherence was essential for effective treatment. Within a short period of time, treatment failure was reported in about half of the patients. Treatment failure was defined as increasing viral loads in the blood of patients who had received HAART for 1 year or more (Barlett, 2002). Failure rates were highest in those with advanced disease, those with who received ARV treatments before HAART was instituted, and those with less than optimal adherence to treatment (Palmer, 2003). ARV therapy is associated with significant toxic and adverse effects. Lipodystrophy syndrome (characterised by high serum levels of triglyceride, cholesterol, and glucose in combination with loss of fat in limbs and truncal obesity) was first reported in 1993. This has raised concerns about potential for an elevated risk for cardiovascular disease and diabetes. Some of the physical manifestations of the syndrome (notably "buffalo hump" on the upper back and marked loss of facial fat) were conspicuous and stigmatising. These adverse effects occurred most frequently in patients taking protease inhibitors and certain nucleoside analogues such as stavudine (Palmer, 2003).

29.7 – EFFECT ON BLOOD-BANKING

On 16 July 1982, the CDC reported *Pneumocystis* pneumonia in three persons with haemophilia (CDC, 1982). This raised possibility of contaminated blood supply. Persons with haemophilia are vulnerable to transfusion-related infections, since a single dose of cryoprecipitate contains products from between 1000 and 20,000 donors. By the time a screening test became available in March 1985, HIV had been transmitted to at least 50 per cent of the 16,000 persons with haemophilia in the United States and to additional 12,000 recipients of blood transfusions. The Institute of Medicine, in its investigation, found that limited safety measures had been adopted and that opportunities for more effective interventions had been lost (Sepkowitz, 2001). In the early 1980s, in France and Japan, officials were publicly chastised and prosecuted, and some were jailed for their failure to protect patients with haemophilia from contaminated clotting factor concentrates (Gottlieb, 2001).

The lessons from the AIDS epidemic have profoundly influenced decisions regarding blood-banking procedures. Now ten tests are performed on each unit of donated blood in the United States, as compared with the two tests (for syphilis and hepatitis B surface antigen) that were required in 1981. Guidelines are being developed to prevent possible introduction of the agent of bovine spongiform encephalopathy (BSE) (a prion) into the blood supply (Sepkowitz, 2001).

29.8 – THE GLOBAL RESPONSE

The epidemic required a global response that would combine the resources and technical capacities of the wealthy nations with the needs and capacities of the poor nations. Unfortunately, the response lacked the social and political will to confront the taboos concerning sexual behaviour and drug use that cause stigma, discrimination, and denial; and tackle the social precursors of the epidemic such as gender inequalities, poverty, and human rights abuses. Consequently, the response to the epidemic was slow, inadequate, inconsistent, and often inappropriate (Merson, 2006).

The Global Programme on AIDS (GPA), launched by the WHO in 1987 was unable to gather the necessary political will in donor and affected countries. The effectiveness of GPA was compromised by rivalries with other United Nations Organisations (UNO) and increasing preference of wealthy countries for bilateral aid programs. In 1996, the Joint United Nations Program on HIV/AIDS (UNAIDS) was established to lead an expanded multisectoral global response. UNAIDS, initially co-sponsored by six UN agencies (now ten), and was constrained by its limited resources and strategic conflicts among its partners. Moreover, wealthy nations disengaged form the global response to the pandemic as their own AIDS-related mortality rates declined (Merson, 2006).

Around the turn of the millennium, multiple events contributed to generating the long-needed global response advocated by UNAIDS. Since 1998, the World Bank increased loans for AIDS-related programmes. In 2000, the XIII International Conference in Durban (South Africa) raised global public consciousness about Africa's AIDS-related mortality and the need for affordable ARV drugs. Around the same time, Brazil reported dramatic reduction in AIDS-related mortality as a result of ARV therapy, providing hope for other developing nations (Merson, 2006).

AIDS activists had a central role in driving the agenda for drug research and early access to new medications for critically ill patients (Sepkowitz, 2001). NGOs agitated for low-cost generic ARV drugs and price reductions for brand-name products (Merson, 2006). In 2001, the World Trade Organization (WTO) announced the Doha Agreement, allowing developing nations to buy or manufacture generic ARV drugs (Kaiser Network, 2006). In April of that year, the pharmaceutical industry abandoned its legal efforts to block South Africa from importing cheaper generic ARV drugs (Swarns, 2001).

Politically powerful religious groups that had been reluctant to support condom distribution and other sex-related programmes accepted the need for global treatment, mainly to reduce the number of children being orphaned by the AIDS epidemic. The high sero-prevalence of HIV in some countries and the spread of HIV into Russia, China, and India raised concerns that the AIDS epidemic could threaten the political stability of entire nations, destabilise global political and economic systems, threatening global security (Merson, 2006). On 10 January 2000 the Security Council of the United Nations discussed the possibility that AIDS may threaten the world's security by devastating a country's entire population of young adults. This marked the first time that a medical illness received attention of the Security Council (Sepkowitz, 2001).

In January 2003, US President George W. Bush announced the President's Emergency Plan for AIDS Relief (PEPFAR) and pledged US$15 billion over a period of 5 years. The objective of PEPFAR is to prevent seven million new HIV infections, treat at least two million HIV-infected persons and care for 10 million HIV-infected individuals including AIDS orphans, and vulnerable children. It will focus on 15 countries that are home to 80 per cent of the people requiring ARV treatment (Merson, 2006; FDA, 2006).

In 2003, the WHO announced the '3 by 5' initiative to extend ARV therapy to three million persons in developing countries by 2005. This '3 by 5' initiative fell short of its target but resulted in ARV treatment for 1.3 million individuals, preventing an estimated 250,000 to 350,000 deaths (Merson, 2006).

In 1986, the average interval between a drug application and the granting of approval by the US FDA was 34.1 months; by 1999 it had decreased to 12.6 months (Sepkowitz, 2001). In May 2004, the FDA announced an expedited review process for ARV drugs (FDA, 2006). In 2005, the Indian pharmaceutical Ranbaxy gained approval of the US FDA to manufacture generic ARV drugs for PEPFAR (Kaiser Network, 2006). In 2005 alone, US$8.3 billion was spent on HIV/AIDS – about 30 times as much as at the time of establishment of UNAIDS (Merson, 2006; UNAIDS, 2006).

29.9 – TWO EPIDEMICS

Since the dawn of the 21st century, there is a widening gap between the rich and poor countries so that a single virus is responsible for two epidemics (Sepkowitz, 2006). In the *high-income* countries, the disease has metamorphosed from being predictably fatal to a chronic manageable condition for individuals in whom the ARV drugs work well (Palmer, 2003) and the focus is on managing the toxic effects of ARV therapy and drug resistance (Sepkowitz, 2006).

In the world's *poorest countries*, access to effective ARV therapy is minimal and the epidemic continues to erode the foundations of society with social and economic consequences that would affect generations. The main concerns are more basic – HIV prevention, diagnosis, access to health care, and palliation (Palmer, 2003). The intertwined epidemics of HIV and tuberculosis continue to devastate. One-third of the increase in cases of tuberculosis since 2001 can be attributed to the HIV epidemic (Sepkowitz, 2006). Despite the recent gains, it is estimated that only one in ten Africans and one in seven Asians in need of ARV treatment were receiving it in mid-2005 (UNAIDS/WHO, 2005).

HIV infection in children is becoming increasingly rare in the developed world because ARV treatment usually prevents transmission of the virus from mother to child. However, the infection is widespread among children in poor countries of the world where HIV-infected mothers and their newborns are not treated (Steinbrook & Drazen, 2001).

29.10 – STRATEGIES FOR PREVENTION

Each mode of transmission requires a different prevention strategy and each raises different sets of complex issues. For example, an HIV-positive mother who is advised not to breastfeed may face social stigma (Sepkowitz, 2001). Prevention programmes must involve the community and be evidence-based (not moralistic), locally planned, and linked to efforts to reduce stigma and elevate the status of women (Merson, 2006). Though behavioural prevention strategies have been found effective, HIV prevention services currently reach less than 10 per cent of the persons at risk. It is estimated that expanding these strategies worldwide would avert more than half of the new HIV infections projected to occur by 2015 and save US$24 billion in treatment costs (Stover *et al.*, 2006; Merson, 2006).

Viral heterogeneity, uncertainty about how to achieve optimal immunogenicity, lack of practical animal model and ethical dilemmas involved in conducting trials are among the difficulties faced by scientists in developing a vaccine against HIV (Sepkowitz, 2001). Until an effective preventive HIV vaccine is available, HIV prevention strategies must rely on other options. Two effective interventions have been shown to limit the spread of HIV – sex education (including condom promotion) and treatment of drug abuse (including provision of clean needles). Implementation of these programmes continues to be constrained by personal, social, and political barriers in almost all countries and governments. Reluctance to implement effective control measures is responsible for the continued spread of HIV rather than the lack of an effective vaccine or remedy (Sepkowitz, 2001).

There is a need for research on new approaches in prevention. Adult male circumcision, pre-exposure ARV prophylaxis in high-risk groups, acyclovir treatment for herpes simplex type 2, and microbicidal agents seem promising (Merson, 2006). The effect of adult male circumcision on HIV transmission remains controversial despite a report from South Africa, where a 61 per cent reduction in rate of new infection was observed in the circumcision group, as compared with the control (observation) group. The study was conducted on more than 3,000 men and the results were controlled for differences in condom use, sexual behaviour, and health-care seeking behaviour (Auvert *et al.*, 2005).

29.11 – THE FUTURE

Preventing new infections, developing an effective vaccine, strengthening health-care systems of poor countries are among the several challenges for the future. It is also necessary to bring about legal and social reforms to address gender inequalities and the potential for vulnerability, discrimination, and stigma related to HIV infection (Steinbrook & Drazen, 2001). The current debate focuses on whether legal provisions in some countries impede the public health response to the HIV/AIDS epidemic by outlawing certain types of sexual behaviour.

In the United States, approximately one million individuals were infected with HIV in 2003 and an estimated 164,000 to 312,000 of them were unaware of their infection (Glynn & Rhodes, 2005). Most of the 40,000 new HIV infections that occur annually in the United States may be due to contact with these undiagnosed persons.

In September 2006, the CDC, Atlanta, recommended that HIV screening be made routine part of medical care for all patients between the ages of 13 and 64 and to improve diagnosis of HIV infection among pregnant women. Patients are to be specifically informed that they have the opportunity to decline HIV testing. CDC has recommended that written consent for HIV testing be no longer required though HIV testing must be voluntary and undertaken only with the informed consent of the patient (www.boston.com, 2006). The availability of a rapid oral test has simplified mass screening. However, it is feared that routine testing would erode patient confidentiality by eroding legal safeguards (Sepkowitz, 2006).

Eradication of HIV by continuous ARV treatment is highly unlikely with the available drugs due to the very long half-life and latency of some HIV-infected cells of the immune system. ARV therapy has become increasingly complex and clinicians must confront numerous issues and dilemmas. Due to adverse effects of ARV treatment, clinicians have become more cautious in advocating early treatment in contrast to the "hit hard and early" approach initially adopted with HAART (Palmer, 2003; Sepkowitz, 2006). The current American, British, and Australian guidelines for starting ARV therapy are much more conservative than those released in 1997 (Palmer, 2003). Till date, there is no established remedy for common side effects of ARV, such as lipodystrophy, lipoatrophy, diabetes, glucose intolerance, insulin resistance, and dyslipidemia (Sepkowitz, 2006).

Ensuring adherence to free ARV therapy does not guarantee adherence. National ARV programmes need to develop adherence measurement and monitoring systems which are built into the national treatment protocols (Nwokike, 2005). The introduction of once-daily dosing regimens has made ARV treatment more convenient for HIV-infected individuals. Though enfuvirtide is effective against drug resistant strains of HIV, twice-daily injections are difficult to sustain for some individuals. Tenofovir, with or without emtricitabine, reduces infection rate in high-risk groups but its prophylactic use may accelerate emergence of drug resistance. Tenofovir has been misused as a 'party' pill by uninfected individuals who planned to engage in high-risk activity (Sepkowitz, 2006). The prevalence of resistance to tenofovir has increased from about 5 per cent before 1996 to at least 15 per cent between 1996 and 2003 (Masquelier *et al.*, 2005).

For controlling the growing epidemic, some countries may have to adapt the legal and institutional environment to facilitate effective HIV prevention strategies such as condom promotion and discussions on sex-related behaviour and sexual activity. Even in the developed countries many physicians had to overcome a prudish reluctance to discuss sexual behaviour with their patients or inquire about their sexual orientation and specific practices. The AIDS epidemic has tested character, beliefs, and values and many health care personnel have matured into compassionate and caring health professionals (Gottlieb, 2001). Activism by AIDS patients has also influenced advocates for patients with diseases such as breast cancer, Parkinsonism, Alzheimer's disease, and juvenile diabetes (Sepkowitz, 2001).

In future, an arsenal of effective ARV drugs may be available along with new vaccines and improved prevention measures (Sepkowitz, 2006). Many pharmaceutical companies have reduced the prices of ARV drugs. Unequal access to ARV treatment poses questions of social justice particularly in poor nations where people can not afford to buy drugs even at reduced prices and ultimately, unequal access may also influence adherence to treatment (Whyte *et al.*, 2005). Extending financial support to poor nations might cost several billion dollars a year. But for the world's wealthiest countries, these costs are not too high. However, international donor support for persons living with HIV/AIDS in sub-Saharan Africa has been insignificant. It was about US$110 million per year from 1996 to 1998 – less than US$5 per HIV infected person in 1998 (Attaran & Sachs, 2001). In order to bring the epidemic under control, the rich nations of the world must contribute the billions of dollars needed for HIV prevention and treatment in poor countries. Controlling the HIV/AIDS epidemic in the poor nations is in the long-term interests of the rich nations (Steinbrook & Drazen, 2001).

Efficient international mechanisms should ensure the funds are used for the intended purposes. This will require high level of cooperation among governments, NGOs, and industry (Steinbrook & Drazen, 2001). As the epidemic becomes chronic, international funding must be partly used to strengthen health care delivery systems in low-income nations. The infrastructure developed in response to the HIV/AIDS epidemic could be utilised for controlling newly emerging and other neglected diseases in the developing countries (Gottlieb, 2001).

REFERENCES

Arno P.S., and Feiden K.L., 1992, Against the odds: the story of AIDS drug development, politics and profits. New York: HarperCollins.

Attaran A, and Sachs T, 2001, Defining and refining international donor support for combating the AIDS pandemic. Lancet 357: 57–61.

Auvert R., *et al.*, 2005, Randomized, controlled intervention trial of male circumcision for reduction of HIV infection risk: the ANRS 1265 trial. PLoS Med 2: e298.

Barlett J.G., 2002, The Johns Hopkins Hospital 2002 Guide to medical care of patients with HIV infection. 10th edn., Baltimore: Lippincot Williams & Wilkins. Cited in Palmer, 2003.

Boston.com, 2006, CDC's revised recommendations for HIV testing of adults, adolescents, and pregnant women in health care settings. September 22. www.boston.com/yourlife/health/

Centers for Disease Control (CDC), 1981a, *Pneumocystis* pneumonia – Los Angeles. Morb Mort Wkly Rep 30: 250–252.

Centers for Disease Control (CDC), 1981b, Kaposi's sarcoma and *Pneumocystis* pneumonia among homosexual men – New York City and California. Morb Mort Wkly Rep 30: 305–308.

Centers for Disease Control (CDC), 1982a, *Pneumocystis carinii* pneumonia among persons with haemophilia. Morb Mort Wkly Rep 31: 353–354, 360–361.

Centers for Disease Control (CDC), 1982b, Acquired Immune Deficiency Syndrome (AIDS): precautions for clinical and laboratory staffs. Morb Mort Wkly Rep 31: 377–380.

Centers for Disease Control (CDC), 1984, Summary – cases specified notifiable diseases, United States. Morb Mort Wkly Rep 33: 4–5.

Connor E.M., *et al.*, 1994, Reduction of maternal infant transmission of human immunodeficiency virus type 1 with zidovudine treatment. N Engl J Med 331: 1173–1180.

Duesberg P., and Rasnick D., 1998, The AIDS dilemma: drug diseases blamed on a passenger virus. Genetica 104: 85–132.

Food and Drug Administration, 2006, FDA News. Washington DC: US Department of Health and Human Services. www.fda.gov. 12 July.

Glynn M., and Rhodes P., 2005, Estimated HIV prevalence in the United States at the end of 2003. Presented at the National HIV Prevention Conference, Atlanta, 12–15 June. www.aegis.com/conferences/nhivpc/2005/T1-B1101.

Gottlieb M.S., 2001, AIDS – Past and future. N Engl J Med 344: 1788–1790.

Gottlieb M.S., *et al.*, 1981, *Pneumocystis carinii* pneumonia and mucosal candidiasis in previously healthy homosexual men: evidence of a new acquired cellular immunodeficiency. N Engl J Med 305: 1425–1431.

Kaiser Network, 2006, Global HIV/AIDS Timeline. www.kff.org/hivaids/timeline

Leonidas J.R., and Hyppolite N., 1983, Haiti and the acquired immunodeficiency syndrome. Ann Int Med 98: 1020–1021.

Masquelier B., *et al.*, 2005, Prevalence of transmitted HIV-1 drug resistance and the role of resistance algorithms data from seroconverters in the CASCADE collaboration from 1987 to 2003. J AIDS 40: 505–511.

Masur H., *et al.*, 1981, An outbreak of community-acquired *Pneumocystis carinii* pneumonia: initial manifestation of cellular immune dysfunction. N Engl J Med 305: 1431–1438.

Merson M.H., 2006, The HIV-AIDS pandemic at 25 – The global response. N Engl J Med 354: 2421–2417.

New Mexico AIDS Education and Training Center, 2006, Fact Sheet 158. AIDS myths and misunderstandings. University of New Mexico Health Sciences Center. www.aidsinfonet.org Revised 18 April 2006.

New Mexico AIDS Education and Training Center, 2006, Fact Sheet 206. How to spot HIV/AIDS fraud. University of New Mexico Health Sciences Center. www.aidsinfonet.org Revised 10 August 2006.

Nwokike J.L., 2005, Baseline data and predictors of adherence in patients on anti-retroviral therapy in Maun General Hospital, Botswana. Essential Drugs Monitor 34: 12–13.

Pallela F.J. Jr., *et al.*, 1998, Declining morbidity and mortality among patients with advanced human immunodeficiency virus infection. N Engl J Med 238: 853–860.

Palmer C., 2003, HIV treatments and highly active anti-retroviral therapy. Austr Prescr 26 (3): 59–61.

Sepkowitz K.A., 2001, AIDS – The first 20 years. N Engl J Med 344: 1764–1772.

Sepkowitz K.A., 2006, One disease, two epidemics – AIDS at 25. N Engl J Med 354: 2411–2414.

Shafer R.W., *et al.*, 1989, Successful prophylaxis of *Pneumocystis carinii* pneumonia with trimetho-prim – sulfamethoxazole in AIDS patients with previous allergic reactions. J AIDS 2: 389–393.

Shilts R., 1987, And the band played on – politics, people and the AIDS epidemic. New York: St. Martin's Press.

Siegal F.P., *et al.*, 1981, Severe acquired immunodeficiency in male homosexuals manifested by chronic perianal ulcerative herpes simplex lesions. N Engl J Med 305: 1439–1444.

Steinbrook R., and Drazen J.M., 2001, AIDS – Will the next 20 years be different (Editorial). N Engl J Med 344: 1781–1782.

Stewart G.T., *et al.*, 2000, The Durban Declaration is not accepted by all. Nature 407: 286.

Stover J., *et al.*, 2006, The global impact of scaling up HIV/AIDS prevention programs in low- and middle-income countries. Science 311: 1474–1476.

Swarns R.L., 2001, Drug makers drop South Africa suit over AIDS medicine. New York Times. April 20: A1. Cited in: Steinbrook & Drazen, 2001.

UNAIDS, 2006, 2006 Report on the global AIDS epidemic. Geneva: UNAIDS. www.unaids.org

UNAIDS/WHO, 2005, AIDS epidemic update. Geneva: UNAIDS-WHO, p 5.

Website: www.hiv.bg/redribbon.history.english.htm. Accessed on 20 September 2006.

Website: www.lifepositive.com/aids.html. Accessed on 20 September 2006.

Whyte S.R., *et al.*, 2005, Accessing retroviral drugs: dilemmas for families and health workers. Essent Drugs Monit 34: 14–15.

INDEX